AF809895

Assessment of Materials in Ageing Offshore Structures

John V. Sharp · John Wintle

Assessment of Materials in Ageing Offshore Structures

Springer

John V. Sharp
Cranfield University
Cranfield, UK

John Wintle
University of Strathclyde
Glasgow, UK

ISBN 978-3-031-84488-1 ISBN 978-3-031-84489-8 (eBook)
https://doi.org/10.1007/978-3-031-84489-8

© The Editor(s) (if applicable) and The Author(s), under exclusive license to Springer Nature Switzerland AG 2025

This work is subject to copyright. All rights are solely and exclusively licensed by the Publisher, whether the whole or part of the material is concerned, specifically the rights of translation, reprinting, reuse of illustrations, recitation, broadcasting, reproduction on microfilms or in any other physical way, and transmission or information storage and retrieval, electronic adaptation, computer software, or by similar or dissimilar methodology now known or hereafter developed.
The use of general descriptive names, registered names, trademarks, service marks, etc. in this publication does not imply, even in the absence of a specific statement, that such names are exempt from the relevant protective laws and regulations and therefore free for general use.
The publisher, the authors and the editors are safe to assume that the advice and information in this book are believed to be true and accurate at the date of publication. Neither the publisher nor the authors or the editors give a warranty, expressed or implied, with respect to the material contained herein or for any errors or omissions that may have been made. The publisher remains neutral with regard to jurisdictional claims in published maps and institutional affiliations.

This Springer imprint is published by the registered company Springer Nature Switzerland AG
The registered company address is: Gewerbestrasse 11, 6330 Cham, Switzerland

If disposing of this product, please recycle the paper.

Preface

The materials used to fabricate the structures deployed offshore in the oil and gas and wind energy sectors age and deteriorate and experience damage with time and their exposure to a harsh operating environment. The principal deterioration mechanisms include, corrosion, hydrogen embrittlement, fatigue cracking, environmental cracking and damage from impacts and severe stress events such as storms, collisions and dropped loads. Significant research and development work has been undertaken since the early 1970's on different deterioration and damage mechanisms, the evaluation of these in terms of their form, rate and effects on the structure, and the need and methods to inspect and repair. This book aims at giving the reader an overview of this research and development work in addition to providing current practice in these areas, both to inform the reader about the existence of this work and to avoid unnecessary repetition of research and development. As many older structures from the 1970's and 1980's remain in use and are showing the effects of ageing, yet are still needed for production in the life extension phase, awareness and knowledge of their main material deterioration mechanisms are important today to ensure that the structures remain safe.

Since early 1980 the first author of this book, John V. Sharp, has been active in the majority of these research programmes through his role initially in managing the relevant UK Department of Energy research work and later as Head of Offshore Research for the UK Health and Safety Executive (HSE). The second author, John Wintle, has over 35 years' experience working on ageing and life extension of safety critical structures in the offshore and nuclear sectors. As a Visiting Professor of Mechanical Engineering at the University of Strathclyde and Consultant Engineer at TWI Ltd. and Fellow of the Welding Institute John Wintle has prepared industry guidance for the Health and Safety Executive, the Energy Institute and other organisations. Both authors have specific relevant expertise and have published a number of papers on material science and ageing and life extension.

The authors have had the benefit of working closely with Gerhard Ersdal, Visiting Professor at the University of Stavanger, in preparing this book and his contributions have made significant improvements to the final text, particularly in the areas of fatigue and microstructure of materials. The authors would also like to thank

Dr Henryk Pisarski for carefully reviewing sections of this document and providing many valuable comments. In addition, the authors appreciate the helpful and patient staff at Springer.

The opinions expressed in this book are those of the authors and they should not be construed as reflecting the views of the organisations the authors represent. Further, the text in this book should not be viewed as recommended practice, but rather as an overview of important issues that are involved in the assessment of ageing materials.

East Hendred, Oxfordshire, UK
Penarth, Vale of Glamorgan, UK
Microsoft Teams, December 2024

John V. Sharp
John Wintle

Competing Interests The authors have no competing interests to declare that are relevant to the content of this manuscript.

Authors' Note

This book is intended as a review of the research and practice associated with the assessment of materials in ageing offshore structures for the purposes of justifying continued service and life extension where required. The data and methods described in this book may not necessarily be applicable for the assessment for individual structures and may not be relevant to the regulations in specific geographic regions. Any assessment of materials for the purposes of gaining approval for life extension should be performed in accordance with the regional regulatory requirements for life extension and in accordance with the recognised technical standards and recommended codes of practice cited in these regulations. This book should, however, provide professional knowledge on the topic to facilitate increased awareness and understanding of the condition and threats to materials in ageing offshore structures particularly during their life extension phase.

The contents of this work are intended to further understanding and discussion only. The book is not intended and should not be seen as a recommended practice for life extension of materials in ageing offshore structures. The publisher and the authors make no representations or warranties with respect to the accuracy or completeness of the contents of this work and specifically disclaim all warranties, including without limitation any implied warranties of fitness for a particular purpose. In view of ongoing research, equipment modifications, changes in governmental regulations and the constant flow of information relating to life extension of ageing offshore structures, the reader is urged to review, evaluate and verify the information provided in this book before it is used in a particular application or context. The fact that an organisation or Website is referred to in this book as a citation and/or a potential source of further information does not mean that the authors or the publisher endorses the information the organisation or Website may provide or recommendations it may make. Further, readers should be aware that Internet Websites listed in this work may have changed or disappeared between when this work was written and when it is read.

Contents

Definitions and Abbreviations

ACI	American Concrete Institute
API	American Petroleum Institute
ASTM	American Society for testing and materials
AWS	American Welding Society
BS	British Standard
C_3A	Tricalcium aluminate
CE	Carbon equivalent
COD	Crack opening displacement
CSE	Copper/copper sulphate electrode
CTOD	Crack tip opening displacement
CVI	Close visual inspection
EAC	Early age cycling
ECA	Engineering critical assessment
EI	Embrittlement index
ERF	Environmental reduction factor
FCAW	Flux cored arc welding
FPSO	Floating production, storage, and offloading installations
FPU	Floating production unit
GGBFS	Ground granulated blast furnace slag
GoM	Gulf of Mexico
GVI	General visual inspection
HAC	High alumina cement
HAZ	Heat-affected zone
HE	Hydrogen embrittlement
HSLA	High strength low alloy
HV	Vickers hardness
ICCP	Impressed current cathodic protection
ISO	International Standards Organisation
ITZ	Interfacial transition zone
Kth	Threshold stress intensity factor
LBZ	Local brittle zone

LWA	Lightweight aggregate concrete
MIC	Microbiologically induced corrosion
MIG	Metal inert gas (welding)
MMA	Manual metal arc (welding)
NDE	Non-destructive examination
NDTT	Nil ductility transition temperature
PFA	Pulverised fuel ash
PWHT	Post weld heat treatment
Q&T	Quench and tempered
SAW	Submerged arc welding
SCF	Stress concentration factor
SMYS	Specified minimum yield strength
SRB	Sulphate reducing bacteria
SSRT	Slow strain rate testing
STRA	Short transverse reduction in area
TIG	Tungsten inert gas
TLP	Tension leg platform
TMCP	Thermo-mechanically controlled processing
UKCS	United Kingdom Continental shelf
UKOSRP	United Kingdom Offshore Streels Research Programme
UPV	Ultrasonic pulse velocity
UT	Ultrasonic testing
UTS	Ultimate tensile strength
YR	Yield ratio

Chapter 1
Introduction to Ageing of Materials

1.1 Why Ageing[1] of Materials Matters

The offshore energy production sector and many other industries rely on structures that are ageing and will continue to do so for the foreseeable future. This is particularly the case for many offshore oil and gas installations, but also for many early offshore wind turbines that are now clearly experiencing ageing. The key question asked is how long can materials and the structure where they are used perform their function safely and reliably without expensive intervention?

In a political climate where environmental concerns are a priority, there is a need to ensure that existing structures and materials can continue to be used, rather than be demolished and replaced with new. The continued use and re-use of existing structures is therefore of considerable importance and will be beneficial both for the economy and the environment, taking into account the CO_2 footprint of the total world-wide manufacturing of steel and concrete (7–10% of the annual CO_2 emissions).

Materials used in offshore structures are susceptible to degradation and deterioration processes as they operate in a hostile seawater environment, which over time can severely affect their properties and performance. The materials are specified at the design stage to last for the planned design life, with allowances for anticipated degradation and deterioration. However, since the structures in many cases are needed for further use beyond their design life in a life extended period, the effects of degradation and other forms of ageing of materials over this period need to be taken fully into account.

The continuing good performance of materials used in structures is essential for their safety and reliability, particularly during the extended life of the structures of offshore installations. This book has examined in depth the performance of the two

---

[1] It should be mentioned that there is an alternative meaning of ageing in the manufacturing of steel, but this is not what is meant by ageing of materials in this book. This alternative meaning of the ageing of steel can be found described in texts on metallurgy.

© The Author(s), under exclusive license to Springer Nature Switzerland AG 2025

J. V. Sharp and J. Wintle, *Assessment of Materials in Ageing Offshore Structures*,
https://doi.org/10.1007/978-3-031-84489-8_1

widely used materials, namely steel and concrete and has also reviewed how ageing affects other materials found on offshore structures. It will be useful for the practising materials engineer whose role is to assess the integrity of ageing structures over their life and to students of offshore engineering wishing to understand how and why ageing of materials is important for the offshore energy sector.

1.2 Forms of Ageing

Degradation of materials is a form of physical ageing and can be considered in two different categories, namely, changes to the material properties (intrinsic changes) and changes to the shape or continuity of the component such as wall thinning, cracking and other geometric changes (extrinsic changes).

Typical intrinsic changes to material properties are hydrogen embrittlement and hardening for ferritic materials and chloride ingress, carbonation and sulphate attack to cementitious materials like concrete. Intrinsic material changes are due to changes at an atomic or microstructure scale, which leads to changes in the material properties (for example yield strength, Youngs modulus, ductility and fracture toughness). Such changes may for metallic materials be a result of ingress of atoms into the material or its surface. Another intrinsic change could be loading the material to an extent that causes permanent deformation. The ingress of atoms can fill vacancies or create reactions that cause debonding, reorganisation of the atomic structure due to plastic deformation, the formation and growth of microcracks and the coalescence of voids.

Extrinsic changes include mechanisms which in different ways removes or changes the material at the surfaces, or which create new surfaces. As already mentioned, these include wall thinning, cracking and other geometric changes. For concrete structures, spalling and delamination are typical examples of such mechanisms. Examples of how such extrinsic changes are relevant for the strength of the structure include:

- Thickness loss—reducing the section area of structural elements and details,
- Geometric change—introducing eccentricities in structural elements and details,
- Discontinuities—macro-scale cracks and sharp transitions in the structure that will act as crack initiators or as locations for harmful substances to enter the material (e.g. hydrogen into steel material and chlorides into cementitious materials).

Ageing of materials generally implies that a material has physically changed from its original state from such intrinsic or extrinsic degradation. However, in evaluating the materials in an ageing structure, ageing can also imply that the material used in the original structure is of lesser quality than we presently expect. Technological changes since the structure was built may lead to the material and joining methods in the original structure not being according to present standards and procedures (outdated).

Technological change may be a problem if, for example, the ductility and fracture toughness of the material are less than expected in present standards. This will require consideration of the effect of the actual ductility and fracture toughness in the structure and to what extent this is acceptable. Even worse, the original material may be incompatible with current consumables for joining methods, which would be a possible issue in situations where repairs or replacement are needed.

It is relatively straightforward to understand that these physical and technological changes, discussed so far, can directly change the way that the safety and reliability of the structure is regarded. However, in the assessment of materials in an ageing structure there are other aspects of ageing that also need to be considered, including:

- *changes to material information* which is the loss and gain of information about the materials in the structure. This includes the loss of information specified in design and the joining methods used in fabrication. The gain of information arises mainly from the additional information from inspections.
- *changes to knowledge and safety requirements* which include experience and research findings that alter our understanding of the behaviour of a material and the improvements in codes and standards, often as a result of this improved knowledge but also as a result of possible stricter societal requirements for structures.

These four aspects of materials ageing are illustrated in Fig. 1.1. It is indicated in this figure that physical and technological changes may directly modify the safety of the structure where the material is used. In contrast, changes in information about the structure (for example loss of material certificates) and changes in knowledge do not change the safety level directly but may influence our evaluation of the safety of the structure. Similarly, Fig. 1.1 indicates that physical and information changes are specific to the structure being assessed, while technological changes and changes to knowledge and safety requirements have a more general effect on all structures of the same vintage.

Physical ageing of materials can be defined as the differences observed in property tests of the material when new and after several years in service. Technological ageing of materials can similarly be defined as the differences observed between properties of the original material and present-day materials of the same type (manufactured using current practice and standards). The original material is here defined as the material used when the structure was fabricated or constructed. Hence, technological developments are captured in, for example, the steel manufacturing and welding processes that improve the properties of the fabricated structure.

1.3 Robust Materials

A fundamental necessity in ensuring materials that are fit for purpose is the ensure their robustness. Robust materials in this context can tolerate to a reasonable extent the environment, operational use and loading to which they are exposed while still being able to perform their full function (fit for purpose), taking into account the normal variability in the properties of the materials. This would also include being

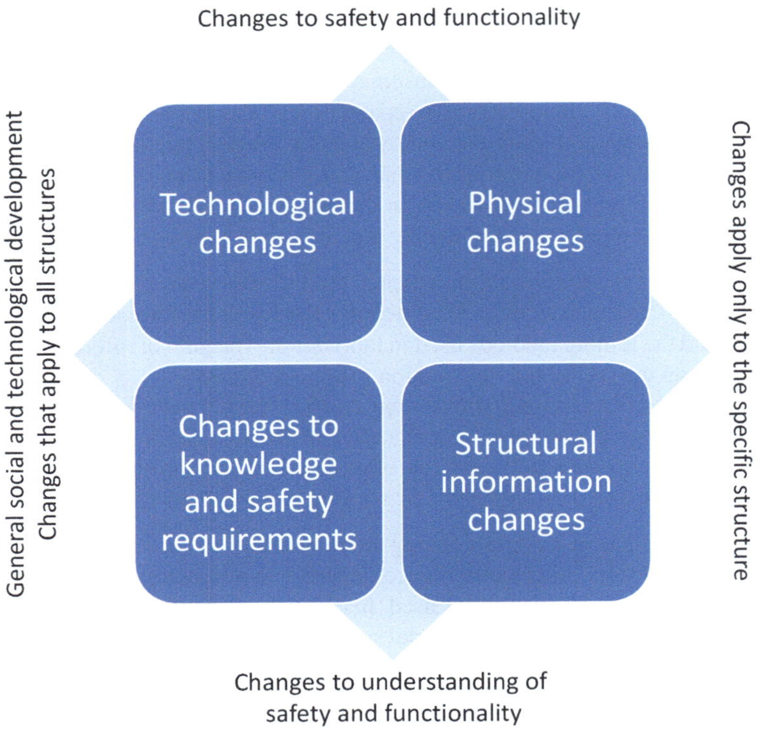

Fig. 1.1 The four main elements of ageing

able to tolerate some level of accidental damage and an amount of physical degradation due to ageing mechanisms. It is also assumed that a robust material will be able to accommodate to some extent unknown or unforeseen circumstances. Robust materials will, in addition, imply that there is a minimum need for intervention (e.g. re-coating, re-painting, replacement) in the service life of the system.

Achieving robust materials in the design life requires having an adequate materials selection process at the design stage. The most probable reasons for failures to select sufficiently robust materials were studied by Drugli et al (2003), who concluded:

1. The testing and evaluation of the materials before they were selected had not been sufficient.
2. The probability for scatter in the material quality had not been sufficiently evaluated.
3. The designers of the systems had insufficient knowledge of the loads and environments to which the materials would be exposed.
4. The designers did not predict the range of worst-case service conditions of the systems.

These reasons for insufficient robustness in materials seems reasonable and, hence, an evaluation of the level of quality in the original material selection would be highly useful in a life extension assessment. It should also be noted that in the early days of the rapid expansion of the offshore oil and gas industry, the material selection process may not have met the criteria for robust materials.

Robust materials selection requires a complete understanding and quantification of all possible sources of environmental impact on materials in all parts of the facility during its life. However, experience has shown that even if a design has been fully qualified, failures have occurred in unexpected modes (e.g. the brittle material failure of Sea Gem, as further discussed in Ersdal et al. (2019)). The root cause of such failure could be a result of gaps in knowledge, unforeseen events and circumstances. The threats from gaps in knowledge, unforeseen events and circumstances related to materials may increase with time and particularly during an extended life of the structure.

When assessing materials in an ageing structure for life extension, the robustness requirements for the materials will still be there. However, for obvious reasons the evaluation cannot be about robust material selection. Rather it more directly has to encompass the robustness of the material that is there in the structure under evaluation. This topic will be discussed particularly in Chap. 10 of this book.

1.4 Failure of Materials and Structures

All materials are composed of atoms, particles and components held together by some form of bonds and essentially all structural materials will include flaws. These flaws may be from an atomic scale to a macro-scale and may include pores, cracks, inclusions and cavities in grain boundaries. These flaws will increase the local stresses and increase the likelihood of the material breaking.

At an atomic level, the atomic structure at flaws does not have the same bonds as those in the main bulk of the material, which means the defects can more easily move under stress. In addition, they can more readily react with other chemical elements to form new bonds. Understanding how a simple defect is affected by a single degradation mechanism such as increased stress, temperature or chemical reaction is in itself challenging, but a component in an offshore structure might have many such interactions occurring in a complex environment.

Some flaws at an atomic level are also useful for the materials. While elasticity is related to the relative movement of the atoms or stretching and compression of the bonds between the atoms, plasticity is related to the permanent movement of atoms (slip) where these form new bonds with different atoms (in for example lines of dislocations). The ability of metallic material to undergo plastic deformations determines its level of ductility.

Failure or material damage normally implies debonding between the atoms or particles. Materials will fail in different ways depending on whether the stresses are

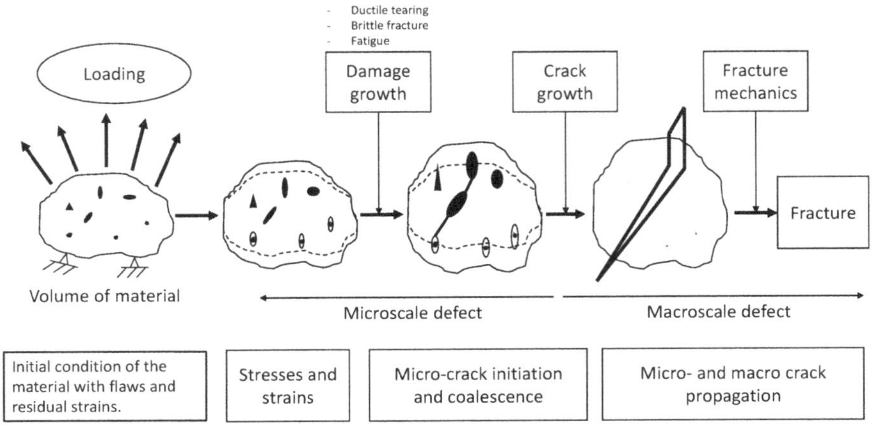

Fig. 1.2 Material failure from micro- to macroscale (inspired by Lemaitre and Chaboche 1990)

in tension, shear or compression. The failure modes of materials include rupture, fracture and crushing, which are determined by the yield stress, ultimate stress, ductility and fracture toughness. In addition, the geometry of the structural part (length and eccentricity), section properties, material properties and boundary conditions will influence its strength (including the buckling failure mode). All these factors can in principle be changed when the structure is put to use. When an initial flaw or defect experiences stresses or chemical attack during operational use, many complex interactions can occur.

Materials will also fail in different ways depending on whether the material is brittle or ductile. Brittle fracture may occur by cleavage or by intergranular fracture and is very sensitive to the presence of flaws, which can result in a rather rapid and catastrophic failure of the material. A ductile failure is a type of failure characterized by extensive plastic deformation or necking prior to the actual failure of the material. In ductile failure, there is an absorption of large amounts of energy and slow propagation before the failure occurs. Materials that contain flaws (which includes most structural materials) will experience nucleation and coalescence of these to form a macroscopic flaw. At a certain stage these voids will initiate cracks that propagates until final separation (rupture or fracture) occurs.

The process of material failure is illustrated in Fig. 1.2.

1.5 Assessment of Materials in Offshore Structures for Life Extension

As a result of the failure of materials, the essence of the assessment of materials in existing structures could be described as:

(1) establishing an understanding of the expected flaws and defects that were present in the original material (after construction or fabrication),
(2) identifying possible new defects that could have occurred as a result of the loading or the environment the structure has been placed in and,
(3) how these have grown and developed during the years of use.

However, in addition there is also a need to establish the properties of the material, such as what yield strength and elongation properties the base material had (ductility) and the expected fracture toughness (the material's ability to tolerate cracks).

As the defects are of such importance, large parts of this book will cover this topic, including both defects in materials after construction and fabrication in addition to defects that may be introduced during operation and use, as a result of the loading or the environment the materials have been placed in. Furthermore, the book will discuss how these defects can be expected to have developed during the years of use. For obvious reasons fatigue crack growth will be important here.

In addition, the material may have had external degradation due to corrosion, wear and tear which may have reduced the amount of material (wall thickness) and possibly also have introduced geometric imperfections to the material.

The testing and inspections that can be performed to establish as much as possible information about the actual condition of the material in the structure at present (the amount of defects in the material, the material properties and the external degradation—material loss). However, since all inspection and testing methods are limited and that only a limited part of the structures can be tested or inspected, it is also important to know the limitations of the inspection and testing methods and their accuracy and ability to detect defects, material property changes and material loss.

Finally, an important part of the assessment of materials in an ageing structure is to provide predictions of future degradation, based on knowledge about the material and history of the use in this specific structure. These evaluations are intended to help estimate a safe life for the structure.

The structural analysis and code checks may often require certain properties of the material. These include the yield strength, ductility, fracture toughness, amount and size of defects in the material, the wall thicknesses and the geometric imperfections of the material and the structural elements. An assessment of the material in an ageing structure should check if these requirements are met. However, most likely the task will rather be to evaluate if the deviations from present material and fabrication standards are acceptable, in many cases in cooperation with the structural engineer. Examples of such evaluations could include:

1. Will the structural element buckle with the added geometric imperfection due to wall thinning?
2. Will the structure be strong enough with reduced wall thickness or reduced yield strength?
3. What will the crack growth rate be from the defects detected (or assumed) and when will this become unacceptable?
4. Will the critical crack size change as a result of reduced fracture toughness and does this reduce the life of the structure?

As stated earlier, life extension is a period of operation beyond a stage where the original standards where applicable. These standards will be based on previous experience and knowledge from a limited time period, which ideally should have been the limit for the original design life of the material. Therefore, materials that were selected during design as being sufficiently robust for the originally limited design life may no longer be robust for the period of life extension. In the final stages of the life of a material, this lack of material robustness is likely become more evident.

As part of assessing the material and its robustness for an existing structure, it is necessary to know:

1. The original material in the structure after construction or fabrication to establish the baseline for subsequent changes (what was there).
2. The present requirements for materials (production, construction and fabrication) to know if the original material meets the requirement to for example strength, ductility, defects, fabrication tolerances and fracture toughness that are required in present structural standards (what is required).
3. The actual condition of the materials in the structure after many years of service, the as-present condition (what is there).
4. The known condition of the structure based on the information available from for example inspection and testing (what we know about what is there).

The difference between 1 and 2 in the above list will be a result of technology changes and changes to knowledge and requirements, as discussed in Sect. 1.1. A general understanding of these changes can be useful in more effectively assessing the material. The difference between 1 and 3 will be the physical changes and the difference between 3 and 4 will be the material information changes. These issues are illustrated in Fig. 1.3, along with indications of the ageing changes mentioned in Fig. 1.1.

This book aims to provide information about each of these steps linked to the overall requirement to the assessment of materials in ageing structures. Hence, the book includes an overview of materials used during the history of offshore structures for energy production, mainly in the North Sea and Norwegian Sea. Further, the book includes information about the degradation mechanisms for these materials in an ocean environment (physical changes). In addition, the book describes the techniques and methods to establish an understanding of the current condition by inspection and testing. Furthermore, the book aims at helping the engineer assess an existing material in the context of current requirements (but will not attempt to present a detailed description of modern material requirements), in addition to describing the development to models and methods to evaluate materials (such as the changes in S–N curves for fatigue calculations). Finally, the book aims to give guidance for the assessment of materials in ageing structures associated with life extension, i.e. to determine whether the materials are adequate.

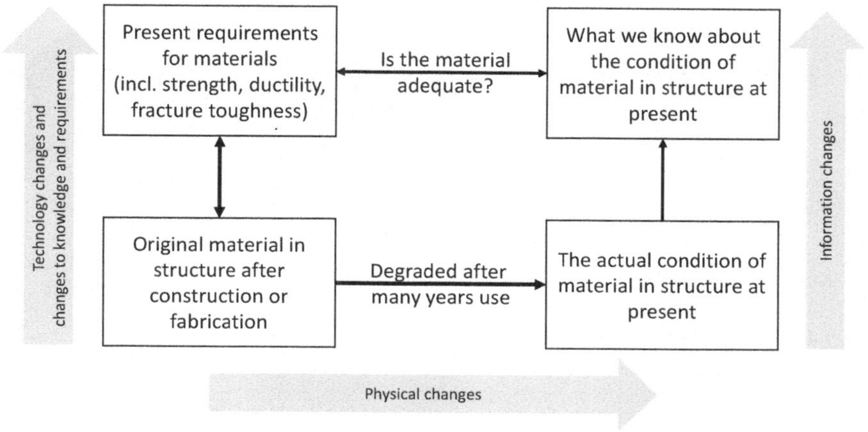

Fig. 1.3 The process for assessment of ageing materials

1.6 The Role of the Material Engineer in Assessment for Life Extension

In an assessment for life extension, the material engineer plays a key role in evaluations of most systems in an offshore asset. The material engineer will need to support the different disciplines with information on several different aspects related to the present state, future degradation and durability of the material. Particularly, the material engineer needs to establish conditions in which the original materials will behave differently from what is expected of materials according to present standards. For example, structural engineers will need information about any significant changes in the parent material, welds or other joining material that could influence the capacity of the structure and would need to be included in an analysis. These changes could include yield strength, Youngs modulus, ductility, elongation limits and fracture toughness of the material. In addition, the structural engineer will typically include the following changes to a structure in an assessment analysis:

- material loss effects from current and potential future corrosion, wear and tear,
- possible eccentricities introduced by unsymmetric deterioration,
- strength loss and remaining fatigue life of cracked components.

Further, the material engineer will need to review the results of inspections performed and history of degradation to establish an understanding of the present as-is status of the material. The as-is status of the material and history of degradation should form the basis for the predictions of degradation in the life extension period. The key issue is to determine the effect these degradations will have and make sure this information is used as input in the assessment for life extension.

An estimate of the expected future corrosion in the life extension phase needs to be specified by a materials specialist. Similarly, if a fatigue assessment beyond

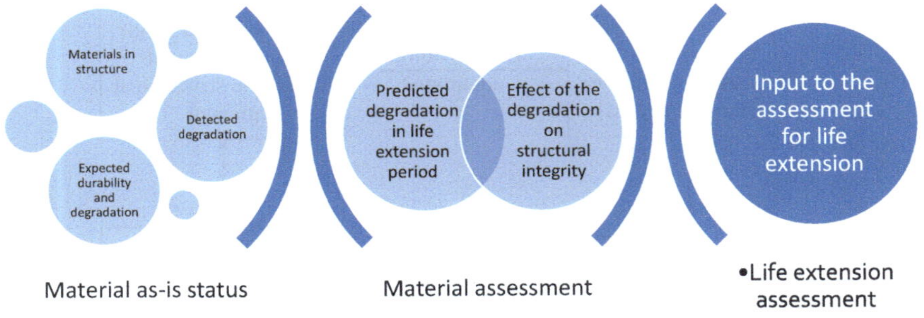

Fig. 1.4 Overview of the assessment of materials in a life extension process

SN-fatigue analyses or simple fracture mechanics is required, an expert in this field is often necessary (for example for complex fracture mechanics analyses and fracture assessments).

In summary, as shown in Fig. 1.4, the material engineer needs to clearly describe the materials in the structure, their limitations and any degradation. This would include to describe:

- the materials and joining methods selected at the design stage,
- the actual materials used in the structure following fabrication (preferably including the material certificates and fabrication procedures),
- any differences between materials as specified prior to fabrication and those received and used in fabrication,
- any known weaknesses in this type of material from this vintage,
- any anomalies and systematic problems with the material observed during operation,
- the general quality of the material.

1.7 Meaning of Materials in This Book

Within this book the word "material" or its plural "materials" has a wider meaning than its conventional usage. "Materials" covers not only the basic fabric of the structure (e.g. steel or concrete), but includes how that fabric has been formed, fabricated, joined, assembled, treated and protected. Steel materials therefore cover processes such as casting and forging, welding processes and procedures, weld consumables, bolting, heat treatment and protective coatings such as paint. Cementitious materials therefore cover reinforcement and pre-tensioning and grouting. In addition, the word "materials" also includes the local faults and defects and particular stress raising features that arise because of its forming and fabrication such as welding cracks and discontinuities in profile as at weld toes.

The performance of materials on an offshore structure in-service depends not only on the condition of its original fabric but also on the environments and physical conditions to which it is subjected. For offshore structures, the environments include mud, sand, immersion in seawater, splash zone or air. Their effect can depend on the degree to which the underlying material is covered by surface coatings such as paint or marine fouling or protected from corrosion by means of a cathodic electrical potential. The physical conditions associated with materials and upon which their properties and performance depend include their temperature and the level and distribution of stress within the material.

Associated with a material are its properties in relation to its environment and physical conditions. For steel, its properties include engineering characteristics such as yield strength, ultimate tensile strength, elongation properties, ductility and fracture toughness. The relevant properties for cementitious materials are different and include compressive and tensile strength and permeability.

Ultimately a material may be deemed to have failed when it ceases to perform the function to which it is required (e.g. load carrying). Materials can fail through a range of mechanisms and causes. For steels used in offshore structures, such failure mechanisms include excessive plasticity and exhaustion of ductility (rupture), brittle and ductile fracture, fatigue cracking, hydrogen embrittlement and hydrogen cracking and stress corrosion cracking. Concrete materials can fail due to compressive and tensile failure (crushing and cracking), corrosion of the reinforcement, loss of prestress of tendons as well as the loss of material through spalling.

This book therefore focuses on materials topics as defined above with regard to their origin, ageing and life extension. While the stress within a material is important to its behaviour, the source and distribution of that stress within the structure is a separate topic covered elsewhere. Fatigue is considered in this book only in consideration of the material properties that influence cracking (such as SN-curves, fracture toughness and crack-growth parameters). The fracture mechanics assessment of remaining life and the crack-growth analysis that needs to be performed are not specifically covered (see Ersdal et al. 2019 for more on such analyses).

1.8 About the Book

This book is specifically about the materials used in offshore structures for energy production (oil, gas and wind) and how they deteriorate with time. The book is primarily about steel and cementitious structural materials but includes a brief account about other structural materials used offshore. The book does not include materials used in other applications, such as pipelines, risers, drilling and subsea equipment, cables, topside processing equipment and piping. The book is primarily aimed at offshore structures for oil and gas production, but some parts are relevant also to other types of offshore energy production, such as wind.

The available literature on structural materials has been primarily aimed at the design of new structures, with almost no published information on the performance of materials in existing ageing structures. Although some books review degradation of individual offshore materials, none cover in depth the assessment of ageing materials for life extension or include the range of materials to be covered in this book. Very little up to date literature is available on the changes in the material properties with time and their effect on performance and strength in later life. This book is intended to help bridge these gaps.

This book will provide an important extension to the topic of materials ageing outlined in the previous book (Ersdal et al. 2019) entitled "Ageing and Life Extension of Offshore Structures". This included an introduction to materials degradation but was limited in the types of materials covered and the depth in discussion of the relevant properties. These two books will form an extensive overview on ageing of offshore structures and materials. In addition, another book by two of the authors (Sharp and Ersdal 2021) entitled "Underwater Inspection and Repair of offshore structures" is also relevant and covers more details on structural inspection and repair.

This book will consider the main materials used in current offshore structures together with materials used in joining such as welding and bolting. It will discuss the processes by which ageing occurs and its effects on the materials and the structure and how these can be mitigated. The development and performance of joining processes and materials used for offshore applications will also be covered, including thicker and higher strength steels. Potential future uses will also be briefly addressed.

Students need a proper understanding of materials and their ageing processes to allow them to contribute more effectively in ensuring continued use of existing structures. This book will be an essential reference for those who want a knowledge of the materials used in offshore oil & gas structures and wind turbines and how they degrade in an offshore environment. In addition, this book will aim to help practising material and structural engineers in their task of assessing the suitability of ageing materials for life extension and providing the necessary input for continued safe use of these ageing materials. This topic will remain important for many years as companies and governments increasingly want to gain more value from their investments and defer the cost of decommissioning.

1.9 Bibliographic Notes

Section 1.3 is partly based on Ersdal et al. (2019).

References

Drugli JM, Thaulow C, Ødegård J, Rogne T, Stokke R, Berge S, Berget J (2003) Pre-project: robust material selection in the offshore industry. In: SINTEF

Ersdal G, Sharp JV, Stacey A (2019) Ageing and life extension of offshore structures: the challenge of managing structural integrity. Wiley

Lemaitre J, Chaboche JL (1990) Mechanics of solid materials. Cambridge University Press

Sharp JV, Ersdal G (2021) Underwater inspection and repair for offshore structures. Wiley

Chapter 2
Offshore Structures

2.1 Introduction to Offshore Structures for Oil and Gas Production

Structures have been used in the sea for millennia, such as ships, lighthouses, mooring buoys, piers and jetties. More recently, offshore structures have been used for producing different forms of energy, such as oil, gas, wind, tidal and wave energy. The first offshore structures for energy production were wooden piers installed on the beach off California in 1896. These were followed by steel structures placed in water close to the shore. These initial offshore platforms were simple structures fixed to the seabed. Fixed structures were typically built around the wells for protection and to provide a platform where the wells could be maintained and serviced (Offshore 2004). Design and construction of offshore platforms followed practices from civil and marine engineering, as would have been used for onshore framed structures and ships.

In addition to fixed structures, the oil and gas industry also used floating units for exploration and drilling since its earliest days in the Gulf of Mexico and Middle East. The first floating production was from around the same time, using steel barges on which drilling rigs were installed. These barges were ballasted to rest on the bottom for drilling. When the wells were completed, the barges could be re-floated and towed away to new well sites.

The actual birth of the offshore technology (Claus et al. 1992) occurred in mid-1940s when two steel platforms were erected in the Gulf of Mexico. One of these platforms was built 18 miles off the Louisiana coast in 5.5 m of water in 1946 by Magnolia Petroleum (now ExxonMobil). The other platform was built in 1947 in 6.1 m of water, also 18 miles off the Louisiana coast, by Superior Oil.

In the UK, oil and gas production commenced in 1965 when BP's Sea Gem drilling rig found gas in the West Sole field. Sea Gem was a jack-up platform with ten legs which could be lowered to the seabed to fix the platform in position. Sea Gem was converted by BP to a jack-up platform from a barge in 1964. The barge was originally

© The Author(s), under exclusive license to Springer Nature Switzerland AG 2025

J. V. Sharp and J. Wintle, *Assessment of Materials in Ageing Offshore Structures*,
https://doi.org/10.1007/978-3-031-84489-8_2

built in 1952. Unfortunately, soon after entering service in the Southern North Sea in 1965, Sea Gem sank in mid-winter after a brittle fracture of structural components (see Sect. 3.1.3).

Semi-submersible floating platforms have been proven to be useful both as drilling rigs and production platforms in the North Sea. Semi-submersible platforms are kept on station by mooring lines or, in later days, by the use of dynamic positioning systems (DP). One noticeable accident involving a semi-submersible platform was the capsizing and sinking of the flotel Alexander L. Kielland in 1980. The root cause of this accident is normally linked to the fatigue failure of one of the main braces.

Concrete has also been used as a material for the construction of offshore structures. Since the 1970s several so-called gravity-based-structures (GBS) have been installed. The first offshore concrete structure was installed on the Ekofisk field in Norwegian waters in 1973, and the tallest was installed on the Troll field, at a depth of 300 m in Norwegian waters in 1995. In addition to the gravity based concrete structures, concrete has also been used as a material in a semi-submersible platform, a tension leg platform and a few storage tanks. Similar to the steel platforms, many of the earliest concrete structures are now ageing and beyond their original design life.

Developments in the UK sector led to over 300 platforms, including 20 floating installations and 10 concrete platforms. Similarly, in the Norwegian sector development has led to over 100 platforms being installed, of which 15 were concrete and several floating facilities of FPSO, semisubmersible, TLPs and 1 spar. In other parts of the North Sea Denmark's first oil production began in 1972 from the Dan Field. In Germany oil was found in the North Sea in 1981 in the Mittelplate field. Many of the structures used in these fields are now operating beyond their original design lives.

Today, the main areas of offshore oil and gas production are, in addition to the North Sea, include:

- the Gulf of Mexico;
- California (Ventura Basin—Santa Barbara Channel and Los Angeles Basin);
- Newfoundland and Nova Scotia;
- South America—Brazil (Campos and Santos Basins) and Venezuela;
- West Africa (Nigeria and Angola);
- Caspian Sea (Azerbaijan);
- Russia;
- Persian Gulf;
- India;
- South-east Asia;
- Western Australia.

The history of these offshore oil and gas producing regions extends to over 50 years and many structures are being used beyond their original design lives which, typically, range from 20 to 30 years. Improvements in the possible oil recovery from several fields have increased the interest for the continued use of these facilities well beyond their original design life. As noted earlier, another reason for extending the life of platforms is to provide a hub for the increasing use of subsea production "satellites".

The use of existing facilities beyond their original design life will in many cases be economically preferable and a replacement would also require significantly more CO_2 emissions, even if rather large modifications, refurbishments, repairs and inspections have to be performed to achieve life extension. A major concern in this regard is that safety requirements should not be compromised.

In the North Sea, approximately two thirds of the infrastructure can be considered to be 'ageing' and in the life extension phase. The proportion of ageing structures is expected to increase, and there are plans to extend some of these platforms up to another 20 years. This trend is also reflected in many other regions, notably in the Gulf of Mexico, West Africa, the Middle East, etc., and consequently the management of Ageing and Life Extension (ALE) is an essential element of the integrity management process.

2.2 Renewable Energy Production

Offshore wind production has evolved over the years, driven by advancements in technology, growing environmental awareness and the increasing demand for renewable energy sources. The concept of harnessing wind energy offshore was initially explored in the 1970s. Early experimental projects were set up in shallow waters near the coast to test the feasibility and viability of offshore wind turbines. The world's first offshore wind farm, the Vindeby Offshore Wind Farm, was commissioned off the coast of Denmark in 1991. It consisted of 11 turbines with a total capacity of 5 MW. This project marked a significant milestone in the development of offshore wind energy.

In the early 2000s, several European countries, especially those with favourable offshore wind conditions, began investing in larger-scale offshore wind projects. The United Kingdom, Denmark, the Netherlands and Germany emerged as pioneers in offshore wind development. The Burbo Bank Offshore Wind Farm in the United Kingdom, commissioned in 2008, marked the beginning of gigawatt-scale offshore wind projects. The wind farm has a capacity of 90 MW, showcasing the increasing scale of offshore installations. The 2010s witnessed a surge in offshore wind development, driven by technological advancements, cost reductions and increased policy support. Turbine sizes grew and floating wind turbine concepts were explored for deeper waters.

All current designs for offshore wind turbines include a hollow cylindrical steel tower (the mast) that extends from the rotor-nacelle assembly down to the connection point to the substructure (foundation). The connection between the tower and the substructure is normally called a "transition piece". For offshore structures this transition piece is typically placed 15–25 m above the waterline.

The most usual sub-structure types (foundation) for wind energy production are steel monopiles for water depths up to ~ 50 m and jackets in water depths up to

~ 100 m. Monopiles are large hollow cylinders that are driven into the seabed and extend above the sea level. A transition piece normally overlaps the monopile and is connected to the tower by either a grout or a bolted connection. Monopile designs are regarded as fast to design, manufacture and install, but can become impractical for deeper water.

Three or four legged jackets are used for supporting the tower in deeper water. As for offshore oil and gas structures, the jackets are typically piled to the seabed. However, the actual pile solution may differ according to the individual design.

Several concepts have been developed for floating substructures (foundations). These include tension leg platforms, semisubmersibles, barges and spars, similar to those used for offshore oil and gas production. The first floating offshore wind turbine, the Hywind project, was a spar type substructure. This was launched off the coast of Norway in 2009 as a test structure and began generating electricity in 2010. This marked a breakthrough for harnessing wind energy in deeper waters.

The world's first floating offshore wind farm, Hywind Scotland, using a developed version of the Hywind spar, started operating off the coast of Scotland in 2017. It consisted of five floating turbines with a total capacity of 30 MW. The newer Hywind Tampen is a further development of the spar concept with a concrete substructure. Hywind Tampen consists of 11 turbines with a total capacity of 88 MW.

In 2021 the Kincardine wind farm developed off the coast of Scotland in water depths of 60-80 m. This wind farm consisted of 5 semisubmersible units supporting five 9.5 MW wind turbines.

The 2020s have seen a surge in large-scale offshore wind projects worldwide. Countries such as the United States, China, South Korea, and Japan have entered the offshore wind market, contributing to its global expansion. Offshore wind capacity continues to increase globally, with larger and more efficient turbines being deployed. Many countries have set ambitious targets for offshore wind development as part of their strategies to transition to clean and renewable energy sources.

The CP protection system for both fixed and floating substructures (foundations) is normally external anodes similar to offshore oil and gas practice.

The future of offshore wind production is expected to involve continued growth, increased efficiency and ongoing innovation. Floating wind technology, advancements in energy storage and collaborative international efforts are likely to shape the trajectory of offshore wind in the years to come.

2.3 Types of Structures

2.3.1 General

Structures have been designed for use in the sea for ages, for ships, lighthouses, mooring buoys, piers, jetties and there are many more. More recently, offshore structures have been used for producing different forms of energy, such as oil, gas, wind, tidal and wave energy and the designs and materials used to construct these structures are the topic of this book.

The oil and gas industry recognised early that some reservoirs extended under the seabed and sought to extract the product from wells drilled offshore, which required offshore platforms. The first offshore structures for energy production were wooden piers installed on the beach off California in 1896 and in the Gulf of Mexico. These were followed by steel structures placed in water close to the shore. The initial offshore platforms were simple standalone platforms resting on the seabed. Construction of offshore platforms followed practices from civil and marine engineering, as would have been used for onshore framed structures and ships, based on plated or tubular assemblies connected by welded or bolted joints. The widespread use of fixed steel offshore structures for oil and gas production continued with structures of increasing size and complexity mounted on piles in the seabed in ever deeper water.

At present, it is believed that there are more than 20,000 offshore platforms of various kinds installed around the world. The types of platforms that are most used are:

- Fixed platforms:

  Fixed steel jacket structures (mostly pile supported or suction anchor supported)
  Concrete gravity based structures (GBS)
  Jack-ups.

- Bottom supported platforms:

  Tension leg platforms (mostly in steel, but one in concrete exists)
  Guyed towers, compliant towers and articulated towers
  Floating platforms:
  Semisubmersible platforms (mostly in steel, but one in concrete exists)
  Ship shaped platforms and barge shaped platforms (mostly in steel, but a few in concrete have been made)
  Spar platforms.

The types of platforms primarily used in the offshore industry are fixed and floating platforms, which are further described in this Chapter. In addition, some examples of special types of bottom supported platforms exist, but these are not discussed in detail in this book. Examples of offshore structure are shown in Fig. 2.1.

2.3.2 Fixed Platforms

Fixed platforms normally include:

- Steel jacket structures (mostly pile supported or suction anchor supported)
- Concrete gravity based structures
- Jack-ups.

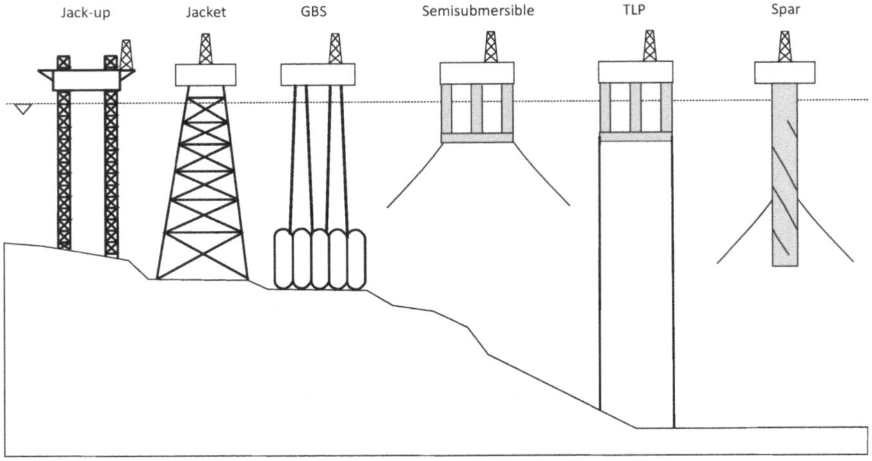

Fig. 2.1 Examples of types of offshore structures

The most commonly used platform type is the steel jacket structure (see Fig. 2.2). Jacket structures are built on steel legs and piled to the seabed, supporting a deck with space for drilling rigs, production facilities and crew quarters. Steel jackets are usually made of tubular steel members. Jacket structures have been used in up to 400 m of water (Bullwinkle Jacket in Gulf of Mexico) but are normally more common in water depths up to 200 m as the weight of jackets in deeper water is becoming very heavy (see Fig. 2.2).

The advantages of offshore jacket platforms include that they: (i) support large deck loads; (ii) allow for being constructed in sections and transported; (iii) are suitable for large field and long-term production (supports a large number of wells); (iv) maintain good stability when piled as foundation; and (v) are not significantly influenced by seafloor scour. Few disadvantages are as follows: (i) cost increases with increase in water depth; (ii) not easily reusable; and (iii) steel structural members are subjected to corrosion and fatigue, causing material degradation.

Steel structures are subject to ageing processes such as fatigue and corrosion and hence life extension is a key issue with regard to the structure itself.

Concrete gravity-based structures (GBS), as shown in (Fig. 2.1), are also a common type of offshore structures and have been successfully used in up to 300 m of water. These structures typically consist of a base with concrete cells for oil storage and typically one to four shafts (also called caissons) that rise to above the sea surface for support of the topside. GBS structures to some degree use the weight of the structure to resist the wave loading on the structure. However, in practice, most offshore gravity-based structures are equipped with a skirt at the bottom that is penetrated into the soil and provides additional support for external loads. GBS platforms are capable of supporting large topsides, both during operation and tow-out, which minimizes the hookup work during installation. In addition, a GBS provides large

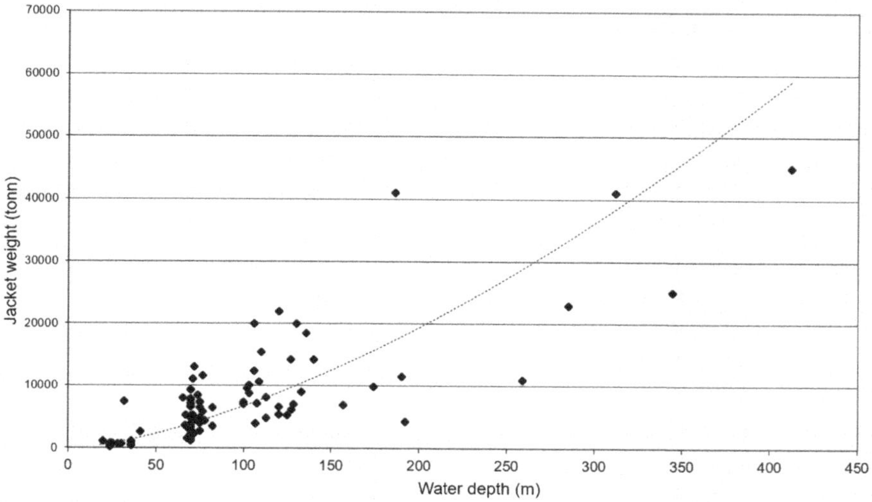

Fig. 2.2 Jacket weight versus water depth for primarily jackets in northern European waters, but also including a few of the most known jackets in GoM (based on publicly available data)

storage spaces for hydrocarbons. Advantages include: (i) constructed and assembled nearshore; (ii) towed to the offshore site as a complete unit; and (iii) use of relatively traditional methods and labour for construction.

Jack-ups (Fig. 2.1) are self-elevating units with a buoyant hull and several legs which, when on location, can be lowered to the seabed and raise the deck above the level of the sea thus creating a more stable facility for drilling and/or production. With its legs in a jacked-up position, it also has the ability to float using its buoyant hull from one field to another. The first jack-up was deployed in mid-1950s. Jack-ups have played an important part in the discovery of oil and gas, but also to a limited extent in the production of oil and gas. At present more than 500 jack-ups are in operation world-wide. Many of these jack-ups have been built with high strength steel in the legs, which may have special ageing issues, see Sect. 3.3.3.2.

Jack-up rigs have been primarily used for exploratory drilling but also for well intervention, production and accommodation. The first offshore exploration in the UK sector was with the Sea Gem jack-up platform with ten legs which could be lowered to the seabed to fix the platform temporarily in position. Sea Gem was an American barge, originally built in 1952 and converted by BP to a jack-up in 1964. Unfortunately, soon after entering service in the Southern North Sea in 1965, Sea Gem sank in mid-winter after a brittle fracture of structural components (see Sect. 3.1.3).

2.3.3 Floating Structures

In addition to fixed structures, the oil and gas industry also has used floating structures since its earliest days in the Gulf of Mexico and Middle East. Floating platforms are in general dependent on their watertight integrity and station keeping in addition to their structural integrity to function. The most used types of floating structures in the offshore industry include:

- Semi-submersible platforms (mostly in steel, but one in concrete exists)
- Tension leg platforms (mostly in steel, but one in concrete exists)
- Ship shaped platforms and barge shaped platforms (mostly in steel, but a few in concrete have been made)
- Spar platforms.

At the time of writing, around 300 floating production platforms are used for production, most of these semisubmersible platforms and ship-shaped platforms (D'souza et al. 2020).

The semisubmersible offshore platforms are the most widely used floating structural type. Semi-submersible units (Fig. 2.1) have hulls, together with columns and pontoons, with sufficient buoyancy to enable the structure to float. Semi-submersible platforms change draft by means of ballasting and de-ballasting (changing the water level in seawater tanks). They are normally anchored to the seabed by mooring systems, usually a combination of chain or wire rope. However, semisubmersibles also can be kept on station by dynamic positioning (DP). This is commonly used for semisubmersibles as drilling units and flotels in deep water. The hull supports a deck on which various facilities for drilling and/or production can be installed.

Semisubmersible offshore platforms are commonly used as mobile offshore drilling units and production units, but are also used as crane vessels, offshore support vessels and as flotels. The earliest in the UK North Sea was the Argyll Floating Production Unit (FPU) in 1975. The semi-submersible drilling rig Sea Quest was one of the first developed for European waters. Semi-submersible drilling rigs were also developed for use in Norwegian waters. A semisubmersible flotel, Alexander L. Kielland, sank in 1980 after a fatigue failure of one of the main braces. The industry learnt lessons from these early experiences and continues to use semi-submersible installations. Many semisubmersible platforms are now in the ageing phase and especially the earlier versions do experience significant cracking (Ersdal et al. 2019).

A tension leg platform (TLP) is similar to the semi-submersible platform but is vertically moored by tension legs which decreases the vertical motion significantly.

A Spar platform is also to some extent similar to a semi-submersible, but consists of a single moored large-diameter vertical cylinder that is supporting the topside. The cylinder is ballasted in the bottom (often by solid and heavy material) to provide stability.

Ship shaped structures and barge shaped structures include floating storage units (FSU), floating storage and offloading units (FSO) and floating production, storage and offloading units (FPSO). Ship shaped structures have the benefit of being able to store oil in the hull.

As oil companies began exploring smaller reservoirs and reservoirs in deeper waters, ship shaped structures were seen as a viable solution. This was particularly the case considering that a fixed installation requires an oil storage facility or access to pipelines, which was not feasible or economic in many cases. The FSO Ifrika in the early 1970s was the first ship shaped platform to be permanently used for production (Paik and Thayamballi 2007).

FPSOs are ship shaped vessels moored in position from a turret on bearings around which the vessel can rotate. Oil is transported from the well-heads by riser pipes that connect to the FPSO through the turret and stored in the FPSO's cargo tanks and at intervals off-loaded onto conventional oil tankers. One of the benefits of an FPSO is that when the production at the field is complete, the risers can be disconnected and the FPSO can be unmoored and taken to a new production site. However, since FPSO's remain on a fixed location for long periods, design against fatigue failure due to a predominant wave direction and the challenges of in-situ inspection have proved to be more critical than in in oil tankers. The integrity of the high strength steel mooring systems was also a key consideration. At the time of writing, approximately 35% of the FPSOs have been purpose built and approximately 65% are converted from conventional oil tankers that had already been in service. The idea of converting conventional oil tankers into FPSOs was to bring the FPSO's into service quickly and economically. Many of the ship-shaped structures are now ageing and, like many ships, they experience various degradation mechanisms.

References

Claus G, Lehmann E, Ostergaard C (1992) Offshore structures—Volume I—conceptual design and hydrodynamics. Springer Verlag

D'souza R, Basu S, Bhaumik T, Sharples M (2020) History of semisubmersible platforms provides guidance for future deepwater projects. Offshore Magazine, 2020. History of semisubmersible platforms provides guidance for future deepwater projects. Offshore (offshore-mag.com)

Ersdal G, Sharp JV, Stacey A (2019) Ageing and life extension of offshore structures: the challenge of managing structural integrity. Wiley, Chichester, UK

Offshore Magazine (2004) https://www.offshore-mag.com/articles/print/volume-64/issue-1/news/special-report/special-anniversary-the-history-of-offshore-developing-the-ep-infrastructure.html

Paik JK, Thayamballi AK (2007) Ship-shaped offshore installations—design, building and operation. Cambridge University Press, Cambridge, UK

Chapter 3
Steel Material Types in Offshore Structures

3.1 Development of Steels for Offshore Structures

3.1.1 Introduction

As outlined in Chap. 2, offshore structures fabricated from steel have been used for producing different forms of energy, such as oil, gas, wind, tidal and wave energy. Construction of the earliest steel offshore structures followed practices from civil and marine engineering based on plated or tubular assemblies connected by welded or bolted joints. For oil and gas production, fixed steel offshore structure for were the main type used. The extensive use of fixed steel offshore structures continued with structures of increasing size and complexity in ever deeper water.

In addition to fixed structures, the oil and gas industry also used floating installations for exploration and drilling since its earliest days in the Gulf of Mexico and Middle East, such as semisubmersible platforms. When oil companies began exploring smaller reservoirs and reservoirs in deeper waters the use of ship-shaped vessels for production became a favourable option. Both semisubmersible and ship-shaped platforms are kept in position using heavy moorings.

As the industry evolved other types of manned and unmanned fixed and mobile installation were developed, such as offloading buoys and accommodation vessels. There was more focus on structural components and appurtenances such as caissons and guides that were essential to safety. With the advent of offshore wind energy since 2000, the offshore wind sector has adopted practices from the oil and gas industry for the design and construction of grouted monopile and tower structures, although there were some important differences in the loading regime and occupancy of wind turbine structures.

© The Author(s), under exclusive license to Springer Nature Switzerland AG 2025

J. V. Sharp and J. Wintle, *Assessment of Materials in Ageing Offshore Structures*,
https://doi.org/10.1007/978-3-031-84489-8_3

3.1.2 Steels in Early UK Platforms (Pre-forties in 1973)[1]

Following the discovery of gas in 1966 in the West Sole field in the UK Southern North Sea many small steel platforms were constructed to extract this gas based on the designs and materials of platforms operating in the Gulf of Mexico. The water depths in these early fields were shallow, typically 20–25 m. Between 1966 and 1974 approximately 30 platforms were installed in the West Sole, Hewitt, Leman and Viking fields for the production of gas, see Fig. 3.1. All of these gas platforms were small in weight, typically between 400 and 1000 tonnes and able to be constructed with limited thicknesses of the steel (e.g. 12.5 mm).

The main criteria for the selection of structural steels for the early offshore installations were the provision of sufficient strength and workability and especially weldability for a relatively reasonable cost. These early structures were typically fabricated from ASTM specified semi-killed balanced carbon manganese steels of relatively low strength and section thicknesses, which had proven to be sufficiently ductile in the calm temperate waters of the Gulf of Mexico and Middle East.

Tolerances of material composition and properties for steels at the time were broad and allowed a wide range of material variations to be accepted for use. The early structures may have been designed with a large margin on design stresses, based on the knowledge at the time. This margin included allowances for variations in materials' performance and construction quality. This may have resulted in many structures using larger quantities of material, often with a significant weight penalty compared to more modern designs.

Fatigue and fracture performance had not previously been a major concern for structures in the Gulf of Mexico and the Persian Gulf where the environmental conditions are relatively warm and benign, as discussed in Chap. 12 (West Sole WE case study). Since the susceptibility to brittle fracture is a function of temperature, stress, thickness and crack size, brittle fracture of these early structures was not considered to be a significant threat. This was because of an assumed absence of welding defects from properly qualified welding procedures that were subject to non-destructive testing (Cotton 1979). The structures were designed to operate

---

[1] The development of steels and welding for deep water northern North Sea structures was reviewed by Professor Harry Cotton in a landmark paper to accompany his presentation of the fifteenth John Player Lecture to the Institution of Mechanical Engineers in 1979. Professor Cotton was Chief Materials Engineer at British Petroleum during the period of construction of the Forties structures in the early 1970s. His recognition of the requirements for steels and welding needed for structural safety of these massive deep-water structures and his demands on the steel industry and welding sector to meet these requirements were instrumental in advancing the technology for the next generation of structures. In his paper Professor Cotton (1979) reviews the development of steel and welding up to the time of construction of the first Forties structures and during the 1970s. Steel making had progressively improved throughout the nineteenth and first part of the twentieth centuries to the point where steel could be made cheaply and with tensile properties sufficiently high and consistent so that it could be used with confidence for major bridges, boilers and pressure vessels. Its tensile strength properties could not, however, be fully utilised until the mid-1950s when electric arc and oxy acetylene manual welding had developed to a point where welding could supplant rivetted construction.

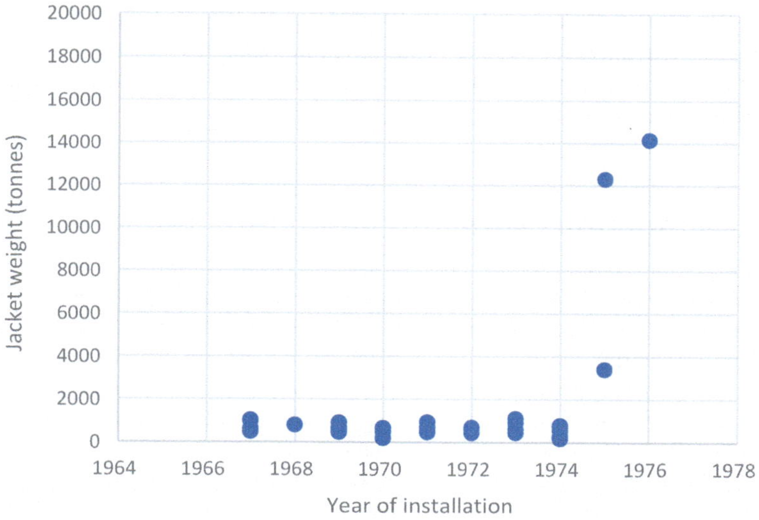

Fig. 3.1 Graph of jacket weight with year of installation of UK offshore structures showing the increase in weight of the Forties platforms installed in 1975 and 1976 (Oilfield Publications Ltd. 1985)

at relatively low stress and the thinner sections limited the size of defects. Consequently, there was no general requirement for impact testing of these materials to demonstrate sufficient toughness and resistance to ductile tearing. The need to give fatigue and fracture performance more attention, especially given that the minimum design temperature for the North Sea was − 10 °C and of a more hostile environment, led to developments in welding and joint design to minimise stress concentrations, avoid crack-like defects and ensure weld material of sufficient toughness.

The development of fixed installations in harsher conditions of the Southern North Sea (from the mid 1960s) and in the colder Northern North Sea (from the early 1970s onwards) drove the oil and gas industry to place greater focus on the composition and properties of structural steels than had hitherto been the case. Particularly, the development of the Forties field from 1975 in much deepwater waters (100–120 m) led to the requirement for larger platforms and thicker section steels. The Forties platforms had weights in the range 12,000 to 14,000 tonnes, a factor of ~ 18 times the average Southern gas platform weight (Fig. 3.1). As a result, there was a requirement for more consistent composition and material properties throughout the plate wall thickness than had been generally available.

Even though these steels were more expensive to produce, the greater emphasis on achieving and maintaining high levels of structural strength in extreme conditions, typical of the Northern North Sea, were major requirements that made the investment worthwhile. The availability and use of thicker steels that were weldable (by controlling their carbon equivalent value) and with assured through thickness properties were reflected in developments in the design codes at that time. In addition, the severe cyclic environmental loading from wind and sea water at lower temperatures led to fatigue cracking and fracture of the structure being recognised as threats that needed to be addressed through design and the selection of materials and welds with appropriate fracture resistant properties.

3.1.3 The Sea Gem Collapse

The Sea Gem was a converted barge originating from the Gulf of Mexico before becoming the UK's first jack-up installation to explore the Southern North Sea. However, the materials and construction practices of Sea Gem which had been shown to be successful in sub-tropical climates were to prove to be demonstrably insufficient for safe operation in the 0 to 10 °C range. The installation collapsed on 27th December 1965 and of the 32 men on board 13 lost their lives.

The report of the UK Court of Inquiry into the Sea Gem's collapse (Ministry of Power 1967) found striking similarities with the collapse of the King's Bridge in Melbourne in 1962 where fatigue and weld under-bead hydrogen cracks had combined to initiate catastrophic brittle fracture. The fracture initiation site in Sea Gem could be attributed to the sharp radius at the spade end of tie-bars associated with the jacking mechanism located at each leg. The tie-bars had been flame-cut and weld repairs had been carried out on the radius section without post weld heat treatment. Small surface cracks in the gas cut edges of the tie bars had extended under the action of cyclic loads into the embrittled heat affected zones (HAZ) of welds. Furthermore, tests conducted subsequently on the tie bars showed that the material had poor Charpy impact energy. Some of girth welds of the tubular legs of the jack-up rig also fractured in a brittle manner. These had a wall thickness of 75 mm and were in the as-welded condition.

The air temperature at the time was 3 °C which was only 7 °C below those of structures in more sub-tropical climates, yet this was sufficient to cause a significant change in the material behaviour and structural integrity performance. Consequently, the Inquiry recommended that in future, a minimum value Charpy toughness of 35 ft lb (47.5 J) at 0 °C should be specified for all safety critical steel in offshore structures operating in the UK sector of the North Sea. In addition, the Inquiry resolved that plates more than 40 mm thick should not be used unless the welds could be post weld heat treated (PWHT) in the stress relieving temperature range of 580 to 620 °C, a decision that proved unpopular giving rise to adverse comment (Cotton 1979).

The Court of Inquiry also recommended that there should be more general statutory requirements for the UK sector covering the safety of offshore installations. Such requirements were subsequently introduced into the Offshore Installations (Construction and Survey) Regulations SI 289:1972 (Department of Energy 1972). These introduced the requirement that offshore installations in UK waters needed to have a 'Certificate of Fitness' issued by a relevant 'Certification Authority'. To support the legislation and provide more definition of what fitness might entail, the Department of Energy issued Guidance Notes as a basis for the Certification of Offshore Structures (Department of Energy 1974).

The limitation of plate thickness did however necessitate extra steelwork to provide adequate strength and stiffness causing severe restraint on the closure welds and resulting in high weld shrinkage stresses. The possibility of weld metal fabrication cracks leading to fracture was controlled through the specification of low hydrogen basic-coated electrodes for all welded joints, full penetration welds between the legs and main braces, combined with 100% non-destructive testing with severe restrictions on the size of flaws that could be allowed to remain in the as-built structure (Cotton 1979).

3.1.4 The UK DoE Guidance Notes

Following the Sea Gem accident, the first edition of the Department of Energy Guidance Notes (Department of Energy 1974) proposed basic requirements for steel. This included a requirement that 38 mm should be the limiting thickness for highly stressed joints before PHWT was required. This was subsequently increased in the second edition of the Guidance Notes (Department of Energy 1977) to 50 mm for structures in the atmospheric and splash zone and 63 mm for permanently submerged parts following representations from the industry (Cotton 1979). The technical justification for this increase was obscure, i.e. no real justification was made (Pisarski and Harrison 1986).

A later experimental research programme showed that that the revised thickness limits of 50 and 63 mm for welds without PWHT could not be justified in thick section welds representative of those used in offshore platforms (Harrison 1981). This research programme was carried out for the European Iron and Steel Community and the UK Department of Energy under the UK Offshore Steels Research Programme (UKOSRP I), reported later by Department of Energy (1988). This revised thickness limit was mainly because the weld metals had insufficient fracture toughness at the minimum design temperature of – 10 °C in the as-welded condition (Pisarski and Harrison 1978; Bokalrud and Karlsen 1978). Consequently, the third edition of the Guidance Notes (Department of Energy 1986) recommended reduced for PWHT limits on the thickness of 40 mm for highly stresses welded joints and 50 mm for welds in members that were remote from intersections. Exceptions to these requirements for exemption from PWHT could be obtained by demonstrating avoidance of brittle fracture using fracture mechanics testing and codified defect assessment procedure such PD 6493 (British Standards 1980) taking high residual stresses and plane strain conditions into account.

According to Cotton (1979), the industry at that time considered that barges converted to jack-up structures were generally not very suitable for use in the stormy waters of the North Sea because of the requirement for very calm weather during jack-up and jack-down operations. This led to an increased use of semi-submersible drilling rigs, such as the Sea Quest which was the semisubmersible drilling rig that discovered the UK's first North Sea oil field in 1969 and later the first giant oil field named Forties in 1970.

3.1.5 The Development of 50D Steels

By the late 1960s, also the issues of lamellar tearing had become evident in welding of thick section plates where delamination could occur in the heat affected zone parallel to the surface being welded. The understanding gained from research into lamellar tearing (Farrar and Dolby 1972) found that lamellar tearing was due to the presence of undesirable manganese sulphide inclusions in the steel.

This led to the development and specification of a new type of Z quality (50D) steel with a modified material composition and having enhanced through thickness tensile properties in terms of a minimum reduction of area in a tensile test specimen. This type of steel had higher strength than had been previously used and was the basis for most offshore tubular structures from the mid-1970s onwards This steel was first used in quantities of thousands of tons in all critical areas of the Forties platforms in the Northern North Sea (including nodal joints) and no cases of lamellar tearing occurred (Cotton 1979). The 50 in 50D indicated a minimum yield strength of 50,000 psi (~ 355 MPa) and the D designation indicated that the material had been impact tested to assure the required ductility. This steel and later medium strength steels are described later in this chapter.

BS 4360 Grade 50D (British Standards 1969, 1986, 1990) is a weldable structural steel and was chosen by many North Sea operators as basis for their steel specification because of its strength, weldability and availability. Nevertheless, many operators specified supplementary property requirements to the basic specification. These included, amongst other things, control of carbon equivalent, sulphur content, more stringent Charpy toughness and through thickness ductility requirements. Consequently, operators often had different specifications, sometimes referred to as "50D modified" but still based on BS4360 50D. This arrangement worked to the advantage of steel makers who could charge more for the supplementary requirements. It took many years before many of the supplementary requirements were incorporated into national steel standards.

3.1.6 Later Developments

By the late 1980s steel was being produced to a fine grain size, by being aluminium treated with fine aluminium nitrides. Another factor was the control of residual elements that could cause embrittlement by segregating to grain boundaries causing an increase in intergranular fractures and a reduction in fracture resistance. Residual elements also increased the likelihood of stress corrosion cracking, hydrogen embrittlement and corrosion fatigue. The production of steels to meet demands of the offshore industry had become a complex process based on a close understanding of the material properties.

The need for assured through thickness properties led to requirements for through thickness tensile, Charpy and fracture toughness testing in different orientations. In parallel with welding, test methods and procedures for determining the fracture toughness of parent steels, weld metal and heat affected zone materials were developed. These requirements were in terms of minimum values of Charpy impact energy and, for highly stressed welds in primary structure, minimum values for Crack-tip Opening Displacement. It is worth noting that in the early 1970s, crack tip opening displacement (CTOD) was simply referred to as Crack Opening Displacement (COD).

In parallel, procedures for assessing fabrication and in-service defects in welds and other metallic structures evolved from the simple methods originally based on the TWI COD Design Curve (Burdekin and Dawes 1971; TWI 1978).

These procedures were subsequently codified in the BSI document PD 6493 (British Standards 1980) and subsequently developed for application to offshore structures (Harrison et al. 1979) and validated using wide plate tests (Kamath 1981). Later, these have been developed and incorporated in the modern standard BS 7910 (British Standards 2019). The evolution of fracture toughness testing and defect assessment procedures is treated in more detail in Sect. 5.5.

In addition to these medium strength steels, some designers specified additional requirements for materials to reflect the more exacting design and service conditions experienced offshore. Higher strength steels with yield strengths up to 650 MPa were required for specific purposes such as for the racks of jack up legs and mooring chains, which are designed to carry high loads. The higher yield strength and operating stress of these steels made them vulnerable to hydrogen cracking under certain conditions. Table 3.1 provides an overview of steel strengths traditionally used in offshore structures and their application.

Table 3.1 Use of medium, high and higher strength steels in offshore structures (HSE 1997)

Strength (SMYS) MPa	Steel grade	Process route	Application area
350	X52	Normalised TMCP	Structures in general
450	X65	Q&T TMCP	Structures in general
550	X80	Q&T TMCP	Structures and mooring
550		Q&T	Jack ups and mooring
650	X100	Q&T	Jack ups and mooring
690		Q&T	Jack ups and mooring
750		Q&T	Jack ups and mooring
850		Q&T	Jack ups and mooring

Higher and high strength steels and their challenges will be described in Sects. 3.4 and 3.5.

3.2 Microstructure, Composition and Manufacturing of Structural Steels

3.2.1 Microstructure and Composition

Generally, the strength of a steel is determined by its microstructure which results from its chemical composition, alloying elements, thermal history (e.g. cooling rates, heat treatment and further post-treatments) and deformation processes it undergoes during its production. Depending on how the steel is produced, the microstructure can consist of different constituents corresponding to different atomic configurations, such as face and body centred cubic, forming grains of different sizes and shapes with boundaries between them. Typical steel microstructures are of ferrite, cementite, austenite, pearlite, bainite and martensite and combinations of these. Details about these can be found in standard textbooks such as Callister and Rethwisch (2018).

Weld metal microstructures are determined primarily by the chemical composition, the amount of non-metallic inclusions present in the microstructure which affect phase nucleation, and by the cooling rate. Alloy design aims to maximise the amount of acicular ferrite present and to minimise the effects of undesirable microstructures such as coarse grain size, grain boundary ferrite and coarse martensite/austenite/carbide constituents (MAC).

The understanding of properties and how materials actually fail require some understanding of materials from the atomic- to meso-scale, see Fig. 3.2.

The atoms of a metal are packed together in a regular structure called the crystalline lattice, which is a repeating number of identical unit cells. There are several different ways the atoms of a metal can pack together, depending on the different types of unit cell. The normal unit cells of metals include face-centred cubic (fcc), body-centred cubic (bcc) as shown in Fig. 3.2 and hexagonal-close packed (hcp).

The strength of a material is, theoretically, obtained by the atomic cohesion bounds in the material, determined by the lattice structure of the atoms. The cohesion strength of iron can be calculated, using Griffith's formula to break the atomic bonds, to 30,000 MPa (Gordon 1991). This far exceeds the strength we expect steel and particularly iron to have because in practice the lattice structure will contain numerous microstructural defects of different types (for example, vacancies, substitutional, interstitial) and which reduce the strength of the material from its theoretical maximum. As we know, stress is increased by cracks, and this also include crack-like defects at the atomic level. The "crack" in the atomic structure due to lack of bonds in a few atoms in a line in the lattice, say 2 microns long (c) and of one Ångström unit tip radius (R), will according to Inglis's stress concentration factor (Inglis 1913) give a $SCF = \left(1 + 2\sqrt{c/R}\right) = 200$ (Gordon 1991), which would reduce the theoretical

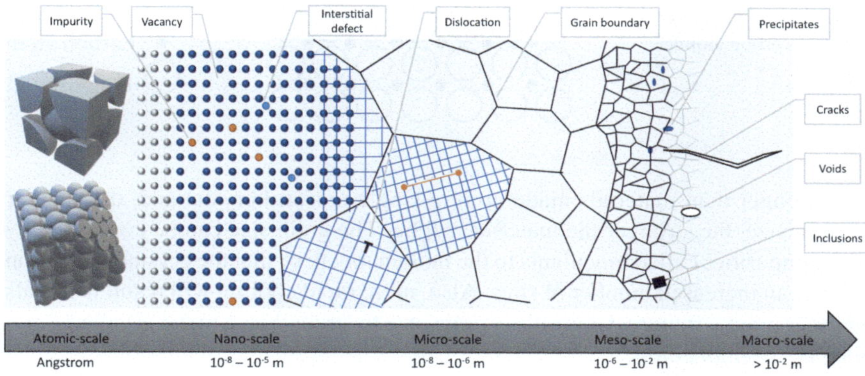

Fig. 3.2 Types of defects in a material. –Point defects (0D): Vacancy (missing atoms), Impurity (foreign atoms), Frenkel defects, Schottky defects; –Line defects (1D): Dislocations (Edge, mixed, screw), Disclinations (random, structured), –Surface/planar defects (2D): Grain boundary, twin boundary, stacking faults, –Volume defects (3D): Twins, precipitate, voids and cracks, porosity, inclusions

strength based on cohesion to a value of 150 MPa, which is roughly the strength of wrought iron.

Lattices also contain 1-D line-like defects (dislocations). A dislocation occurs where a number of atoms are offset from the usual position in the lattice (edge, screw and mixed dislocations). These dislocated lattices are a stable configuration, but when sufficient stress is applied to the lattice, the atomic bonds break free and re-forms. Hence, the process is irreversible, and the dislocation will not return to its original position when the applied stress is removed. This motion of dislocations at the atomic level is the underlying mechanism behind *plastic deformation* in metals. In contrast, *elastic deformation* is caused by the stretching of atomic bonds and is completely reversed when the stress is removed.

Materials that contain many dislocations have improved strength, as dislocations can get tangled and prevent each other from moving through the lattice. The motion of dislocations through the lattice is also affected by how the atoms pack together and the easiest way for dislocations to move is along planes where the atoms are close to each other, as those bonds most easily break and re-form. This corresponds to different planes depending on the structure of the unit cell.

The making of metals includes heating them to their liquid phase and cooling them down in the wanted shape. As the metal cools down, the atoms group together and a lattice structure begins to form in several different locations at the same time. Each of these lattices will have their own random orientation. When the metal is completely solidified, the result will be multiple grains with lattices that are oriented in different directions. The grains are separated by grain boundaries, a 2-D surface or planar defect.

Since each grain has specific planes along which slip will occur, the presence of grains impedes the motion of dislocations. Hence, polycrystalline materials tend

Fig. 3.3 Metal lattice with interstitial carbon atoms

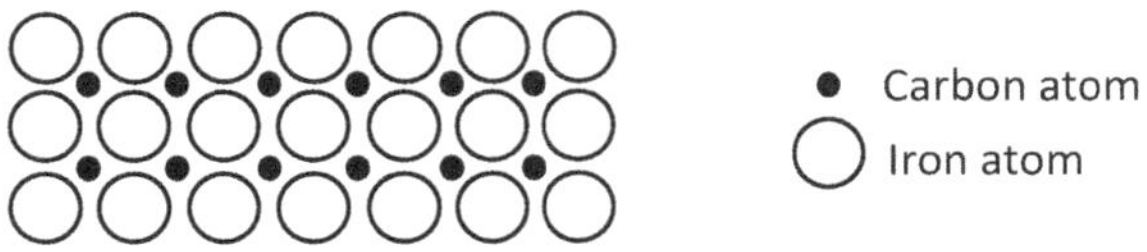

to be stronger than materials made up of a single uniform crystal and, the smaller the grain size, the stronger the material will be. The grain size can be controlled by adding impurities called inoculants to the molten metal, so that the crystal nucleation occurs at an increased number of sites. Also, more rapid cooling will result in metals with finer grain structure. Controlling grain size to strengthen a metal is called *grain boundary strengthening*.

The metal can also be strengthened by plastically deformation (cold rolling or forging), called *work hardening*. This increases the number of dislocations, and so increases the strength of the material. However, such treatment significantly reduces the ductility of the material.

Metals can in their molten form be mixed with small quantities of other metallic and non-metallic elements to improve the material properties. Such metals are called alloys. Alloys are typically split into ferrous and non-ferrous categories, depending on whether the base metal of the alloy is iron. For example, pure iron has only limited engineering usefulness, but the addition of a small quantity of carbon lifts iron into the position of a unique structural material known as steel. The added carbon atoms between the atoms of iron enable the material to flow and yield allowing plastic strain without rupturing of the material (Fig. 3.3).

Steels are separated into a few different categories, depending on the amount of carbon and other alloying elements. Low-carbon (mild) steel contains up to 0.25% carbon and is not particularly high strength but is ductile and relatively low-cost. Medium-carbon steel contains between 0.25 and 0.6% carbon and high-carbon steel contains between 0.6 and 2% carbon. The more carbon the stronger the steel. In addition, they can be more easily strengthened using different heat treatment methods like quenching and tempering. Cast iron contains between 2 and 4% carbon. The additional carbon lowers the melting point of the alloy, making it good for casting. The increase of carbon in general makes the steel more brittle, and this is particularly true for cast iron which tends to be brittle. Strengthening of the material by alloying is called *solid solution strengthening*.

The alloying elements will not always be able to fully dissolve into the lattice of the base metal. In some cases, these alloying elements may separate out and produce a distinct homogeneous phase within the microstructure of a different composition, a 3-D defect called precipitates. There are several different ways the second phase particles can be incorporated into the material and they can significantly affect the properties of the material. The boundaries between phases impede the motion of dislocations and as a result increases the strength of the material. Heat treatment can be used to intentionally produce second phase particles with the goal of strengthening the material. Such treatment is called *precipitation hardening*.

All these processes influence the properties of materials important in structures, including strength (yield strength and ultimate tensile strength), elasticity, ductility, hardness, density and fracture toughness. In addition, the processes influence properties such as weldability and machinability, important for the manufacturing and fabrication of structures.

Precipitates (2nd phase material) has already been mentioned as a 3D volume defect. These are different material (e.g. carbide grain in steel) in the metal lattice and can strengthen the material. Other 3D volume defect can form during the manufacturing process and are harmful to the material. These includes:

- Pores/voids and porosity can be a result of gas entrapment during melting and pouring,
- Inclusions of foreign elements like slag, oxide particles or sand entrapment,
- Cracks due to uneven heating or cooling, thermal mismatch, constrained expansion or contraction which all lead to stress development that may lead to cracks.

All these defects will lead to stress concentrations and crack initiators with the potential for further damage to the material. As a result, large portions of this chapter on steel and welding will explain the possible defects in different types of steel from different time periods and what these do to the properties of the steel.

Joining processes such as welding will add to the complexity of the material by additional heat treatment as part of the welding process and the possibility of added flaws (mostly 3D volume defects) as part of the welding process. Section 4.7 will further provide information on the welding process and Chap. 5 will describe typical defects in welded steel material after fabrication.

3.2.2 Manufacturing Routes for Steels Used in Offshore Structures

Weldable structural steels can be classified as as-rolled, normalised, control rolled and quenched and tempered with progressively higher strength. The minimum yield strength values specified in standards at that time were in the range of 200 MPa for ordinary as-rolled material to 690 MPa for quenched and tempered (UEG 1985). Structural steels for offshore structures were typically supplied to UK, other European, American, or other international standards for steel products for general application. These standards have been updated over time and assessors of ageing structures should refer to the edition relevant at the time of the fabrication of the structure.

Different manufacturing routes are used to achieve the required properties of steels used in offshore structures. Heavy ingots are transformed into forged plates through a unique combination of hot forging and hot rolling to ensure consistency of properties throughout the thickness, dimensional tolerances, and good surface condition. The plates are then heat treated to obtain steels of different strengths.

(a) Normalising

Normalising consists of heating a suitable steel to a temperature typically in the range 830–950 °C (at or above the hardening temperature of hardening steels) and then cooling in air. Heating is usually carried out in air, so subsequent machining or surface finishing is required to remove scale or decarburised layers. The objective of the treatment is to counter the effects of prior processes, such as for example rolling, by refining the existing non-uniform structure into one with a fine grain size which enhances machinability/formability or, in certain product forms, meets final mechanical property requirements.

(b) Thermo-mechanically controlled processing (TMCP).

In this process route, the rough steel section (i.e. the slab) is heated to a temperature typically used for hot working (about 1200 °C). Initial hot rolling ('roughing') is carried out as standard but the final thickness reduction or 'finishing pass' is carried out at a lower temperature. Plastic deformation at this lower temperature promotes finer grain sizes and retards precipitation. The final hot working may continue down to a critical temperature for the transformation from austenite to ferrite. TMCP requires heavy rolling equipment capable of deforming the steel at lower working temperatures. The optimum precipitate size and dispersion is obtained when the finish rolling temperature is around 775 °C.

The cooling brings the steel to the transformation temperature range, where the austenite to ferrite transformation results in fine ferrite grains and fine dispersed precipitates. For some TMCP steels, this last stage of cooling, during which transformation is completed, is accelerated by water cooling, to give a finer grain size. Accelerated cooling can sometimes result in bainite formation as well as, or instead of, ferrite formation. The strength of a TMCP steel is generally higher than that for normalised steel of the same composition.

(c) Quenched and Tempering route

Most conventional structures use only moderate strength steel produced by the normalised or TMCP routes. For higher strength levels there are processing thickness restrictions to TMCP steels and normalising cannot produce the strength levels required in the necessary section thicknesses. Quenching and tempering are therefore the standard production route for thicker higher strength structural steel.

Quenching and tempering consist of a two-stage heat-treatment process. Stage one includes hardening, in which the steel is austenitized to approximately 900 °C and then quickly cooled with water or oil. Upon being rapidly cooled, a portion of austenite (dependent on alloy composition) will transform to martensite, a hard, brittle crystalline structure. The quenched hardness of the steel depends on its chemical composition and quenching method.

Stage two consists of tempering the steel to obtain the desired material properties. It is done by heating the metal to some temperature below the critical point for a certain period of time, then allowing it to cool in still air. The required temperature and time determine the amount of hardness removed, and these depend on both the specific composition of the steel and on the desired properties in the finished product. In carbon steels, tempering alters the size and distribution of carbides in the martensite, forming a microstructure called "tempered martensite" Quenching and tempering achieves a fine-grained and homogeneous microstructure. Quenched and tempered steel is characterized by high strength and good ductility but involves more processes than either normalising or using the TMCP route.

In summary, most conventional offshore structures use only moderate strength steel produced by the normalised or thermo-mechanically processed routes. At higher strength levels, there are thickness restrictions in the production processing of these steels because normalising cannot produce the strength levels required. Quenching and tempering (QT) are therefore the standard production route for higher strength structural steels.

Heat-treated carbon and alloy steels may be normalized or quenched and tempered to improve the mechanical properties. Normalizing produces essentially the same ferrite–pearlite microstructure as that of carbon and high strength low alloy (HSLA) steels, except that the heat treatment produces a finer grain size. This makes the steel tougher and more uniform throughout its length. Quenching and tempering produce a martensite or martensite–bainite microstructure in both carbon and alloy steels with decreased grain size (ferrite and/or bainite) to give increased strength and increased toughness. This is usually achieved by microalloying with Nb, V or Al and by some form of thermomechanical processing.

3.3 Medium Strength Steel Materials (SMYS < 420 MPa)

3.3.1 B 4360 Grade 50D and Equivalent European Specifications

From the 1970s, the standard specifications of thick structural carbon manganese steel for North Sea structures were the UK specification BS4360 Grade 50D (British Standards 1969) and the equivalent Norwegian specification NS12101 Grade St52-3N. Standards such as BS 4360 specified the compositional chemistry (including carbon equivalent), heat treatment condition, tensile properties and Charpy toughness. However, there was no specific requirement for steels with resistance to fatigue cracks.

In the mid 1990s, BS 4360 Grade 50D (British Standards 1990) was superseded by the European standard EN 10025–4 (EN 2019), with Grade S355 being equivalent to Grade 50D and S420 Grade also used for applications requiring higher strength. This steel is the standard material for many oil and gas structures today and increasingly for offshore wind turbine towers.

Grade 50D was a medium tensile strength, low carbon manganese steel that could be supplied in plate thicknesses up to 75 mm with a minimum UTS of 490 MPa and a minimum yield strength of 355 MPa. Grade 50D and St52-3N were required to have 27 J impact resistance at sub-zero temperatures. The D designation meant that it had been impact tested to achieve sufficient impact energy and ductility even for thick sections, initially at – 10 °C, and by 1990 at – 20 °C, see Table 3.2.

The compositional specification for steel to BS 4360 grade 50D, NS12101 St52-3 and equivalent specifications required by DNV and the Norwegian Standard NS 12101 are also shown in Table 3.3.

Table 3.2 Yield and ultimate tensile strength of medium strength structural steels

Specification	SMYS (MPa)	UTS (MPa)	Charpy V (J)
NS12101:1972 Grade St52-3N	353	510–608	27.5 J@ – 20 °C
BS4360:1969 Grade 50D	355	490–640	27 J@ – 10 °C
BS4360:1990 Grade 50D	355	490–640	27 J@ – 20 °C
EN 10025-4: 2019 Grade S355	355	370–530	40 J@ – 20 °C
EN 10025-4: 2019 Grade S420	420	~ 520	

Table 3.3 Chemical composition of steels used in offshore construction (Walker 1987)

Date	C% (max)	Si%	Mn% (max)	S% (max)	P% (max)	Nb (max)	Al	Ni% (max)
BS4360: 1969 Grade 50D	0.22	0.1–0.55	1.6	0.05	0.05	0.1		
Typical supply to fabricators in 1972/73	0.17	0.4	1.3	0.2 (sic)	0.025	0.05	0.028	
NS12101: 1972 St52-3	0.2	0.55	1.5	0.045	0.045			
DNV Rules: 1977 Special	0.2	0.5	1.6	0.03	0.04			0.4
DNV Rules: 1977 Primary	0.22	0.5	1.6	0.04	0.04			0.4
DNV Rules: 1977 Secondary	0.24	0.5	1.6	0.05	0.05			0.4
Typical supply to fabricators in 1986	0.12	0.4	1.5	0.002	0.012	0.03	0.035	0.25
BS4360:1990 Grade 50D	0.2	0.15	1.5	0.04	0.04	0.003/ 0.1		
EN 10025-4: 2019 Grade S355	0.23	0.05	1.6	0.05	0.05			

The table shows that the specified composition of Grade 50D did not change significantly between 1969 and 1990. In addition, the compositions of comparable steels required by DNV and the Norwegian Standard NS 12101 are broadly similar to BS 4360 50D but allow increasing carbon content for less critical applications.

The need to improve weldability resulted in steelmakers developing steels with lower carbon content (to around 0.12%). In normalised steels this reduction in carbon was compensated for by an increase in elements that promote solid solution hardening such as manganese, niobium, vanadium, nickel and copper. BS 4360 50D is therefore readily weldable with a carbon equivalent not exceeding 0.41 (see Sect. 4.6).

A feature of steelmaking practice for higher strength steels led by 1987 to the control of residual elements that could cause embrittlement by segregating to grain boundaries. These could cause a marked increase in the tendency for intergranular fracture along prior austenite grain boundaries creating a potentially serious reduction in fracture resistance and exacerbating the risk of stress corrosion cracking, hydrogen embrittlement and corrosion fatigue. The possibility of embrittlement was kept in balance by the careful use of other alloying elements such as Molybdenum, Titanium and rare earth metals. The need to improve the resistance to lamellar tearing and impact toughness led by 1987 to a large reduction in the sulphur content (to 0.002%).

It should be noted that oil companies involved in the development of North Sea platforms, used steel standards such as BS4360 and St52 only as a basis for specifying requirements for steels used in the primary structures. Each oil company would have their own specification. Typically, they would have more stringent requirements on microalloying additions, carbon equivalent, sulphur content, and Charpy toughness. Therefore, assessors of ageing structures should examine the original specification which could have had better properties and performance than the standard grades, although the significance of any differences should be investigated.

3.3.2 American Specifications for Steel Plate

Historically, American specifications for steel plate used offshore have been based on ASTM and API standards. As platforms designed to these specifications were usually intended for use in the normally benign and temperate waters of the Gulf of Mexico or the Middle East, the main requirement was to guarantee tensile strength rather than fatigue and fracture resistance. Structures fabricated to these early American specifications and standards may therefore not have been impact tested and have guaranteed minimum impact properties. As a result, structures constructed from material specified to early American standards could be less tolerant to cracks and impacts than those constructed to material meeting later US, UK or Norwegian specifications.

Occasionally, American specifications were used for the early platforms in the Southern North Sea. An example is the BP West Sole platform, see Appendix 1. A detailed study (BP 1979) was made to confirm the original materials specifications following its recovery to shore. The study showed that the main tubular members to the template were specified to ASTM 252 Grade 2 (ASTM 2019) for piling steel pipe, while the brace members were specified to meet API 5LB refinery piping steel (API 2019). BP specified St 52 Grade 3 German steel for the piles that were subject to the highest bending stress.

American materials specifications for offshore structures have changed over the years in line with UK and Norwegian practice. There are four main API specifications, namely API 2H, API 2Y, API 2W and API 2MT1 with different strength options (API 2001, 2006a; b, c). The primary use of these steels is in tubular joints, stiffened plate construction and other intersections where portions of the plates will be subjected to tension in the through-thickness (Z) direction. They are intended primarily for use in offshore structures in critical positions that require assured impact toughness and resistance to lamellar tearing. The API 2H specification covers intermediate strength (Grades 42 and 50 (ksi) minimum yield) standard normalised steel plates up to 4 in. thick.

The API specifications 2W and 2Y (API 2006b, c) cover companion steels providing similar or better mechanical properties (Grades 50 and 60) than 2H with the advantage of allowing lower welding preheats. These steels are intended for more specialist applications where higher strength and ease of welding are required. This improvement results from a reduction in the maximum allowed chemical composition, particularly carbon (0.16%), and is made possible by the method for manufacturing the steel using heat treatment and/or processing. API 2W is produced by thermo-mechanical control processing (TMCP), while the higher strength API 2Y is produced by quench and tempering.

A further steel used in offshore structures is API 2MT1 Grade 50 (API 2001). This steel is intended primarily for use in Class "B" applications, as defined in API RP 2A (API 2014). API 2MT1 steel plates are suitable for use in positions of offshore structures which must resist impact and fatigue loading but are not subjected to substantial through-thickness (Z-direction) stresses. They do not therefore have the guaranteed through thickness properties of API 2H, 2W and 2Y (API 2006a; b, c). Table 3.4 compares the chemical compositions of these steel specifications.

API 2H, 2W and 2Y all have a minimum, mid-thickness, transverse Charpy V-notch impact toughness of 30 ft-lbf [41 J] at – 40 °F [– 40 °C] and are therefore suitable for use in colder conditions. When assessing the fracture toughness based on a correlation with Charpy V notch energy, a correction needs to be made for the effect of thickness. In addition, cold-formed material has less ductility due to work hardening than that of the original flat plate, especially in those areas aged by the attachment welding of stubs or braces and therefore may exhibit lower fracture toughness.

With the potential use of lower pre-heats with these higher strength steels, in 1998 API introduced a route for prequalification of heat affected zone (HAZ) toughness within API RP 2Z (API 2020). The purpose of the recommended practice was to provide the purchaser with information which can be used to minimize the amount of time and testing necessary to prepare and certify fabrication welding procedures (and to assure that the steel to be supplied is inherently suitable for welding). Particular attention is given to fracture toughness and resistance to cracking of the HAZ.

Table 3.4 Comparison of the chemical compositions of steel specifications

Steel	C	Si	Mn	S	P	Nb	Al	N_2	Mo	Ni	SMYS
API 2H Grade 50 (API 2006a)	0.22	0.05–0.45	1.15–1.60	0.015	0.03	0.01–0.04	0.02–0.06	0.012			350
API 2W Grade 50 (API 2006b)	0.16	0.05–0.50	1.15–1.60	0.010	0.03	0.03	0.02–0.06	0.012	0.08–0.15	0.75–1.0	420
API 2Y Grade 50 (API 2006c)	0.16	0.05–0.50	1.15–1.60	0.010	0.03				0.15	1.0	450

API RP 2Z (API 2020) requires that the supplier specifies target fracture toughness values for the HAZ in terms of CTOD. Furthermore, API RP2Z requires the supplier carries out post-test metallography to ensure that a minimum amount of grain coarsened HAZ is sampled because, in general, this region of the HAZ has the lowest fracture toughness. The RP was developed in conjunction with and is intended primarily for use with API Specifications 2W and 2Y, but it can be used as a supplement to other material specifications (e.g. API Specification 2H). Older structures will not have the assurance about HAZ toughness provided by API RP2Z and therefore this aspect should receive close consideration when a structure is being assessed for continued service or life extension.

3.3.3 ISO Standards

The ISO 19902 (ISO 2020) is the current international standard for the design and manufacture of fixed offshore steel structures (jackets). This standard was first published in 2007 and the latest edition in 2020. It includes requirements for different steels that could be used in offshore structures. Structural steels are grouped into five groups according to specified minimum yield strength (SMYS).

ISO 19902 Materials requirements are given in Table 3.5. Appendix 20 of ISO 19902 (ISO 2020) correlates steel groups and toughness classes for steel plates to US specifications (e.g. ASTM and API) and also EN and NORSOK specifications.

Group II, with a SMYS between 280 and 395 MPa, covers steels comparable with the forementioned BS4360 Grade 50D. As indicated the carbon equivalent of group II steel is not to exceed 0.43, while the equivalent value for BS4360 was 0.41. Although these are relatively similar, it is also believed that present manufacturing processes are able to deliver more ductile and weldable steels, allowing for this higher carbon equivalent value.

ISO 19902 states that consideration shall be given to the selection of steels with adequate notch toughness characteristics suitable for the conditions of service. The

Table 3.5 Specified minimum yield strength for five groups

Group	SMYS range MPa	Carbon equivalent
I	220–275	< 0.45
II	280–395	< 0.43% for all toughness classes
III	400–455	ns
IV	460–495	ns
V	Above 500	ns
VI	Reserved for possible higher strength structural steels in the future	

Note ns not specified

toughness classes and Charpy impact temperatures are listed in Table 3.6 and the description of the steel classes is given in Table 3.7.

ISO 19902 states that the designer shall select either a material category (MC) or a design class (DC) approach for determining the steel group and toughness class. The material category approach is based on three levels (MC1 to MC3) depending on the exposure level for the structure with MC1 relating to exposure level 1 structures and correspondingly for the other levels. The required toughness class is determined based on the type of component, the material category (MC) of the structure and the

Table 3.6 Minimum toughness requirements for structural steels

Steel group	SMYS range	Charpy toughness Joules	Toughness classes and Charpy impact temperatures			
			No testing	CVI test at LAST	CV2 test at 30 °C below LAST	CV2Z and CV2ZX
I	220–275	20	x	x	x	
II	280–395	35	x	x	x	x
III	400–455	45		x	x	x
IV	460–495	60		x	x	x
V	Above 500	60			x	x

Note LAST is lowest anticipated service temperature

Table 3.7 Description of steel classes

Steel class	Description and conditions of service
No testing	Secondary structures only
CVI	Suitable for application in welded structures at service temperatures above 0 °C, and for which no formal requirement is made for impact testing
CV2	Suitable for use where thickness, cold work, restraint, stress concentration, impact loading, and/or lack of redundancy indicate the need for improved notch toughness
CV2Z	Should have a large margin between impact testing temperature and lowest anticipated service temperature as prescribed in Table 18.2-2. Such steels should be considered for major primary structures and other failure critical or non-redundant applications, particularly when the need for reliability is compounded by the existence of factors such as high stresses; residual stresses; degree of cold work from fabrication; freezing temperatures; fatigue loading; and impact loading with due regard to the thickness of the sections in question
CV2ZX	Should have through thickness (short transverse direction) ductility, for resistance to lamellar tearing caused by tensile stress in the direction of thickness, in addition to the other characteristics of class CV2 steel. Through-thickness ductility shall be demonstrated by either having a minimum reduction in area of 30% in a tension test conducted on a specimen cut from the through-thickness direction, in accordance with an internationally recognized standard, or by specifying a low sulphur content

Table 3.8 Design class, component complexity and consequences of failure

Design class	Component complexity	Consequences of failure
DC 1	High	Applicable for joints and members where failure will have substantial consequences and the structure possesses limited residual strength
DC 2	Low	
DC 3	High	Applicable for joints and members where failure will be without substantial consequences due to residual strength
DC 4	Low	
DC 5	Any	Applicable for joints and members where failure will be without substantial consequences

Notes High joint complexity means joints where the geometry of connected elements and weld type leads to high restraint and to a triaxial stress pattern. "Substantial consequences" in this context means that failure of the joint or member will entail either, danger of loss of human life, significant pollution or major financial consequences

steel group based on SMYS. For typical North Sea structures, the relevant exposure level is I being a manned non-evacuated platform. The basis for Design Classes is given in Table 3.8 which relates class to the component complexity and consequences of failure.

Chapter 20 of ISO 19902 concerns welding, including the requirements for welding procedures and their qualification, welder qualification, production welding, and welding inspection for fixed steel offshore structures.

3.4 Higher Strength Steel (SMYS > 420 MPa)

3.4.1 Introduction

While the specification of Grade 50D steel was common, higher strength steels were used with advantage. The use of higher strength steels enables thinner sections which reduces weight and weld volumes. This smaller thickness has a significant effect on the required quantity of weld metal and welding time and can also minimise the need for post-weld heat-treatment (PWHT). However, higher strength steel was less used for welded nodes due to its perceived lack of fatigue properties compared to normal grade steel (such as grade 50D). The reduced weight of higher strength steel facilitated towing and installation, with the reduction in the volume of welding required had major economic benefits.

Quenching and tempering simple carbon manganese steels with small additions of micro alloying elements had been found to result in steels of considerably higher strength with good notch ductility and weldability and plates up to 55 mm thick in Grades 60 or 65 became available.

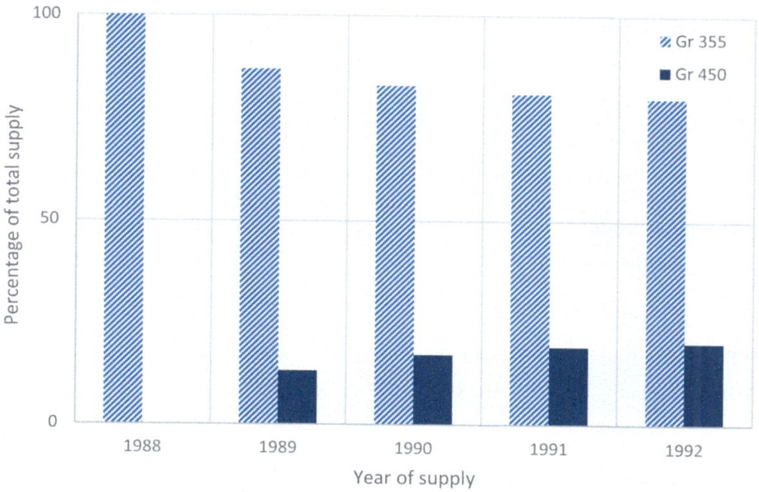

Fig. 3.4 Increasing use of grade 450 MPa steel over the period 1988 to 1992 (based on Billingham et al, 1995)

An increasing amount of higher strength steel with yield strengths of 420–460 MPa became in use, primarily after 1988. A survey (Billingham et al. 1995) undertaken for structures being designed in the period 1988 to 1995 showed that the use of 450 MPa steels increased from zero in 1988 to approximately 20% by 1992, as shown in Fig. 3.4. The survey indicated that this increase in use of higher strength steels was primarily where weight was a consideration for lift reasons. Further, the survey indicates that the use of higher strength steel increased to 40% by 1994. Figure 3.4 shows that the dominant material used was grade 355 steels.

These higher strength steels with yield strengths of 420–460 MPa correspond to St42 and St46 in Europe, Grade 55E in BS 4360 in UK (British Standards 1990) and API 2W and API 2Y in US.

Table 3.9 shows some specific platforms where such higher strength steel was used. The amounts of such steel used vary from 100 tonnes in the Shell Gannet structure to over 10,000 tonnes in several other structures.

The chemical compositions of UK grade 50D and 55E are shown in Table 3.10. It can be seen that UK Grade 55E has a higher carbon content and hence a higher carbon equivalent (CE) value.

3.4.2 Fatigue Capacity of Higher Strength Steels

As already noted, there was a perceived lack of fatigue endurance properties of higher strength steels compared to normal grade steel (such as grade 50D). In early editions of the Department of Energy Guidance Notes (Department of Energy 1974, 1977,

Table 3.9 Amounts of higher strength steel used in some offshore platforms (MTD 1995)

Year	Owner/project	Tonnage of steel 420 < YS < 460	Proportion of total tonnage %
1992	BP/Bruce	1600	18
	BP/Unity	1000	10
	Shell/Gannet	100	1
1993	Enterprise/Nelson	8000	30
	Amerada Hess/Scott 2 jackets	25,000	50% grade 420
	Chevron/Alba	15,000	15
	Phillips/Judy	9000	40
1994	Total/Dunbar	10,000	40
	Statoil/Sleipner 2 jackets	30,000	95% grade 420
1995	Phillips/Ekofisk 2 jackets	33,000	95% grade 430

Table 3.10 Chemical compositions of grades 50D and 55E steels for thickness 20/25 mm (MTD 1995)

Grade	C	Mn	Si	S	P	Nb	Al	CE	Typical strength (MPa)
BS4360: 50D	0.10	1.35	0.42	0.016	0.015	0.028	0.022	0.43	360
BS4360: 55E	0.22	1.6	< 0.6	0.04 max	0.04	0.003–0.10	–	0.49	450

1986) a caution was introduced concerning fatigue for steels with yield strengths greater then 400 MPa, due to the lack of data in seawater. In 1995 it was also reported by MTD (1995) that there was limited fatigue data on grade 450 steels. This absence of data limited the application of higher strength steels in fatigue sensitive areas.

New data for higher strength steels has later been established and was reviewed in the background to the new fatigue guidance for steel joints and connections in offshore structures (HSE 1999), which supported the fatigue section in the fourth edition of the Guidance Notes (HSE 1994). These data included fatigue tests in seawater on higher strengths steels. Environmental reduction factors (ERF) for both freely corroding and cathodically protected steels were introduced as a comparison of the fatigue performance in air. This showed that the ERF's were similar for all steels with a yield strength of up to 540 MPa. The Review Panel considered that this supported allowing the fatigue guidelines to be applied to steels with yield strengths up to 500 MPa. However, it was noted that this change was restricted to welded plates as at that time only two tubular joints with yield strengths greater than

400 MPa had been tested. For structures fabricated with higher strength steels in nodes it was recommended that environmental factors should be determined using an approved test programme incorporating the correct environment, CP level and seawater temperature. An alternative approach was to use fracture mechanics to derive environmental factors.

More recent data has shown that the basic nodal fatigue S–N curve in ISO 19902 applies to all structural steels (including tubular nodes) with yield strengths less than 500 MPa steels (ISO 2020). If higher strength steel was used in the nodes, it is likely that the original fatigue calculations were on the safe side compared to what we now believe to be the capacity. However, a detailed evaluation for each case is for obvious reasons recommended.

3.4.3 Capacity of Higher Strength Steel Joints

Higher strength steels generally have different stress–strain behaviour compared to medium strength steels (350 MPa). This is because the composition of higher strength steels has been selected specifically to increase the yield strength and these have much less influence in subsequent stress–strain behaviour. One measure to illustrate this difference is the yield ratio ($YR = SMYS/UTS$) which is the ratio of the specified minimum yield strength (SMYS) to ultimate tensile strength (UTS). This ratio generally increases as the strength of the steel increases. Most higher strength steels (450 MPa) have yield ratios close to or greater than 0.85, which is in excess of the required 0.67 or 0.7 in current codes and standards that apply to tubular joints. Figure 3.5 shows examples of two stress–strain relationships for a Grade 50D steel (350 MPa) and a higher strength steel (450 MPa). The change in the stress–strain behaviour and the related strain hardening is evident illustrated by the change in the YR from 0.7 to 0.85.

Current design equations are based mainly on test data from medium strength steels, where some degree of strain hardening is present. Limited strain hardening could lead to failure at a stress level of $1/0.9 = 1.1$ of the yield stress, compared to the often-assumed $1/0.67 = 1.5$ originally in codes. This limited strain hardening could have significant implications for the capacity of tubular joints in service, as constrained nodes may not be sufficiently ductile to develop plastic behaviour prior to local failures (rupture), which is required for obtaining their full strength. These restrictions in structural design codes reflect this by not allowing to have a yield strength value greater than 0.85 times the UTS. In their report, Cranfield in their report for HSE (2003) considered that the code restrictions for nodal connections may be conservative in respect of high strength steels.

The Panel developing the ISO 19902 standard re-examined the test data on joint strength and developed some improved design equations. However, recognising the lack of data on higher strength steel joints, the ISO Panel concluded that these equations should be limited to steels with yield strengths less than 500 MPa and the YR should not exceed 0.85 (ISO 2020). It was also argued that individual test results

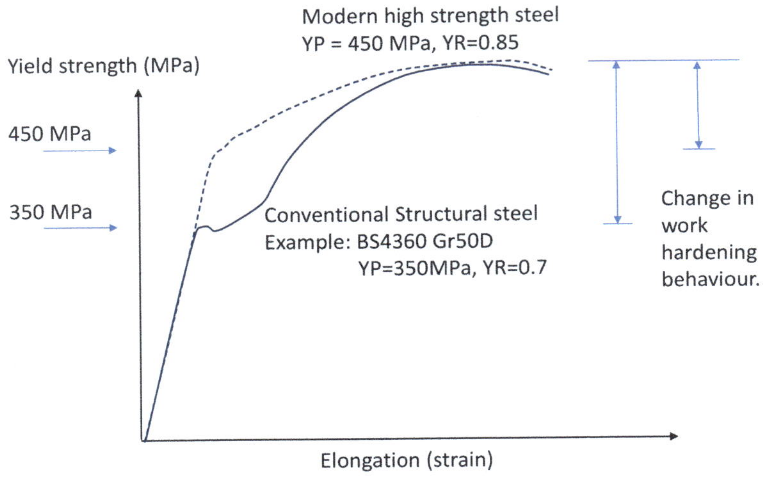

Fig. 3.5 Change in stress–strain characteristics between higher strength and medium strength steels (based on MTD 1995)

of 0.90 from single test coupons could be acceptable. The use of higher strength steel in practice therefore generally requires the provision of test data, which is costly and time consuming.

In life extension, consideration of the YR for higher strength steels should be taken into account, noting the possible limitations in steels with specifically high YR values.

3.5 High Strength Steel (SMYS > 500 MPa)

3.5.1 Introduction

High strength steels with yield strengths higher than 500 MPa have been used for many years in mooring lines, prestressing tendons in concrete structures and in the fabrication of mobile jack-up installations.

There is less information on high strength steels compared to for example 50D steels on the long-term performance under the severe environmental conditions to which structures on the UK Continental Shelf (UKCS) are subjected and hence a certain amount of uncertainty is associated with their use. Particular concerns are the greater susceptibility of higher strength steels to hydrogen cracking. This can be enhanced considerably when bacteria producing hydrogen sulphide are present (see Sect. 6.2).

There is also limited detailed information on the fatigue performance of these materials, especially when welded. These concerns were heightened by the discovery in the late 1980s of cracking within the spud can region of jack-ups operating on the UKCS. This was believed to be due to hydrogen cracking caused partly by a sour environment and partly by excessively negative cathodic potentials rather than corrosion fatigue (HSE 1993).

High strength steel is more vulnerable to hydrogen embrittlement and cracking. This was exemplified by the detection of cracking in jack-ups in the late 1980s led to a significant research programme to understand the effects of hydrogen cracking, particularly in situations offshore where cathodic protection potentials are strongly negative.

This led to specific Guidance on limiting such potentials for high strength steels in vulnerable situations. The recommendations at the time were that the CP level should be limited to a maximum negative value of − 850 mV Ag/AgCl. Special measures (e.g. voltage limiting diodes) were recommended to keep potentials within acceptable limits. In addition, steels for use offshore should be assessed for their vulnerability to hydrogen cracking. In the supporting document (HSE 1993), testing of potential materials, using a slow strain rate with a range of CP values is recommended. It is also stated that all steels in the strength range above 650 MPa should be examined for the possibility of hydrogen damage in service, both in the parent material and in the weldments. Regular in-service inspections were also recommended, to detect any other deterioration.

DNV (2015) provides limited guidance for such steels. DNV-OS-C101 states that the use of steels with yield stress greater the 550 MPa should be subject to special considerations for applications where anaerobic environmental conditions (e.g. stagnant water, active mud and in the presence of hydrogen sulphide) may exist. Anaerobic conditions are defined as a concentration of SRB's in the order of magnitude of > 103 SRB/ml. The susceptibility of high strength steel to hydrogen induced stress cracking should be particularly considered when used for critical applications such as jack-up legs and spud cans.

3.5.2 *High Strength Steel in Jack-Ups*

3.5.2.1 Introduction

Historically, the primary application of high strength steels in offshore structures has been in jack-ups. There has been a significant increase in the size of jack-up rigs, enabling them to work in greater water depths and during harsher weather conditions than earlier designs. As noted earlier, several jack-ups are now being used for production, without the ability to inspect in dry dock in sheltered waters on a regular basis.

The principal application of high strength steels in jack-ups is in the fabrication of the legs, rack, pinions and spud cans. Steels with nominal yield strengths in the range 460–700 MPa have commonly been used. In general, each lattice leg is composed of three or four longitudinal chord members which may contain a rack plate for elevating the hull and a series of horizontal and diagonal tubular braces which connect the chords and form a truss. Supplementary braces (span breakers) are frequently used between main brace mid-points to increase buckling resistance.

The thickness of the rack plate varies typically between 150 and 250 mm. The chord shell cans are usually fabricated from plate with a wall thickness between 35 and 80 mm and a diameter in the range 800–1200 mm. Weldability and good toughness and ductility are important material considerations in this application and the steel maker provides this by careful control of alloy composition and by processing using the quenching and tempering route.

3.5.2.2 Steels for Chord Racks

In the 1980'ties, the majority of high strength steels used in jack-up legs were quenched and tempered Ni–Cr–Mo–V steels manufactured in France, Japan and the USA. These were characterised by high hardenability in thick sections and good toughness properties for a wide range of strength levels (Stacey et al. 1996).

(a) Type HT80 Steels for Chords and Racks

The main European and Japanese producers manufacture steel plate for chords and racks based on differing grades of either:

- A514 high strength quenched and tempered alloy steel plate suitable for welding, or
- A517 high strength quenched and tempered alloy steel.

The actual product specifications are usually modifications of the basic ASTM specifications and may vary with time to achieve better weldability or reflect advances in steelmaking practice. Both the F and Q grades are Ni–Cr–Mo–V steels and can be described by the designation HT80.

(b) French Chord Rack Steels

Creusot-Loire Industrie (CLi) used to be a major producer of rack and chord steel in Europe. Their products were specified by a majority of designers in Europe and the Far East. The basic chord and rack steels from Creusot Loire have been known as 1.5 DFO (early 1980s), Superelso 700 CR, Superelso 700 TR and (modified) A517 FCR (early 1990s). These were quenched and tempered Ni–Cr–Mo steels with a minimum yield stress of 100 ksi (690 MPa), as shown in Table 3.11.

According to the manufacturers, the steels are all covered by the ASTM designation A517 (ASTM 2023) Grade Q. A517 FCR is a later designation for Superelso steel, for this steel the chemistry has been altered with a reduced chromium content and additions of vanadium which were not present in Superelso 700 and 1.5 DFO. The A517 FCR grade is applicable to both chords and racks and is closest to the basic A517 Grade Q specification.

There are also differences in the Superelso 700 chemistry of the CR grade for racks and the TR grade for chords. Because of the decreased thickness of the latter,

Table 3.11 Yield strength of French chord rack steels (Stacey et al. 1996)

Location	Grade	Sy (ksi)	Fy (MPa)
Chord rack	A517 F (mod)	100	690
	1.5 DFO	100	690
Chord shell	Superelso 700	100	690
	Centrishore VA/VB	95/100	655/690
	Centrishore V	100	690
	Nicuage 60 (Dillinger)	85	590

it is possible to achieve the mechanical properties with a leaner chemistry showing a reduced nickel level and a matching increase in (cheaper) manganese.

French foundries have also supplied Centrishore centrifugal castings for tubular chords (CFEM designs) and AFNOR 18 NCD 10M cast nodes (MSC CJ-46 design), although the compositions vary both from each other and from plate products. AFNOR specification 18 NCD 10M is a quenched and tempered 2.5% Ni–Cr–Mo steel. The minimum yield stress is 100 ksi in both cases. In addition to having a different thermo-mechanical history, both the Centrishore V and AFNOR 18 NCD 10M cast steels have compositions which differ significantly from the wrought products considered above. Neither of these steels is considered to be represented by a generic type.

(c) Japanese Chord Rack Steels

Major producers of Japanese steel for chords and racks include Nippon Steel and Sumitomo with a range of 100 ksi quenched and tempered steels, i.e. Wel-ten 80 and Sumiten 80 respectively—the '80' designation refers to the 80 kg/mm^2 (800 MPa) tensile strength. These are both Ni–Cr–Mo–V steels similar to the French A517 FCR steel but also contain minor additions of copper and boron. Steels of this type are often known as HT80 type, although this can refer to any 80 kg/mm^2 quenched and tempered steel. The manufacturers also refer to them as ASTM A514 Grade F (modified).

The levels of most alloying additions in these steels have been increased to ensure adequate strength in thick sections. This is particularly noticeable with the nickel content which has a specification range of 1.2–3.5%. The leaner end of the composition range is used for braces (1.2–1.7%) whilst the higher concentration of 1.5–3.5% is required to generate sufficient strength in the rack. Greater amounts of manganese, nickel and chromium are allowed to compensate for a tighter restriction on carbon.

(d) North American Chord Rack Steels

The principal source of US steel for jack-up units has been the Le Tourneau Steel company which supplied steel for the legs on Le Tourneau designs which were built both in the USA and in Singapore. The arrangements of the chord and rack differ from many other types of jack-up. The relevant materials are N-30 HN for the rack and flat plates of N-20 BHT for the chord.

The N-30 HN rack steel appears to be similar to AISI 4340 (SAE 2022) engineering steel but has a lower carbon equivalent. However, even though the carbon

Table 3.12 Yield strengths of Le Tourneau chord rack steels

Location	Grade	Yield stress	
		(ksi)	(MPa)
Chord rack	N-30-HN	90	620
Chord plates	N-20 BHT	70	482

content is lower than that of 4340 steel, it is still significantly higher than in the corresponding French or Japanese materials, which may indicate poorer weldability.

The N-20 BHT chord steel is a medium to high strength low allow steel produced through a quenching and tempering route. Table 3.12 shows that the 480 MPa yield strength of this steel is much lower than that of several other types of tubular chord steel (690 MPa).

3.5.2.3 High Strength Steel for Jack-Up Bracing

The braces of most jack-up units are manufactured from pipe rather than formed plates. Therefore, the material specifications tend to be derived from the appropriate pipe specifications, typically API 5LU Specification which is for ultra-high test heat treated line pipe (American Petroleum Institute 1991). These specifications provide considerable latitude in the requirements for the composition. In the case of higher strength grades of API 5L this may be by agreement. Therefore, in some areas the API specifications define mechanical and dimensional requirements rather than that for a material.

In practice, the strength of most braces lies in the range 450–690 MPa (X65 and X100 indicating 450 to 690 MPa yield strength), although regular 345 MPa structural steels have been used for diagonal braces in some cases.

The chemistry of the highest strength grades is similar to that for the chord and rack steels. However, it is often a little leaner as with a thinner wall thickness the strength is easier to achieve.

A particular noteworthy steel is the Le Tourneau Grade MM structural tubing which is used for braces on all Le Tourneau units. This is a manganese-molybdenum steel with a yield strength of 590 MPa.

3.5.3 Research Projects on High Strength Steels

In the 1980s high strength steels had been available for many years and used offshore initially in Jack-up platforms and pipelines. For example, the X70 grade steels with a yield strength of 480 MPa had been commonly used subsea. Such steels are ferrite pearlite steels which are micro-alloyed to produce a fine grain size to provide both strength and resistance to low temperature brittle fracture. Low carbon contents lead to good weldability and toughness. Higher strength steels with yield strengths of around 700 MPa had been used in jack-ups. High strength steels had also been used in mooring applications, both in chains and also for tethers in tension leg platforms where their high strength and good resistance to fatigue could be exploited. However, in this application the steels are not joined by welding but by screwed connectors.

In the early years of using high strength steels, it was recognised that there were a number of concerns which are listed below (MTD 1995).

- Restrictions in the available codes particularly for the higher strength steels (> 700 MPa yield), including yield strength and the yield /ultimate ratio,
- Weldability and the ability of fabrication yards to weld these materials. What level of weld metal overmatch should be recommended,
- Reductions in the properties of HAZ's following welding, given that as the heats input increases HAZ toughness may decrease,
- Fatigue performance as results indicate that this in the welded condition is no better than that for lower strength steels, hence limiting their applicability in fatigue sensitive applications,
- Hydrogen embrittlement, particularly when over-protected by cathodic protection systems.

This lack of data at the time led to research programmes in the UK and abroad being initiated. For example, the Marine Technology Directorate (MTD 1995) funded a number of research projects on high strength steels, with particular emphasis on their fatigue and fracture performance via a two-year Managed Programme which included a series of projects being carried out at Cranfield Institute of Technology and UMIST (Manchester). The UK Health & Safety Executive (HSE) and industry were sponsors.

The MTD's Managed Programme, co-ordinated by Cranfield University, was one of the more significant projects on high strength steels. This consisted of several phases over the period 1985–1995 (MTD 1995). It included a number of projects concerned with the performance of high strength steels offshore with yield strengths in the range 350–690 MPa and the development and properties of the associated weld metals.

These projects looked at the fracture, fatigue and corrosion fatigue behaviour, and the embrittlement tendency of both steel and heat-affected zone (HAZ) material in relation to microstructural changes produced by welding and the micro-alloying elements and composition of the material. Certain projects addressed a wide range of heat inputs to embrace fabrication and welding repair under adverse conditions, backed up by controlled laboratory simulations of the most crucial microstructures.

A range of the latest lower carbon weldable high strength low alloy (HSLA) steels was studied, with yield strengths nominally between 450 and 700 MPa. They were manufactured, in Europe and Japan, by controlled rolling, quenching and tempering (Q&T), and accelerated cooling, and included castings as well as plates and beams.

It was concluded that thermo-mechanically controlled processing (TMCP) high strength steels could compete favourably with Q&T materials, particularly at the lower end of the strength and thickness range. Accelerated cooled steels could also compete favourably with modern quenched and tempered steels at least up to 500 MPa yield strength.

From a review of international structural steel design codes at the time, a yield-ratio limit of 0.85 was widely specified for structural steel work, and a lower fraction of 0.66 was associated with tubular joint design. Both provisions limited the weight saving advantages that could be taken with high strength steels where YR values of 0.8 to 0.95 are typical. No account was taken of specific material properties in its formulation. The changes in stress–strain behaviour had been correlated with advances in steel processing and alloying. The increase in YR value with increasing strength had been shown to depend on factors such as production route, composition, thickness and steel plate manufacturer. Despite this trend, several manufacturers could consistently deliver high quality, tough, weldable steels at least up to 500 MPa strength level while still satisfying the current material specification requirement of YR < 0.85.

Weldability trials using CTS test procedures confirmed that high strength steels had excellent weldability, with critical heat inputs being < 0.8 kJ/mm for tests with hydrogen levels of 4.7 ml/100 g. Both maximum and average HAZ hardness decreased with increasing heat input, and values of maximum hardness were generally < 350 Hv. It was found that there was little correlation between performance and published predictive documents such as BS5135 (British Standards 1984).

The Charpy V notch impact properties of the HAZ regions of high strength micro-alloyed steels generally decreased as the welding heat input increased. However, several steels, including a titanium oxide strengthened steel, retained excellent properties even at the highest heat inputs used (6 kJ/mm). The most beneficial HAZ microstructural features were a fine bainite grain size and a low MAC content.

A number of Grade 460 and 550 SAW and FCAW weld consumables had good levels of impact toughness when welded at low heat input levels of 1.0 and 0.7 kJ/mm, values being well above the 45J and 55J at −40 °C specified in BS7190 (British Standards 2019). This was associated with the fine-grained nature of the microstructures brought about by the relatively fast cooling rates, the higher oxygen content providing an even distribution of small non-metallic inclusions.

The research showed that high strength steels suffered from accelerated corrosion fatigue crack growth in seawater when tested under conditions of high negative potential (−1100 mV sec), crack growth rates being between 2 and 10 times faster than when protected at − 800 mV. Secondary cracking played a significant role in slowing down crack propagation, such cracks commonly being associated with microstructural factors such as substructure boundaries. The proportion of cleavage type failure increased as the potential decreases, strongly indicating that hydrogen embrittlement played a significant role in the failure process.

There was generally little difference in corrosion fatigue performance between the various micro-alloyed high strength steel compositions tested. Prolonged exposure to cathodic hydrogen charging led to accelerated crack growth compared with tests pre-charged for much shorter periods (8 weeks cf 2 weeks). This was thought to be because more time was available for hydrogen diffusion to the crack tip region.

This result indicated that in-service performance could be inferior to short exposure laboratory test results, and that care must be taken in comparing results from different data sets. A pre-charging time of 8 weeks was recommended for future test programmes. It was found that the absorption of damaging hydrogen by cathodically protected steel was considerably enhanced by the presence of active SRB in the local environment.

As a result of this research significantly more information became available about the mechanical properties of Grade 420 and 460 steels as they became more widely used offshore. However, it was concluded that there was less information available on the higher grade materials (Grade 550 and above).

Also noticeable, there was a joint industry project completed by Det Norske Veritas on the fracture toughness properties of two Grade 690F jack-up steels, such as steel of Nippon Steel Welten 780 and Creusot Loire A 517 F type, with plate thicknesses of up to 250 mm (DNV 1995).

3.6 Steel for Ship-Shaped Floating Installations

As previously mentioned, since the 1970s ship-shaped vessels have been developed for use as floating production, storage, and offloading installations (FPSO's) with a rapid increase in their number during the 1990s. According to a worldwide survey of FPSOs in 2022, approximately 37% of the FPSOs had been purpose built and approximately 63% converted from conventional oil tankers that had already been in service (Offshore 2022). The idea of the converting conventional oil tankers into FPSOs was to bring the FPSO's into service quickly and economically.

The suitability of a steel for use in commercial shipping is decided by the Classification Society appointed by the ship owner in agreement with the fabricator. Individual Classification Societies have their own standard specifications for steels used in vessels under their classification, which have developed over many years, and which are separate from those of national standards bodies.

The failure by fracture of the Liberty ships and others in the 1940s drew attention to the possibility of cracks initiating at low temperatures from defects associated with welds and propagating into ship plate where the fracture could run a long way before arresting. During the post war period the Classification Societies independently developed requirements for different grades of ship steels to minimise the likelihood of brittle fractures in welded ships. An example of an early specification of composition of ship steel from the 1948 American Bureau of Shipping Rules (SSC 1962) is shown in Table 3.13. It can be noted that there are no specified limits for the content of carbon, manganese and silicon for Class in steel not exceeding 0.5″ which meant that the properties could be highly variable.

Most steel traditionally used for oil tankers, merchant vessels and other ship-shaped floating installations can be classified as low strength carbon manganese mild steel without other alloying elements. The steels used for ships can have moderate

Table 3.13 Composition of ship steel from the 1948 American Bureau of Shipping Rules (SSC 1962)

	Class A	Class B	Class C*
	Plates not exceeding ½ inch and all shapes	Plates over ½ inch but not exceeding 1 inch	Plates over 1 inch
Carbon, max (%)	–	0.23	0.25
Manganese (%)	–	0.6–0.9	0.6–0.9
Phosphorus, max (%)	0.04	0.04	0.04
Sulphur, max (%)	0.05	0.05	0.05
Silicon (%)	–	–	0.15–0.3

*plate steels produced to the requirements of Class C shall be made with fine grade practice.

levels of sulphur, phosphorous and silicon. The yield strength is typically in the range of 250–270 MPa and UTS in the range of 465–500 MPa depending on thickness.

Table 3.14 shows mechanical properties and compositions of some typical ship steels tested that were tested in 1974 (SSC 1975). Plates were obtained from several steelmakers and shipyards. The material type refers to ABS Grades A and B reflecting their differing impact test requirements. It can be noted that the carbon content for Grade A steel varies between 0.16 and 0.21% and that for Grade B from 0.1 to 0.2%. Despite these large variations there is no consistent effect on the mechanical properties which are determined mainly by the steel making practice.

The lack of a common approach for the definition of steel selection led to considerable inconvenience in the steelmaking and ship building industries. Thus, in 1959, the Classification Societies introduced uniform requirements for different Grades of steel based on their Charpy impact energy defining four Grades of normal strength steel as follows. For Grade A steel there is no requirement of the steel maker to undertake or document a Charpy or any other fracture test that would guarantee any minimum level of fracture resistance. Grades B to E have a requirement for 27 J Charpy impact energy at decreasing temperatures. For North Sea structures, a design temperature of – 10 °C would be typical.

Grade A	No Charpy requirement
Grade B	27 J at 0 °C
Grade D	27 J at – 10 °C
Grade E	27 J at – 40 °C

By its nature without control testing, different batches Grade A steel can have vastly varying fracture properties depending on the composition and manufacturing route. As most merchant shipping operates in temperate waters above 0 °C where the risk of fracture is relatively low the shipping industry has historically used and continues to use Grade A ship steel plate as its main material contained of large quantities of Grade A steel. This leads to considerable uncertainty about its fracture

Table 3.14 Details of plates (SSC 1975)

Material type	Thickness (mm)	Plate code	%C	%Mn	%P	%S	%Si	YS (MPa)	UTS (MPa)	NDTT (°C)	CV at NDTT (J)	CV at upper shelf (J)
A (*)	50	U-11	0.19	0.42	0.004	0.025	0.23	270	390	+ 10	18	108
A (*)	19	U-13	0.21	1.02	0.007	0.019	0.01	270	440	− 7	27	117
A	25	U-23	0.16	0.57	0.003	0.026	0.22	260	430	+ 4	–	–
A	20	U-25	0.16	0.68	0.007	0.023	0.06	270	410	− 7	–	–
A	25	U-31	0.19	1.41	0.004	0.017	0.26	325	410	− 7	–	–
B (*)	25	U-14	0.20	1.00	0.005	0.023	0.008	225	450	− 1	19	128
B	25	U-20	0.12	0.87	0.007	0.025	0.15	255	430	− 7	–	–
B (*)	25	U-21	0.10	1.01	0.005	0.019	0.19	250	430	− 1	19	185
B	22	U-26	0.15	1.06	0.005	0.027	0.04	280	455	− 7	–	–
B	25	U-33	0.17	0.80	0.004	0.030	0.17	280	428	− 7	–	–
B	25	U-34	0.15	0.93	0.006	0.013	0.01	270	390	− 1	–	–

*plate steels produced to the requirements of Class C shall be made with fine grade practice. Fine grade practice would require slower cooling rates to produce a finer grade microstructure.

performance. The implications of this for vessels used as FPSOs are considered in more detail in Sect. 1.10.6 on fracture toughness later in this book.

A feature of merchant ships including oil tankers is the high proportion of Grade A ship plate used in their construction. According to one study (Randerson and Morrison 1989) the proportion of Grade A ship plate in large ships (including most of the oil tankers converted to FPSOs) was 80–85% and up to 95% in intermediate and smaller ships. The remaining material is usually Grade B or Grade D with the higher strength grades AH32/36 and DH32/36 used in more highly stressed areas or crack arrest strakes.

3.7 Castings and Forgings

3.7.1 Use of Castings

Castings have occasionally been used to replace the conventional welded node joints between tubulars in steel-jacket platforms. Cast steel nodes were first installed in offshore platforms in the North Sea in 1984 with their use in the Conoco Viking installation. These were four X joints manufactured by River Don Castings (latterly Sheffield Forgemasters) which were installed on the Conoco Victor JD platform (Herden et al. 1986).

The advantage of cast nodes is that they can provide significant improvements in fatigue life and design flexibility as well as all-round strength in offshore components. However, they are also complex to manufacture and inspect and a major defect can result in scrappage and remanufacture. Therefore, they have only been used in specialist applications.

Since Viking, castings have been used for nodal joints and other offshore components particularly when fatigue life is an issue. In particular, castings have been used for joining and anchor shackles. In 2006, cast steel specialist Kohlswa Gjuteri delivered 85 joining and anchor shackles to Statoil's Volve development in the North Sea (Offshore 2006). Each unit weighed around 1 tonne. The shackles, which were required to have good tensile strength and impact resistance, were made from offshore grades of hardened cast steel. The shackles were ordered by chain-makers Ramnäs to form part of its anchor chain delivery to APL. The latter supplied the submerged turret loading (STL) system through which oil would be transported from the Volve jack up production platform to a floating storage unit.

Other large cast offshore components included two 110-tonne trunnion nodes measuring 4.6 m × 4.4 m × 3.7 m were supplied by Vulcan (a subsidiary of Sheffield Forgemasters International) which were, used for lifting Statoil's 9000-tonne Valemon jacket into place in the Norwegian sector of the North Sea, in 2012 (Offshore Engineer 2012).

The Canadian experience of cast steel nodes in tubular construction is presented in the paper by Oliveira et al. (2006). The authors noted a resurgence of interest in steel castings for onshore tubular structures in road and railway bridges that are susceptible to fatigue. Castings can provide viable solutions to tubular connection design problems, particularly under seismic and fatigue loadings and describe Canadian experience with cast steel nodes and relevant university research projects.

In 2013, Vulcan SFM (subsidiary of Sheffield Forgemasters) had an order for engineered steel castings as part of what was then the world's largest floating spar gas production platform for the Aasta Hansteen deepwater gas field off Norway (Foundry Management and Technology 2024). The spar was a type of floating oil platform used in very deep water, usually as an alternative to conventional platforms. It consisted of a large-diameter (50 m), single vertical cylinder (198 m) supporting a deck weighted at the bottom, and the whole structure was anchored to the seabed at a water depth of ~ 1200 m. The platform was located approximately 186 miles off the Norwegian coast and was the first spar platform built with storage capacity.

The order included 12 cast nodes, the largest of which weighed 117 metric tonnes. These steel castings formed high-strength, fatigue-resistant nodal joints connecting the main leg and tubular braces that formed the lower, truss section of the spar platform connecting the main leg and tubular braces. The strength was developed through heat treatment involving water quenching. The castings were a specially developed grade of steel offering a balance of strength, toughness, and weldability.

3.7.2 Defects in Castings

Research was needed on cast joints for offshore use following their introduction in the mid 1980s. A joint industry funded research project, which included HSE funding, was undertaken in the late 1980s aimed at providing a better understanding of the behaviour of cast joints and the provision of design guidance which was intended to extend their application in offshore structures (Ma and Sharp 1997). Four areas were investigated in the project which included:

- Material properties of cast steels,
- Stress concentration factors,
- Fatigue S–N relationship and
- Defects in cast nodes.

Of particular interest to this book is the research on materials properties and defects. As part of the programme a series of laboratory tests on small scale cruciform specimens and real sized cast nodes was conducted. In the materials part of the study information from foundries and other relevant companies was obtained. In general, it was found that the cast steel material at that time was compatible with BS 7191 Grade 355 or 355 EMZ EM steel with similar performance to standard wrought steels used offshore (British Standards 1989). The work on defects is important for the fatigue performance of cast joints. Defects in castings were be classified as shown in Table 3.15.

Table 3.15 Defect types in castings (UEG 1985)

Defect type	Description	Cause
Blow-holes	Smooth rounded holes visible on the surface	High moisture, low permeability, excessive fine grains and gas producing ingredients. Another cause could be extra hard ramming or improper venting
Pin-holes	Very small holes visible on the surface on the casting, after cleaning by shot blasting	Sand containing too much moisture or gas producing materials
Gas-holes	Holes that appear when a surface is removed by machining or sectioning	Metal in poor condition, moist sand may aggravate the effect
Hot-tears	Shrinkage cracks or pulls formed by the casting pulling itself apart whilst cooling in the mould. Cracks have ragged edges and with a discoloured, oxidised appearance	Insufficient feeding of the casting

For inspection purposes a cast steel node is usually divided into differing zones, based on the applied stress level (Ma and Sharp 1997). For an X node there are four different zones, Fig. 3.6, which are as shown below:

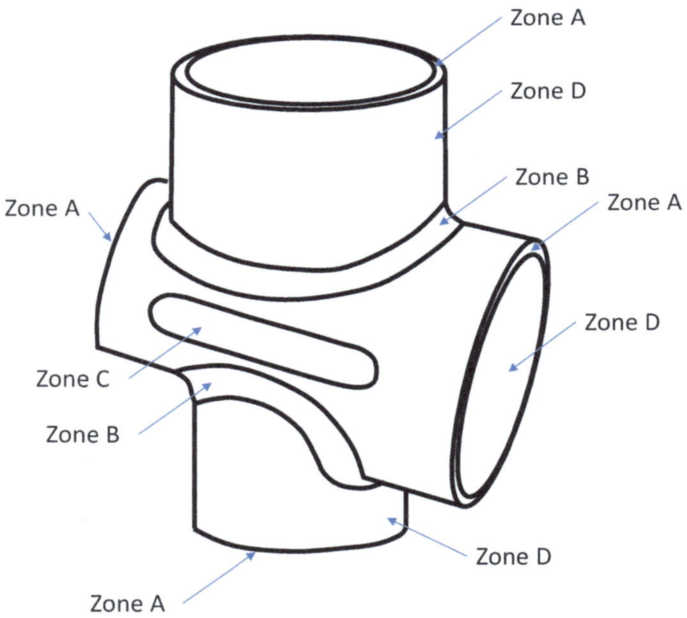

Fig. 3.6 Casting—showing zones A to D

- Zone A—weld preparation to a distance twice the section thickness away from the preparation
- Zone B—band around radii and a further 50 mm width along the parallel portion
- Zone C—high stress region
- Zone D—remaining area.

It was stated that the acceptance levels for defects at the time of the JIP were subject to agreement between the manufacturer, client and the certifying authority, A common acceptance level for defects found in zones A, B, C was that the area of defects in these zones should not be larger than 250 mm^2. It was also noted that defects due to hot-tears, which are usually planar, are normally not allowed.

In the experimental nodes cast for the JIP detailed examination by ultrasonic inspection showed that defects in one of the nodes constituted non-planar shrinkage cracks (varying in length from 10 to 150 mm and with widths varying from 5 to 20 mm. The larger defects were found in zone D which has a higher acceptance level. In a second node the defects were all non-planar slag inclusions with lengths ranging from 10 to 35 mm, all within the acceptance criteria.

As part of the Joint Industry project a large-scale X node was sectioned and samples taken from the node by mechanical sectioning to reveal the true size and nature of defects. Prior to sectioning the node was subjected to 100% NDE using both radiography and UT. Several defects were found. One small defect found in the high stress region was examined closely by microscopy and it was seen that the shapes of defects due to shrinkage and slag inclusions were similar to each other. The overall shape was volumetric and sharp points existed along the edge of the defect. It was concluded that it might be appropriate to consider both defects as planar defects, although it was suggested that further work was needed to confirm this, in the weld repaired region only very fine defects were found, implying that the weld repairs were adequate, and no structural defects were introduced.

On comparing the defect size obtained by the standard NDE methods used it was seen that radiography gave a better correlation with the defects revealed by sectioning. This may be due to the fact that shrinkage, porosity and slag inclusions were all volumetric in nature. It was found that UT failed to detect clusters of defects and provided less accurate measurements of defect size when compared to radiographic inspection. Both methods failed to report the detection of small defects with sizes up to 0.5 mm in length.

Local weld repairs are often used to remediate and minimise the effect of casting defects following the actual casting. These need to be taken into account particularly in fatigue design. The MMA welding process is typically used with consumables selected to be similar to the base material (UEG 1985). The chemical composition and the mechanical properties of the weld metal should correspond to those of the base metal.

The EEMUA specification for structural castings (EEMUA 1987) to be used offshore sets out the requirements for the design, manufacture and testing of carbon-manganese steel castings for use as structural members of offshore platforms. Chemical and mechanical properties are specified, together with the requirements for fracture toughness, non-destructive testing and weldability testing.

3.7.3 Forgings for Mooring Chains and Other Components

Chains play a crucial role in mooring floating offshore vessels, typically crafted from forged low alloy steel with minimal alloying elements. Initially, chains were of R3 or ORQ quality, boasting yield strengths of approximately 400 MPa and 500 MPa respectively. However, as offshore vessels grew in size and ventured into deeper waters, higher strength chains became imperative. Thus, the R4 and R5 series emerged, offering minimum yield strengths of 580 MPa and 760 MPa respectively (see Table 3.16). These chains can either feature studs or be studless, with studs mechanically fitted in R4 and R5 chains rather than welded.

The chemical composition for grade 4 chain is given in Table 3.17, with standards like DNV-OS-E302 (DNV 2022) emphasizing a minimum Mo content for higher-grade chains. Mechanical properties and toughness are paramount, necessitating specific heat treatments and grain sizes as per DNV-OS-E302. The standard specifies the use of killed steels with a fine grain heat treatment, ensuring a defined austenite grain size.

Grades R4 and R5 should be supplied in quenched and tempered condition, with specific tempering temperatures and cooling requirements.

Material compatibility and degradation issues of mooring chains revolve around corrosion and wear. Chemical composition, microstructure, hardness, and surface

Table 3.16 Required mechanical properties for different chain grades (HSE 2017)

	R5	R4	R3	API ORQ
Ultimate strength MPa min	1000	860	690	641
Yield strength MPa min	760	580	410	–
Reduction in area % min	50	50	50	40
Elongation % min	12	12	17	17
Impact Joules—Design temperature – 20 °C	58	50	40	40
Impact Joules—Impact temperature 0 °C		70	65	65

Table 3.17 Chemical composition of steel (%) for R4 chain (average of three from 3 different manufacturers (HSE 2017)

C	Si	Mn	P	S	Cr	Mo	Cu	Ni	Al
0.22	0.27	1.27	0.001	0.001	1.08	0.30	0.15	0.70	0.17

condition influence these properties, with hardness typically correlating with steel strength. R4 grade chains exhibit a Vickers hardness of approximately 275, while R5 chains measure around 305.

In summary, mooring chains have evolved to meet the demands of larger vessels and deeper waters, with stringent standards dictating their composition, heat treatment, and mechanical properties. Their durability and resistance to corrosion and wear are crucial for ensuring the safety and stability of offshore installations.

In addition to corrosion and wear mooring components are susceptible to fatigue cracks at various locations, including the flash weld and at or near the inter-grip area, as well as erosion and gouging. Other failure mechanisms in mooring lines include excessive tension from severe environmental loading, damage during operation, inspection and dropped objects, abrasion with the seabed and corrosion.

References

American Petroleum Institute (1991) API 5L specification for line pipe, 39th edn. Washington D.C, USA

API (2001) API 2MT1 specification for carbon manganese steel plate with improved toughness for offshore structures. American Petroleum Institute, Washington, US

API (2006a) API 2H specification for carbon manganese steel plate for offshore structures. American Petroleum Institute, Washington, US

API (2006b) API 2W specification for steel plates for offshore structures, produced by thermo-mechanical control processing (TMCP). American Petroleum Institute, Washington, US

API (2006c) API 2Y specification for steel plates, quenched-and-tempered, for offshore structures. American Petroleum Institute, Washington, US

API (2014) API recommended practice 2A-WSD—planning, designing, and constructing fixed offshore platforms—working stress design. American Petroleum Institute, Washington, US

API (2019) API Grade 5LB line pipe. American Petroleum Institute, Washington, US

API (2020) API RP 2Z Preproduction qualification for steel plates for offshore structures, 4th ed. American Petroleum Institute, Washington, US

ASTM (2019) ASTM 252 Standard specification for welded and seamless steel pipe piles

ASTM (2023) ASTM A517/A517M-17 Standard specification for pressure vessel plates, alloy steel, high-strength, quenched and tempered

Billingham J, Healy J, Spurrier J (1995) Current and potential use of high strength steels in offshore structures, MTD Publication 95/102 (1995). ISBN 1-870-533-24-1

Bokalrud T, Karlsen A (1978) Fracture toughness and significance of defects in heavy offshore structures

British Petroleum (1979) North Sea–West sole field platform WE; Post removal condition report and lifetime performance survey

British Standards (1969) BS 4360 British Standard specification for weldable structural steels. British Standardisation Institution

British Standards (1980), BS PD 6493:1980 Guidance on some methods for the derivation of acceptance levels for defects in fusion welded joints. British Standards

British Standards (1984), BS 5135:1984 Specification for arc welding of carbon and carbon manganese steels. British Standardisation Institution

British Standards (1986) BS 4360 British standard specification for weldable structural steels. British Standardisation Institution

British Standards (1989) BS 7191 Specification for weldable structural steels 1989

British Standards (1990) BS 4360 British Standard specification for weldable structural steels. British Standardisation Institution
British Standards (2019) BS 7910:2019, Guide to methods for assessing the acceptability of flaws in metallic structures. British Standards
Burdekin FM, Dawes MG (1971) Practical use of linear elastic and yielding fracture mechanics with particular reference to pressure vessels. In: Proceedings IMechE conference, London, May 1971
Callister WD, Rethwisch DG (2018) Material science and engineering—an introduction. Wiley Publishing
Cotton HC (1979) Welded steel for offshore construction, 15th John player lecture. Proc Inst Mech Eng 193(26):1979
Department of Energy (1972) SI 289:Offshore installations (construction and survey) regulations 1972
Department of Energy (1974) First edition guidance notes for the design and construction of offshore structures. HMSO London, UK
Department of Energy (1977) Second edition guidance notes for the design and construction of offshore structures. HMSO London, UK
Department of Energy (1986) Third edition guidance notes for the design and construction of offshore structures. HMSO London, UK
Department of Energy (1988) United Kingdom offshore steels research project—phase 1 final report (UKOSRP I), OTH88 282. HMSO London, UK
DNV (1995) Toughness acceptance criteria for rack materials and weldments in jack-ups. Report 95-3056, 1995, Høvik, Norway
DNV (2015) DNV OS C101 Design of steel offshore structures, general
DNV (2022) DNV-OS-E302 offshore mooring chain
EEMUA (1987) EEMUA-150 Mechanical plant and equipment: steel specifications for fixed offshore structures
EN (2019) NS-EN 10025-4:2019 Hot rolled products of structural steels—Part 4: Technical delivery conditions for thermomechanical rolled weldable fine grain structural steels. CEN
Farrar and Dolby (1972) Lamellar tearing in welded steel fabrication. Welding Institute report, 1972
Foundry Management and Technology (2024) https://www.foundrymag.com/molds-cores/article/21927997/vulcan-sfm-casting-steel-nodes-for-offshore-platform, visited September 2024
Gordon JE (1991) The new science of strong materials; or why you don't fall though the floor. Princeton Science Library
Harrison JD (1981) Material selection considerations for offshore steel structures. Welding Institute Research Report 150, 1981
Harrison JD, Dawes MG, Archer GL, Kamath MS (1979) The COD approach and its application to welded structures. ASTM-STP 668:1979
Herden DC, Tebbett IE, Wood T (1986) New data on the fatigue performance of cast steel tubular joints. OTC-5310-MS
HSE (1993) Hydrogen cracking of legs and spudcans on Jack-up drilling rigs—a summary of results of an investigation, written by Abertheny K, Fowler CM, Jacob R and Davey VS, Offshore Technology Report OTH 91 351, HSE Books, Sudbury, Suffolk
HSE (1994) Fourth edition of the guidance notes for the design and construction of offshore structures. HMSO London, UK
HSE (1997) OTO 97 072 data surveys of hydrogen assisted cracking in high strength jack-up steels, prepared by JCE Baker for the Health and Safety Executive (HSE), HSE Books
HSE (1999) OTH 92 390, Background to new fatigue guidance for steel joints and connections in offshore structures. HMSO London, UK
HSE (2003) Review of the performance of high strength steel, Cranfield University, published as Health and Safety Executive Research Report 105

HSE (2017) RR 1096 The effect of wear and corrosion of steel components on the integrity of mooring systems for floating offshore installations Mooring Integrity Joint Industry Project Phase 2, HMSO London, UK

Inglis (1913) Stresses in a plate due to the presence of cracks and sharp corners. Read at the spring meetings of the fifty-fourth session of the Institution of Naval Architects, May 14, 1913

ISO (2020) ISO19902:2020 Petroleum and natural gas industries—fixed steel offshore structures. International Standardisation Organisation

Kamath MS (1981) The COD design curve: an assessment of validity using wide plate tests. Int J Press Ves Piping 9(2):1981

Ma A, Sharp JV (1997) Fatigue design of cast steel nodes in offshore structures based on research data. Proc Inst Civ Engrs Wat Marit Energy 124:112–126

Ministry of Power (1967) Report of the inquiry into the causes of the accident to the drilling rig Sea Gem, HMSO Comnd 3409, October 1967

MTD (1995) Marine technology directorate (MTD)—high strength steels in offshore engineering, MTD Publication 95/100, ISBN 1-870553-21-7

Offshore (2006) Dual-role on Volve for harsh environment jackup, May 2006, . Offshore (offshore-mag.com)

Offshore (2022) 2022 Worldwide survey of floating production storage and offloading (FPSO) units, 2022. Offshore (offshore-mag.com)

Offshore Engineer (2012) The Valemon field, August 2012, The Valemon field (oedigital.com)

Oilfield Publications Ltd (1985) North sea platform guide. Oilfield Publications Ltd., Ledbury, Herefordshire, UK

Oliveira JC, Willibald S, Packer J, Christopoulos C, Verhey T (2006) Cast steel nodes in tubular construction—Canadian experience. In: Proceedings of the 11th international symposium and IIW international conference on tubular structures, pp 523–529, Quebec City, Canada, September 2006

Pisarski HG, Harrison JD (1978) Fracture resistance of materials used in offshore structures. Select seminar: European offshore steels research, Cambridge, November 1978. Published by the Welding Institute on behalf of the Department of Energy

Pisarski HG, Harrison JD (1986), Fracture toughness considerations for offshore structures in UK waters. Metal Construction 18:748

Randerson R, Morrison WB (1989) Materials for ships, British Steel Technical Publication, Swinden Laboratories, Report No PPRSI 197/86/39/D, March 1989

SAE (2022) Steel, bars, forgings, mechanical tubing, and forging stock 0.80Cr–1.8Ni–0.25Mo (0.38–0.43C) (SAE 4340) normalized and tempered AMS6484F. SAE International

SSC (1962) SSC 142 investigation of the notch-toughness properties of ABS ship plate steels. Ship Structure Committee (Ship Structure Committee: List of SSC Reports)

SSC (1975) SSC 248 Fracture toughness characterisation of ship building steels, written by JR Hawthorne and FJ Loss for ship structure committee

Stacey A, Sharp JV, King RN (1996) High strength steels used in offshore installations. In: OMAE conference, Florence, 1996

TWI (1978) COD design curve, assessment of validity using wide plate tests. TWI industrial member report summary 71/1978

UEG (1985) Underwater engineering group (UEG) design of tubular joints for offshore structures, Vol 3, Published by UEG, ISBN, 0860172317, 9780860172314

Walker EF (1987) Steel quality, weldability and toughness, Steel in marine structures. In: Proceedings of the third international ECSC offshore conference on steel in marine structures, Delft June 1987, Edited by CC Noerdhoek and J de Back, Elsevier

Chapter 4
Steel Joining Methods and Joints

4.1 Joints in Steel Offshore Structures

The main process for joining the parts of a steel substructure (supporting a topside) is by welding. Bolting is also used in topside modules for applications and especially where joints may have to be assembled or disassembled. In addition, bolting is favourable where there is a need to avoid hot work on a platform in operation.

The jacket type structures consist of tubular members welded together into a space-frame. The large diameter tubular legs, chords and braces are formed by rolling plate to a curved section and then longitudinally and circumferentially butt welding the sections with full penetration welds to create tubular subassemblies. The tubular subassemblies are connected at welded joints called nodes or nodal joints, which are critical for transferring and distributing the load carried by the structure.

In contrast to tubular structures, the hulls of semisubmersible and ship-shaped substructures primarily consist of rolled and machined butt-welded plates forming a containment boundary. The plates are normally stiffened and having longitudinal girders and transverse frames welded to them, as shown in Fig. 4.1.

In addition to the main structure, welded joints are also used to fabricate the secondary structures (such as attachments, flow-line supports and guides, caissons and topside modules). These welds traditionally have been partial penetration or fillet welds with unfused land which are easier and quicker to fabricate than the full penetration welds used for the primary structure. These secondary structure welds have, however, proved to be a problem due to fatigue cracking.

The development of welding processes, procedures and consumables for offshore structures is treated in Sect. 4.7. Traditional manual metal arc welding using welding rods is still used but has been largely superseded by methods such as automated or semi-automated flux core arc welding (FCAW) and gas shielded FCAW. Submerged arc welding is still used for longer linear strakes between curved plates.

© The Author(s), under exclusive license to Springer Nature Switzerland AG 2025
J. V. Sharp and J. Wintle, *Assessment of Materials in Ageing Offshore Structures*,
https://doi.org/10.1007/978-3-031-84489-8_4

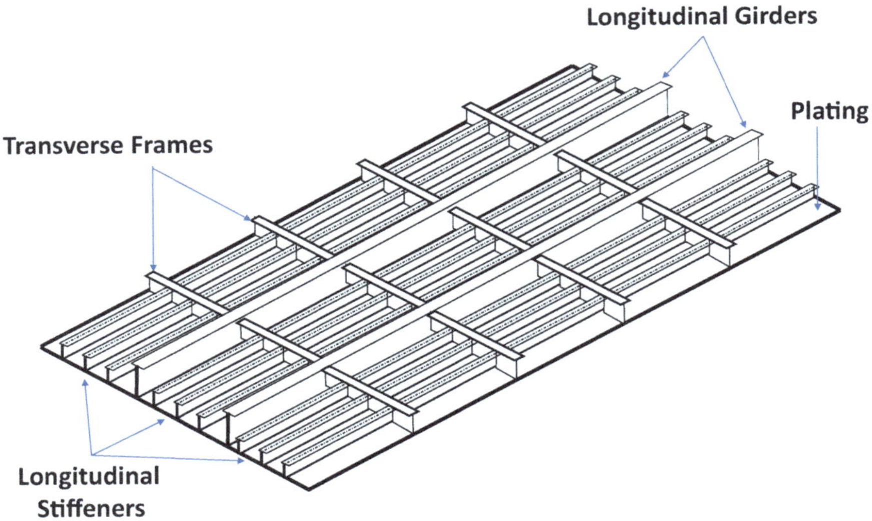

Fig. 4.1 Plated structure with stiffeners, transverse frames and longitudinal girders

By its nature, the process of welding can give rise to a range of types of different fabrication defects. The occurrence of weld fabrication defects cannot be easily foreseen or altogether prevented. Careful choice of the welding procedure can minimise the occurrence of fabrication defects while post weld inspection enables the detection of larger defects which can then be removed or repaired. Because fabrication defects can act as initiators for fatigue cracking, and hence shorten the life of the joint and weaken the structure, considerable efforts have been made to determine their frequency and size distribution, see Sect. 5.3.

Although welding is the pre-dominant joining method, bolted joints are also used. The primary application of bolted joints is mainly in the topside areas and the fixture of equipment and structures to the deck. Bolted joints often found for assembling the clamps that are used for repairing tubular joints.

In the offshore wind sector, large bolted flanged joints are now the main method of connecting the transition piece and tower sections to the monopile of offshore wind turbines. These are subject to significant fatigue loading as a result of wave impacts changes in wind loading (speed and direction).

4.2 Importance of Defects in Welds for Assessment of an Ageing Structure

This section shows how welds and the defects that are associated with them can be of vital importance to structural integrity and safety. Experience is that most structural problems originate from welds joining the structural members. In an assessment of an ageing structure for life extension, it is therefore essential to investigate and understand how the structure under consideration was welded and the strengths and weaknesses of welding processes and procedures which were used. These can have a huge effect on the likelihood and size of fabrication defects remaining in the structure, the susceptibility to fatigue initiation and crack growth in service, and on the resistance to fracture of the joints.

Welding is a process that requires a high degree of operator skill both in developing the procedure and set-up as well as ensuring stable controlled conditions during welding. As a result of the natural variability associated with any manual operation, welds generally contain different types of fabrication flaws that form due to process or metallurgical factors. Welding flaws are characterised by their type, size, density, orientation and position. Examples of flaws include lack of fusion, porosity, inclusions, metallurgical cracks, hydrogen cracks, lamellar tearing, undercut and lack of penetration is shown in Fig. 4.2.

After they are completed, welds are inspected for defects using non-destructive techniques and a judgement is made regarding their quality. Standards specify quality levels for workmanship that welds must satisfy. Despite inspection and quality

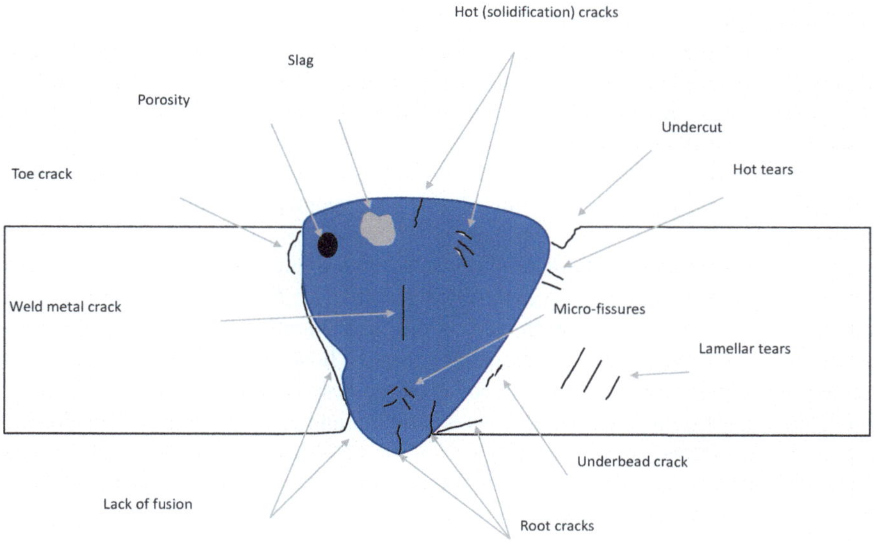

Fig. 4.2 Possible defects in a weld

control, some flaws may remain undetected in the welds, which can be initiators for fatigue and other types of cracks growing in-service that can lead to more widespread damage.

An assessment of an ageing structure for life extension should review the critical primary and secondary structure welds where a failure could lead to significant consequences. It is particularly important to review or inspect single sided, partial penetration welds and underfilled fillet welds where that the as-fabricated weld throat size is less than that specified in the design. More recently, the quality of the welds (at the best fabrication yards) has been improved by automated welding, but the presence of fatigue cracks produced by welding flaws and cyclic loadings outside the design specification should not be discounted.

Fatigue design assessment of welded joints was not common practice for earliest structures operating in the shallow waters of the Southern North Sea. These structures were based on experience from structures in the Gulf of Mexico which were not exposed to significant cyclic loading. However, with time, even these early structures experienced fatigue cracking which necessitated repairs (MTD 1995).

In regions such as the Northern North Sea, the more severe forces from wind and waves give rise to higher fluctuating loads and stresses on the welded joints which render them more susceptible to fatigue cracking. Early structures designed to standards where fatigue assessment was undeveloped may not have a fatigue assessment consistent with modern standards and should have been reanalysed. Later structures were designed following the methods in the Department of Energy Guidance Notes, see for example Department of Energy (1984), where there can be reasonable confidence in their fatigue endurance. Fatigue of welded joints remains, however, a major focus for the integrity and life extension of ageing offshore structures and is considered in more detail in Sect. 6.7.

Welded joints in semi-submersible and ship shaped floating structures are also susceptible to fatigue cracking from the motion of the vessel and the impact of waves. The penetration of cracks through the wall of hulls and ballast tanks resulting in leakage or loss of containment can have very serious consequences. In these types of structures fatigue cracks can initiate at the toe of fillet welds attaching internal stiffening girders and transverse frames and subsequently grow through the wall of the plate.

Some members and joints of offshore structures lie at sea level or in the splash zone above the sea. These pose a significant challenge related to corrosion as the cathodic protection system is ineffective in this area. Members and joints in this area usually have improved corrosion resistant coatings, however, experience has shown that members and joints can still corrode if the coating is damaged. Many designers incorporate additional thickness to these members and joints to compensate for the wall thinning effects of corrosion. This additional thickness is referred to as the "corrosion allowance" and a check on that remaining would be a normal part of an assessment of an ageing structure.

4.3 Formation of a Weld

Welding is a process of joining smaller pieces and components to build up bigger components. The joining is mostly done by creating a molten pool of weld material at the joints and allowing it to fuse with the parent material of the component and then solidify on cooling. Creation of a molten pool is not always desired and thus joining in the solid state is sometimes done but this is not generally relevant to the welding of offshore structures. This discussion will pertain only to metallic welded joints made by creating a molten pool which solidifies to give the final joint.

Welding requires a source of heat to melt the materials. In welding of steel offshore structures this is usually provided by an electric arc, although historically this was by using an oxy acetylene gas flame. The heat input from the welding arc and its travel speed are important variables in determining the properties of the weld and the thermal history of the weld region.

Weld metal is deposited by melting of a weld metal consumable, such as a stick or wire feed, that is fed into the arc. The weld consumable usually has a composition that matches the parent material but normally has other alloying elements to provide the required strength and to mitigate dilution in the fusion with the parent material. The choice of the correct consumable depends on the welding process and requires a qualified welding engineer.

Associated with the welding stick or wire is a flux whose primary role is to stabilise the arc and provide the arc and molten weld pool with a gaseous envelope to prevent the pick-up of oxygen and nitrogen which would lead to inferior mechanical properties of the weld. The stick or wire can be covered or cored with a band of flux but sometimes the flux is fed separately into the weld pool. The arc and molten pool can also be shielded by a welding gas to prevent oxygen from oxidising the molten metal. Typical gases are argon or carbon dioxide or a mixture of the two.

It is important that the welding consumables are dry and free from moisture and contamination. Damp electrodes or flux can result in the release of hydrogen into the weld. This can have a deleterious effect on properties and can give rise to hydrogen cracking. Therefore, welding consumables need to be stored under dry conditions. This is usually in heated cabinets, although these conditions are harder to maintain when welding is on site or in the open.

Melting of the weld metal consumable (stick or wire) and when the molten pool solidifies giving off heat creates time-dependent variations of temperature and temperature gradients in the weld region mainly due to the heat sink effects of the solid metal surrounding the molten pool. In multi-pass welds there is a further phase of reheating and cooling material that has already been through a thermal cycle. Welds may also be pre-heated to reduce the thermal shock and post weld heat treated to reduce residual stress and improve the metallurgy.

4.4 Features of a Weld and Regions of the Heat Affected Zone (HAZ)

The different temperature histories across the weld affect the final microstructures and create distinct metallurgical zones. These zones can have different properties and internal residual stress due to different temperature cycles and cooling rates of the metal during and after welding. In structural integrity assessments it is important to recognise these zones and to determine their individual properties by appropriate testing. Figure 4.3 shows a schematic sketch of the cross section of a butt weld joint with a single-run weld deposit.

Zone (1) in the middle is the weld metal solidified from a molten state and having columnar grains. The zone (3) on the other side of the HAZ is the original parent metal where the temperatures have not reached a magnitude to cause significant metallurgical change. The zone shown in black (2) is the region of the parent metal that has been affected by the heat of welding and is known as the heat affected zone (HAZ). This can subdivide into four different metallurgical sub-regions described below (Pisarski 2023).

The coarse grained HAZ (2i) is close to the fusion boundary. This region has been heated to peak temperatures more than 1200 °C. The original microstructure of the parent material has been destroyed and the new microstructure formed depends on the composition of the parent material and the cooling rate. As the name suggests the effective grain size of the various microstructural constituents of this region is larger than the ferritic grain size of the parent material. The effective grain size within the coarsened region (except very close to the fusion boundary) and the width

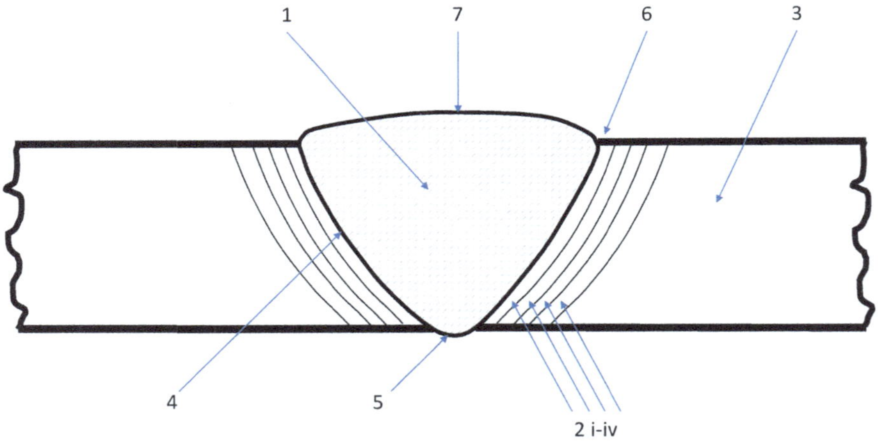

Fig. 4.3 Section of a single pass weld 1 is the weld metal, 2 is the heat affected zone including zones (i) to (iv), 3 is the parent material, 4 is the fusion boundary, 5 is the weld root, 6 is the weld toe and 7 is weld cap)

of the region itself are limited by the presence of precipitates[1] available to pin the grain boundaries and also by the heat input. Generally, in Carbon-Manganese (CMn) and similar micro alloyed steels this coarse grained HAZ (2i) region has the lowest toughness.

The fine grained HAZ (2ii) is further away from the fusion boundary and has been heated to peak temperatures between 900 and 1200 °C. It has a grain size which is usually finer than the parent material. The fine grains mean that it does not normally present any problem with low toughness.

The inter-critical HAZ (2iii) in the next zone and has been heated in the temperature range of between 720 to 900 °C. The microstructure of this region is essentially the same as the parent material except that the carbon rich areas have been transformed to austenite during welding and retransformed on cooling. This inter-critical HAZ (2iii) may show varying degrees of embrittlement depending on the parent material chemistry and heat input.

The subcritical HAZ (2iv) has been heated to no more than 720 °C. In carbon manganese and similar micro-alloyed steels. The temperature may be sufficient to spherodise pearlite, but often the microstructure when seen under the optical microscope is indistinguishable from the parent material (Pisarski and Harrison 1978). The toughness properties of this region can, however, be different from the parent material. The toughness properties of this region can be reduced by for example by strain ageing that occurs if the time was long enough and the temperature were high enough to allow interstitial carbon and nitrogen atoms to diffuse to the dislocations.

Other features of the weld shown in Fig. 4.3 are the weld preparation and fusion boundary, the weld root, the weld toe and the weld cap. The weld preparation (4) is the machined angled profile of the sides of the parent metal to be joined and the fusion boundary is where the weld metal and the parent metal fuse and where mixing of the molten metals can occur. The weld root (5) is at the base of the weld where welding starts and is an area where defects associated with the root such as lack of root fusion occur. The weld toe (6) is where the weld metal fuses to the parent material at the surface and is an area where defects such as undercut occur (see Sect. 5.3).

The weld cap (7) is the profile of the weld at the surface which can increase the stress concentration particularly at the weld toe. This can make the weld toe a site for fatigue crack initiation. The as welded profile at the weld toe can be improved by grinding to remove undercut and reduce the stress concentration thereby making fatigue initiation less likely (see Sect. 6.11.2).

Figure 4.4 shows the same for a multi-pass weld deposit. The shape of the contour of the fusion boundary has changed because of multi pass nature of the successive weld deposits. The HAZ micro-structures are slightly more complicated, and the microstructure can vary not only as one moves away from the fusion face but also parallel to it. Due to the heat input given during the succeeding weld passes, grain refinements take place in the weld deposits and HAZ. In the HAZ this can create fine gained microstructures with high toughness bounding coarse grained material.

---

[1] Precipitates are second phase particles that can form in the microstructure and can include aluminium nitride, see Sect. 3.2.

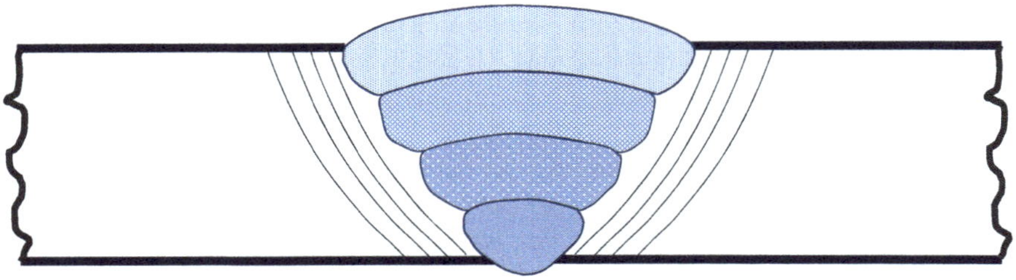

Fig. 4.4 Section of a multi-pass weld

The detailed microstructure will primarily depend on the composition and welding procedure of which the main factors affecting toughness are the arc energy, bead placement and interpass temperature.

Figures 4.3 and 4.4 give only simplified representations of the microstructural states in welded joints and in practice the actual microstructural states may not have clear cut demarcation lines. Microstructural analysis of weld sections is required to determine the different microstructures. Complexities in the microstructures arise in alloy steel systems in which precipitations of secondary phases take place at various temperature regions and at various rates of cooling and heating.

4.5 Properties and Characteristics of a Weld

Welds are usually designed to have material properties that are matched or exceed those of the parent material. Normal practice is to ensure that the yield strength of welds is overmatched (i.e. higher strength) to that of the parent steel by a judicious choice of the welding wire forming the weld metal. A sufficient degree of overmatching is appropriate to allow for higher stresses around volumetric defects whereas too much or too little can lead to problems. Too much overmatching can produce a stress concentration in the heat affected zone and susceptibility to fatigue cracking whereas too little overmatching or undermatching makes the weld metal vulnerable to plastic strain.

In addition to tensile properties the weld metal and heat affected zone regions can have ductility and toughness that is different from the parent material. High heat input and rate of cooling of the weld can lead to coarse grain microstructures in the weld metal or heat affected zone and corresponding low ductility and resistance to fracture. Low ductility and toughness make these regions more susceptible to cracking and fracture if defects and high residual and/or applied stresses are present.

Another aspect of welding is the generation of self-balancing residual stress within the materials. Differentials in temperature and cooling rate together with volumetric changes due to phase transformations at a microstructural level create internal strain incompatibilities giving rise to residual stresses that can reach the yield point of the material. The magnitude of residual stress can be reduced by post weld heat treatment, although the effectiveness of such treatment depends critically on the temperature, thickness and level of constraint and maintaining the required temperature for sufficient time. Although self-balancing, residual stress can locally increase the crack driving force for the material to fracture and increase the susceptibility to hydrogen induced or stress corrosion cracking in-service.

4.6 Development of Welding for Offshore Structures

Processes and procedures for welding and welding consumables have developed significantly over the period of construction of offshore structures but not always aligned with the improvements in parent steels. During the period from 1936 to 1950 many new welding processes evolved together with a wide range of new electrodes suitable for manual metal arc welding. These included solid extruded rutile, rutile/lime and basic low hydrogen electrodes which enabled welds to be produced with tensile properties that approximately matched to the parent plate (Cotton 1979).

Progress in welding generally slowed in the period between 1950 and 1975 and the development of welding failed to match the improvements in the quality and range of parent steels. In particular, the consumables for manual metal arc welding did not improve in respect of their weldability, slag, appearance and mechanical properties, especially for the notch ductility from welding positions that were not flat. A further concern was the volume of weld repairs necessary to achieve a weld of acceptable quality (Cotton 1979).

These were major concerns for the construction of the Forties platforms and other steel platforms when there was no alternative to the use of MMA for nodal joints. Repair welding was not only time consuming but held the possibility that repairs could introduce high residual stresses and further defects. Despite intensive research, a more reliable automatic high speed hand-held welding processes had not evolved by the mid to late 1970s, although the flux cored self-shielded semi-automatic process was starting to be used for positional welding on some structures (Cotton 1979).

The quality of welding achievable has improved since 1980, assisted by advances in equipment, steel and consumable composition and inspection technology backed up by better knowledge and understanding from many research programmes. In addition, generally, an extensive increase in the quality control procedures was implemented for the whole industry in the aftermath of the collapse and sinking of the concrete platform Sleipner A1 in Norway in 1991 (Steen 1993). Although this was a concrete platform the improvements in the quality control also influenced the construction of steel structures.

Welding is a fundamental enabling technology in the fabrication and repair of structures in the offshore industry above, at or below sea level, and without which the offshore industry cannot operate at its present level of sophistication. Yet after the construction of the early platforms welding technology was often taken for granted as a mature and established technology and mostly remained in the background, almost forgotten, until the Alexander L. Kielland disaster in 1980 (Moan et al. 1981). The inquiry found that a fatigue crack had initiated from a defect in a weld attaching a hydrophone to a structural member, and the crack grew by fatigue around the member until it reached a size where the member could no longer carry the load. This grew attention to the relationship between welding and the initiation of fatigue cracks.

Welding has like most technologies developed steadily over time, allowing new benefits in terms of what can be achieved in terms of quality and reliability, and in terms of the process economics of the rate of production. Welding technologies now have a wide range of different processes including process variants. While no fundamentally new arc welding techniques have emerged in recent years, there have been significant developments focused on improving process efficiency and productivity together with the facilitation of increased welding automation.

Welding of offshore structures is specified by international codes and standards, ship class rules, engineering specifications and client specifications depending on the service environment. Standards cover welding, processes, consumables, procedure qualification and welder qualification. According the UEG Design Guide (UEG 1985) typical standards for welding would have historically included:

- AWS D1.1 structural welding code giving detailed guidance on the welding of carbon and low alloy steels in the construction of buildings, bridges and tubular structures (AWS 2020).
- API RP2A (API 1993) on planning designing and constructing fixed offshore structures specifying provisions additional and alternative to those of AWS D1.1.
- DNV Rules for the design construction and inspection of offshore structures which contains detailed requirements for weld procedure qualification tests.
- Norwegian Petroleum Directorate Design Rules requiring weld procedure qualification tests relating to material thickness.
- British Standards Code of Practice 6235 (British Standards 1982) for fixed offshore structures specifying welding procedures in accordance with BS 5135 (British Standards 1984) and welders tested in accordance with BS 4871 (British Standards 1985).
- UK Department of Energy Guidance Notes specifying welding to one or more of the above standards.

Nowadays welding of offshore structures is covered by ISO standards such as ISO 19902 (ISO 2020). These standards place additional requirements for the strength, cracking resistance and toughness of the weld metal. Welding procedures used during the fabrication and repair of offshore structures should take account of the properties and metallurgy of the grade of the parent steel used. These include strength, microstructure, impact energy, hardness of the heat affected zone and numerous others.

The specified welding parameters, such as the joint dimensions, weld geometry, heat input and parent steel composition, can influence the resulting weld microstructures and susceptibility to cracking. After considerable research, Bailey developed a diagram called a nomogram for determining combinations of welding parameters to produce a sound crack-free joint (Bailey 1994). The diagram was based on the carbon equivalent value for the parent steel. In general, a steel with lower carbon equivalent value has improved weldability compared to one with a higher value.

Formulae to determine the carbon equivalent value were originally developed in the 1940's to give a numerical value for a steel composition as an indication of a carbon content which would have an equivalent level of hardenability for that steel. These formulae were later extended to represent the contribution of the composition to the susceptibility to hydrogen cracking. They are also used as compositional characterising parameters for other properties that may be linked to hardness, such as toughness and strength (TWI 2024a).

A commonly quoted carbon equivalent (CE) equation is that first recommended by the International Institute of Welding in 1967 and now incorporated into several codes and standards, including EN 1011-2:2001 (EN 2001) and in a modified form in the American welding specification AWS D1.1 (AWS 2020). The EN 1011-2:2001 (EN 2001) formula is:

$$CE = C + \frac{Mn}{6} + \frac{Cr + Mo + V}{5} + \frac{Ni + Cu}{15}$$

where the levels of each constituent CE, C, Mn, Cr, Mo, V, Ni, Cu are in per cent.

An alternative equation recommended for high strength steels is that in the DNV Offshore Standard B101—Metallic Materials (DNV 2020). In general, steels with reduced carbon contents and limitations on the levels of chromium, nickel and molybdenum have improved weldability, i.e. they have lower CE values.

Most repair procedures include welding which is quite different from joining during fabrication regarding the parameters, the weld sequences as well as the metallurgical conditions and the design features. Ideal welding conditions cannot always be achieved during repair welding. Additionally, poor material conditions of the component segments to be repaired might result from previous service operation.

The factors contributing to failures of welds are well known. They include residual stresses, defects or microstructural imperfections, loss of optimal bases metal microstructures during the weld thermal cycle, toughness changes, dissimilar material properties and the stress raising effects of geometrical discontinuities at the joint. Add to this the welding process variables, joint configurations and the variations in base material and welding consumable composition and we experience an extremely complex environment. Welds of older platforms are more likely to have been fabricated using manual processes and therefore may be more variable in quality and properties than welds made by later automated or semi-automated processes.

4.7 Welding Processes

4.7.1 General

Welding of structures for the offshore industry has and continues to be dominated by the conventional arc welding processes (TWI 2024b). Typical welding processes used in the offshore industry are:

- Manual Metal Arc (MMA) otherwise known as shielded Metal Arc Welding (SMAW),
- Flux Cored Arc Welding (FCAW)—mostly gas shielded,
- Metal Inert/Active Gas (MIG/MAG), otherwise known as Gas Metal Arc Welding (GMAW),
- Submerged Arc Welding (SAW),
- Tungsten Inert Gas (TIG) other known as Gas Tungsten Arc Welding (GTAW).

Of these processes the first four are the most commonly used, while TIG is mostly specified for root welds of high integrity joints. Different welding processes, welding consumables and process combinations are used to meet the fabrication and repair requirements of offshore structures. The ability of some of the welding processes to operate in the open on site is a major advantage.

Research is ongoing in using friction stir welding, laser welding and other welding processes in the fabrication and repair of offshore structures. The main aims are to provide increased production efficiency and increased quality of welding during fabrication and repair. In the future we will see these welding processes used more frequently.

4.7.2 Manual Metal Arc Welding

Manual metal arc welding using stick electrodes and either AC or DC power source is the most commonly used welding process for fabricating offshore structures. It is used wherever the welding is complex and of irregular geometry such as the welds forming the nodal joints between a brace and chords. The operator is required to manipulate and feed the electrode stick as it is melted away to ensure that the arc remains in contact with the work for long enough to obtain good fusion and, at the same time, to manoeuvre the stick along the joint to achieve even deposition of weld metal.

When the stick is deposited there is a break in welding as the operator inserts a new stick discarding the stub of the old and strikes a new arc to resume welding. If molten slag becomes trapped in the weld the operators must chip it away at the start position before commencing with a new electrode. These stop-start positions are often an area where welding defects such as slag inclusion, porosity and lack of fusion can arise.

Hydrogen present as moisture in the electrode coating or the atmosphere can enter the weld during welding and cause the weld to become brittle with low fracture toughness. This combined with residual stress due to differential contraction of the joint during cooling can lead to the generation of hydrogen cracking of the weld, which can be an initiator for fatigue crack growth. Steels and consumables with a high carbon equivalent are susceptible to hydrogen cracking. The problem has now been largely overcome with the development of low carbon equivalent materials and can be controlled in practice by ovens to keep electrodes dry and the specification of sufficient pre-heat and inter-pass temperature during welding. While these measures are now well understood, they were less well applied during the early years of offshore structure fabrication.

The notch ductility of MMA can be highly variable depending on the heat input, welding procedures and electrodes, the position of welding and whether post weld heat treatment is carried out. This variation was investigated in a notable project for BP by the Welding Institute (Cotton 1979) using BS4360 50D grade steel of 53 mm and 63 mm thick (British Standards 1986). The results found that no welds in the as-welded condition could achieve the crack opening displacement COD requirements, while post weld heat treatment showed substantial benefits in terms of the allowable flaws that could remain.

The project established that poor notch ductility was associated with less microstructures that occurred during positional welding. Positional welding could also produce dispersion of the shield gas produced from the flux. Adequate properties COD could only be achieved by remelting to produce recrystallisation. The project also established the benefit of acicular ferritic microstructures which nucleate on non-metallic inclusions within a narrow range of oxygen content.

The success of the manual metal arc process is highly dependent on the training and skill of the operator. The correct specification and oversight of the welding procedure by the Welding Engineer in terms of heat input (voltage and current), deposition rate and other variables and the weld preparation and fit-up is equally important. Standards for qualifying welders and weld procedure specifications for manual metal arc welding have developed and improved over the years so that it is now more reliable. Despite this almost all MMA welds contain cracks and crack like defects as a result of various mechanisms including hydrogen, reheat and solidification cracking.

4.7.3 Automatic Submerged Arc Welding

The automated submerged arc process uses one or more continuous bare electrode wires and a shielding flux fed into the arc in powdered or granulated form. The electrode feed is automatically regulated to ensure a constant current and arc length allowing very high currents and deposition rates. The root of submerged arc welds is normally produced using manual metal arc or a manual gas shielded arc process where better control can be maintained.

The submerged arc process can produce deep and wide welds that are smooth and regular with low levels of inclusions and porosity. The higher currents used allow high production rates and long production runs, while developments such as twin and multi-arc techniques, hot and cold wire additions and cut wire and iron powder additions increased production rates further. Submerged arc welding was typically specified for fabricating long longitudinal seams welds of tubular sections under workshop conditions and also circumferential seam welds if the workpiece could be rotated.

As with MMA, the choice of flux and electrode wire composition can have an important effect on the weld composition and properties such as fracture toughness. Weld metal impact properties could be low due to coarser grain structures in large single run high heat input welds with high rates of cooling, while the heat affect zone could also have poor properties if welding conditions were not properly controlled. Subsequent improvements in welding fluxes have resulted in reducing oxygen content and improved notch ductility.

Submerged arc welds are susceptible to thermal shrinkage cracking on account of their significant width and heat input. They can be subject to hydrogen cracking if the moisture and rust occur during the storage of the parent plate and electrode wire. Being an automated process, submerged arc welds can also be susceptible to extended and large lack of sidewall fusion defects from inaccurate set-up. These are difficult to defect with conventional NDT but can have an important effect on the structural integrity of the joints.

4.7.4 Gas Shielded Arc (TIG/MIG/MAG)

Gas shielded arc processes may be divided between manual TIG (tungsten inert gas) and the semi-automated metal inert/active gas processes (MIG/MAG) and the flux cored arc process (FCAW). In TIG an arc is struck from a non-consumable tungsten electrode, the joint protected by a flow of shielding gas, which can be either insert (argon or helium) or active, and the weld consumable is a fine wire fed in separately at the edge of the weld pool. The process requires a high level of skill from the operator and the deposition rate is relatively low compared with other processes. Therefore, in offshore structures, TIG is only used where a high-quality weld is required with a low probability of significant defects, such for as weld roots or specialist applications.

MIG and MAG are semi-automated processes where the arc is created between a continuously solid drawn wire feed and the workpiece. For ferritic materials shielding gas is typically a mixture of argon with either carbon dioxide (insert) or oxygen (active). As the wire is fed into a hand-held gun at the end of a flexible conduit the operator only needs to position the gun at the start position and traverse the gun along the weld. The processes combine the advantages of automated welding with the flexibility of manual welding and have the advantage over MMA that they are continuous and achieve higher deposition rates.

4.7.5 Self-shielded Flux Core Arc Processes (FCAW)

Flux cored arc welding uses a wire constructed with a flux core providing self-shielding so a separate shielding gas is not needed. It is well established for welding offshore structures, particularly on windy fabrication sites since it is quicker than MMA and more versatile than submerged arc. However, as with all shielded arc processes, operators have been found the transition from a simple plate or pipe test piece to a complex tubular joint challenging and this has limited their applications in the offshore industry. Where these processes were used the skill and management of the welding operators were key important factors determining the quality of the welds produced.

4.8 Welding Consumables

The welding processes described above all involve the addition of weld metal to fuse and fill the gap between the parts to be joined. There are some autogenous welding processes that do not require the addition of a filler metal, for example electron beam or friction stir welding, while tungsten insert gas welding (TIG or TGAW) and oxyfuel welding can also be used without a filler for certain types of joint where there is sufficient material to fuse. However, autogenous welding processes are not generally relevant for construction of offshore oil and gas structures.

Weld metal may in the form of a consumable electrode in the shape of a rod, as in manual metal arc welding, or fed into the joint as a wire as in submerged arc and gas shielded arc processes. In TIG welding the arc is struck from a non-consumable tungsten electrode. In oxy acetylene welding the heat is produced by a combusting flame and a consumable welding rod plied into the joint but this technique was rarely used in welded offshore construction.

During the early development of welding, electrodes were of the same composition as the parent metal. Elements for alloying and arc stabilisation were present in the electrode covering or flux. However, since the 1980s electrodes have included these elements in the composition of the weld metal rather than the flux because there was a tendency for the elements in the flux to burn off and not enter the weld.

There are generally three categories of manual metal arc electrodes: acid (rutile-titania), electrodes with a high cellulose content and low hydrogen. For offshore construction low hydrogen carbon manganese electrodes are still generally used to minimise the risk of hydrogen cracking. When assessing an ageing structure, it is pertinent to determine the type of electrode used so a judgment can be made on the susceptibility to fabrication hydrogen cracks that could develop into fatigue cracks later in life.

From the early 1980's there was increased interest in electrodes with 1.5 to 3% nickel content. These usually result in weld metal with improved fracture toughness particularly in thicker sections (UEG 1985). However, Dawson and Judson (1982) showed that there was also a potential stress corrosion cracking problem associated with excessive tensile strength of high nickel electrodes which could be overcome by adjustment of the other alloying elements.

After welding the strength of the weld metal should generally overmatch the yield strength of the parent material and the associated heat affected zone. If the weld metal under-matches the parent material in strength, then any enforced shrinkage stress, deformation and plastic strain will be concentrated in the relatively small weld metal volume, which increases the likelihood of weld metal cracking.

Welding on offshore structures is regulated by international codes and standards, ship class rules, engineering specifications and client specifications depending on the service environment. These standards can place additional strength, cracking resistance and toughness requirements on the weld metal. Welding procedures used during the fabrication and repair of offshore structures should have considered the factors related to the properties of the parent steel grade used. These include strength, microstructure, impact properties, and hardness of the heat affected zone and numerous others.

In addition to weld metal, other welding consumables include fluxes and cover gases. These perform several important functions:

- To stabilise the arc
- To protect the arc and weld metal from atmospheric contamination
- To deposit a slag on the surface which assists the formation of a weld bead of proper shape.

Holdcroft (1979) showed the ability of the flux to protect the arc was important in preventing nitrogen ingress into the weld metal. This could result in porosity and could also affect the weld metal fracture toughness by lowering cleavage resistance, particularly at the start-stop position.

Correct site storage of welding consumables is essential to avoid the presence of moisture and the risk of hydrogen cracking after welding. Electrodes arrive from manufacturers hermetically sealed and once opened the electrodes should be stored in an oven before being handled in accordance with the manufacturers' recommendations. Typical offshore specifications quote hydrogen contents of less than 10–15 cc/100g for all flux coated electrodes, cores wire and submerged fluxes of the basic type. Rutile electrodes and those with celluloid coatings should not have been used for offshore construction.

When assessing an ageing offshore structure for life extension, the assessment engineer should review the welding processes and welding consumables used for the critical welds. The choice of process, the quality assurance and management of the welding operation, and the selection and storage of consumables can all be factors when assessing the likelihood of cracking and weld failure later in life. Welding of offshore structures has generally improved over the years as more knowledge and experience have become available but historic problems may still be relevant for late life assets.

4.9 Welding of Higher Strength Steels

The use of higher strength steels in offshore structures has continuously increased over time, as described more in Table 3.9. Higher strength steels need additional demands placed on their welding procedures to avoid under-matching and hydrogen cracking. Important aspects to be investigated during an assessment of an ageing structure containing higher strength steel include:

- The strength of the parent material, where challenges increase with increasing strength.
- The composition and choice of the welding consumables.
- Welding procedures, including the use of pre-heat.
- The strength of the resulting weld and level of overmatching compared to the parent material.
- The likelihood of intrinsic deterioration of the weld from, for example, hydrogen embrittlement.

The normal statistical variations in yield strength that occur in steel plates also occur in weld metals for high strength steels. This poses a problem because there is the distinct possibility that the weakest weld metals may undermatch the highest yield strength of the parent material. Significant levels of overmatching can produce problems of hydrogen embrittlement and low toughness of the weld metal. Weld metal specifications therefore often call for 20 to 30% overmatch which can be easily achievable in 450 and 550 grade steels with satisfactory weld property performance (HSE 2003).

A problem arises in producing the necessary combination of strength and toughness in welds and HAZ required to overmatch steel at yield strength levels higher than 550 MPa, for example steel of 620 and 690 grade (du Plessis 2024). Weld metal properties should be reasonably insensitive to variations in welding procedure parameters, such as heat input and inter-pass temperature, especially during manual welding (Ramirez 2007). As the parent material strength increases, greater precautions are needed to ensure that welding procedures are satisfactory to avoid hydrogen embrittlement (HSE 2003).

Weld metals for welding high strength steels have typical alloying elements in the ranges of (Karlsson et al. 2007).

- Carbon 0.04–0.08%
- Manganese 1–2%
- Silicon 0.2–0.5%
- Nickel 1–3%
- along with some additions of Cr, Mo and sometimes Cu.

As the alloying content and strength increase, bainite and martensite gradually become the dominant microstructural components rather than the softer phases associated with strength levels less than 690 MPa. Although well-balanced mixed (martensitic/bainitic/ferritic) microstructures can offer attractive combinations of properties, the microstructure and hence the mechanical properties, tend to become sensitive to the cooling rate of the steels (du Plessis 2024). However, it has been shown it was possible to maintain a minimum impact energy of 35 J at – 60 °C for weld metals with minimum yield strengths between 350 and 690 MPa (Kobelco Welding Today 2003).

Weld metal microstructures are determined primarily by the chemical composition, the number of non-metallic inclusions present in the microstructure which affect phase nucleation, and by the cooling rate. Weld metal alloy design aims to maximise the amount of acicular ferrite present and to minimise the effects of undesirable microstructures such as coarse grain size, grain boundary ferrite and coarse martensite/austenite/carbide constituents (MAC). Welding consumables employ sophisticated alloying techniques, incorporating the optimum balance of deoxidising elements (aluminium, silicon and manganese), to produce a high density of small non-metallic inclusions which are known to act as intragranular nucleation sites for acicular ferrite. The carbon content is generally kept low to aid weldability, so increased strength is achieved through additions of molybdenum in SAW wires and titanium-boron in FCAW wires, with impact toughness improved by nickel additions (HSE 2003).

The need for pre-heating and inter-pass control to avoid hydrogen cracking problems increases with higher strength parent steel. As an example, for 690 grade steels, preheat temperatures of 125 °C are recommended and electrodes and fluxes with very low hydrogen content should be used to prevent hydrogen cracking (Bennett et al. 1992). The degree of pre-heating depends on the thickness of the plate material and for thinner plates (e.g. 40 mm) no pre-heating may be required resulting in significant time and cost savings.

The formation of hard or brittle phases can affect the toughness of the weld region (HAZ and the weld itself) during multi-pass welding. This may influence the ability to withstand exposure to hydrogen cracking. Significant factors during multi-pass welding are:

- the grain size in the grain coarsened HAZ near to the fusion line,
- the microstructural changes that occur in the weld metal during subsequent weld passes.

In general, the Charpy toughness of the coarse grained HAZ decreases with increasing heat input and increasing impurity content (HSE 2003).

To limit coarse grained microstructures in the HAZ, restrictions were often placed on the maximum heat input that can be used (e.g. 3.5 kJ/mm for submerged arc welding). Productivity was not greatly compromised for generally thinner sections with smaller volumes of deposited weld metal (HSE 2003). However, steel that could be welded satisfactorily at higher heat input levels would offer economic advantages particularly for thicker sections (Krancioch 1994).

Published literature in the 1980's indicated that there were weld consumables which could produce the necessary strength properties required in service, even for the 690 grade steels (Bews 2000). However, for the very highest strength steels there was much less experience of weld consumables with suitable fracture toughness properties. Significant pre-heating and inter-pass control were necessary to avoid hydrogen cracking problems (HSE 2003).

Later work at Cranfield University in the 1990's (MTD 1995; Billingham et al. 1998) showed that certain combinations of steels and weld consumables could limit the extent of the coarse grained HAZ even at high heat input levels and could therefore offer the economic advantages of higher productivity.

The 1998 Cranfield report (Billingham et al. 1998) found that consumables up to 550 MPa yield strength showed adequate toughness throughout the weld, with upper shelf values > 150 J and the 50 J impact transition temperatures below – 60 °C. All the welds had low hardness values and showed no indication of hydrogen cracking. Acicular ferrite was the major microstructural feature of the welds and microstructures generally coarsened as heat input increased.

The Cranfield report also showed that 690 MPa yield strength consumables for both SAW and FCAW showed mixed microstructures containing acicular ferrite, martensite and polygonal ferrite. The impact properties were inferior to those of the 550 MPa consumables with upper shelf values ranging from 80 to 100 J and 50 J impact transition temperatures between –50 and – 80 °C. It was concluded at the time that more development work was needed before these consumables could be specified generally for offshore application.

HSE (2003) reported that the residual stresses in restrained high strength steel joints could be less than those encountered in lower strength steels which can have an important influence on subsequent fatigue and fracture behaviour. Bennett et al. (1992) claimed that the residual stresses in a 40 mm thick restrained joint were only approximately one half of those obtained with mild steel and were largely independent of heat input. The explanation of this effect was thought to be partial counterbalancing of the thermal contraction stresses by the martensitic/bainitic transformation which occurred during cooling. The potential benefit of this different behaviour is reflected in smaller limiting defect sizes from an engineering critical assessment than would otherwise be the case.

HSE (2003) details the welding practices and procedures successfully used for welding the Elgin jacket where high strength steels, Superelso 500 and Superelso 600, were used in leg chords (Bews 2000). These materials have specified minimum yield strengths of 450 MPa and 550 MPs and yield ratios of 0.78 and 0.80. Significant use was made of MMA welding during the fabrication because of its fully positional welding on site capability. Oerlikon electrodes, Tenacito 70, were used which gave yield strengths between 490 and 550 MPa and excellent Charpy V-notch toughness values of 130 J at – 40 °C. The welding of the chord to the prefabricated 160 mm thick rack sections was carried out with a minimum preheating temperature of 150 °C followed by a dehydrogenation treatment of 2 h at 200 °C. Very low repair rates (< 1%) were reported for the project.

The assessment of an ageing structure containing high strength steel therefore requires particular attention to be given to the welding consumables and welding procedure that were utilised for the high strength welds. The assessment should determine the degree of over-matching of the yield strength of the weld metal to the parent material and if possible, measure the hardness and microstructures of the weld and HAZ. The toughness of the weld and HAZ and the susceptibility of the weld to hydrogen cracking are the main questions to be answered.

4.10 Welding of Nodal Tubular Joints

The welded joints (nodes) between the tubular members of legs, chords and braces of the main structure are complex that must be able to transfer load and resist cracking in-service. Gulf of Mexico practice was to use point to point construction methods, that is braces were directly welded to legs using a single sided weld on-site. Some of the nodes for early Southern North Sea structures were fabricated in this way.

The harsher environmental conditions and deeper water in the Northern North Sea meant that design and fabrication of joints against fatigue failure and brittle fracture became paramount. To achieve high-quality joints, specialized welding procedures were required enabling a fast deposition rate with minimum defects. The procedures typically used automated multi-pass welding techniques normally including pre-heating and controlled cooling after the welding but generally without post weld heat treatment. This required the fabrication of nodes as sub-assemblies with a stub or can for subsequent butt welding to the tubular members and was a North Sea innovation for Forties platforms (Cotton 1979).

To minimise the possibility of fatigue failure and brittle fracture, section thicknesses had to be increased to keep stresses low at highly loaded locations such where a brace joins a leg or chord. A consequence of the use of a thicker plate and welds meant that the risk of brittle fracture was increased, and post weld heat treatment would have been required where the joint thickness exceeded 38 mm. Since heat treatment of a completed frame was impractical, the option was to fabricate sub-assemblies which incorporated the thicker sections required at joints that could be post weld heat treated if required. These welds were partly fabricated under cover in specialised welding shops and partly on site under temporary cover (tents) to enable controlled conditions to be maintained.

The tubulars and nodes were then moved to the fabrication yard where the final assembly of the structure took place. Here conditions were less controlled typically under temporary and local cover. This welding, critical to integrity, was more difficult since it often meant working at height and under possibly damp conditions. This increased the chance of welding defects occurring, particularly of hydrogen cracking. Where access to the inside of tubulars was not possible, for example in the case of closure welds, welds were single sided which increased the likelihood of lack of root penetration defects.

This put onerous requirements on non-destructive testing teams to find welding flaws in the site welds between tubulars and nodes, especially those that might be located at the weld root. In some structures a cut-out was created in the tubular for a welder to get inside and fabricate a double-sided weld to the node. This practice has proved to be problematic due to defects in the single sided weld created when the cut-out was replaced.

Jacket structures contain different types of nodal joints such as T, X, Y and K joints, see Fig. 4.5. The choice of joint type depends on the load transfer path and structural performance (strength, fatigue endurance, etc.) required for the specific joint. Welding procedures were developed to be appropriate for different configurations of joint (such as X, T, Y and K joints) and were typically welds of full penetration without unfused land. Internal ring stiffeners were sometimes used to add strength to the joints. After the nodes had been fabricated, their ends were tapered so they could be joined to the tubular legs and braces which had thinner wall thicknesses using butt welds which were normally single sided.

The joint of two tubular subassemblies is typically not linear or in one plane and therefore the welding of these joints is challenging and may have to be done using a manual process. The welds between tubulars could be single or double sided depending on whether there is access for welding from one or both sides. However, as shown in Fig. 4.6 access to both sides can be limited by steep angle and limited space between the two members along the circumference (particularly locations 2–3 in Fig. 4.6) and a full penetration weld may not be achievable, Fig. 4.6. Locations with partial penetration welds are particularly vulnerable to fatigue cracking and should be given special attention in an assessment of an ageing structure.

Cast nodes are an alternative product form developed to avoid the defects and discontinuities associated with welding and providing an efficient technique for creating the shape required (see Sect. 3.7). The presence of casting defects, such

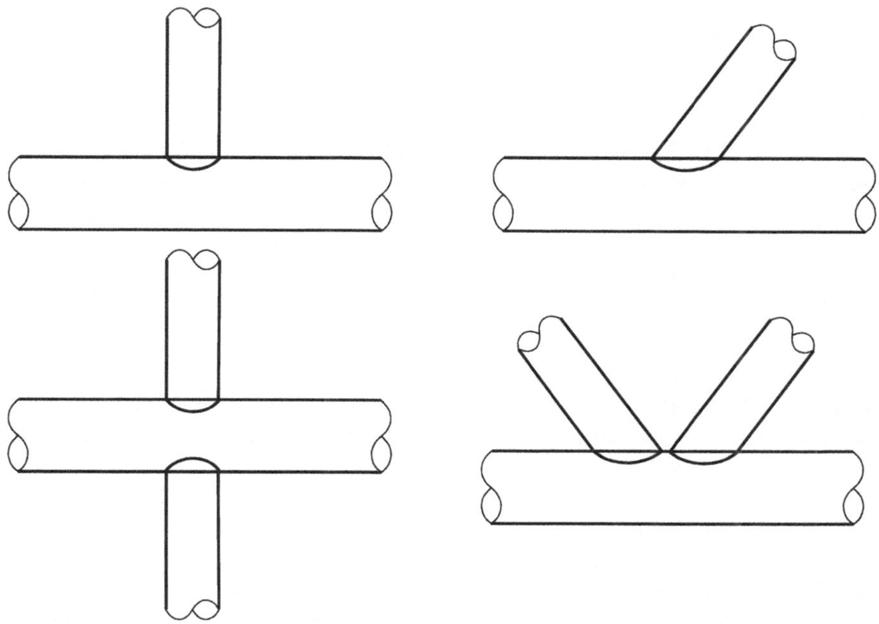

Fig. 4.5 T, Y, X and K-joints

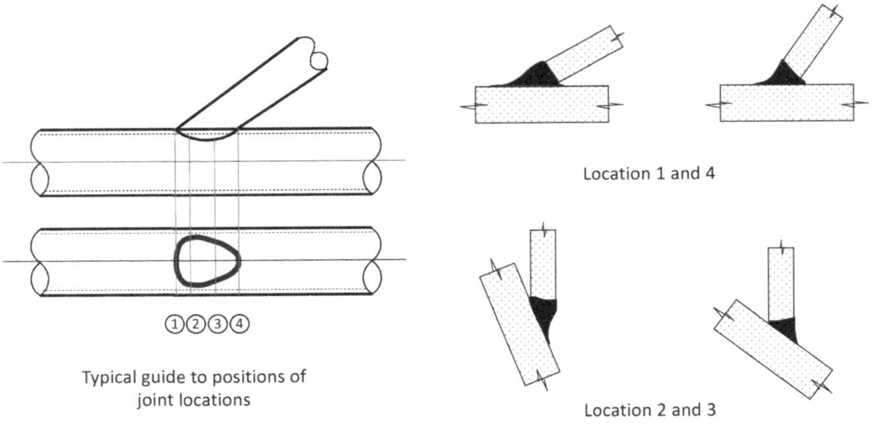

Fig. 4.6 Welds in a tubular joint

as excessive porosity, which might need to be repaired by welding, thereby creating a potential defect and local residual stress, is a significant problem.

4.11 Single Sided Closure Welds

To reduce the likelihood of fatigue cracking, full penetration welds made from both sides supported by detailed NDE during fabrication were generally recommended for platform primary structures. However, where access from the inside of a tubular member is not possible or convenient, single sided welds were commonly used. The inaccessibility of the root of single sided welds and difficulty for inspection and repair creates the possibility of weld root defects that can develop into fatigue cracks later in life.

Single sided welds were typically specified for connecting braces to brace stubs and chords to chord cans, particularly in large jacket installations. The joining of cast nodes to the associated tubular members, such as legs and braces, was normally by single sided welds. Single sided welds were also often used for other thickness transitions which have added complications. They are also found where windows in tubular members have been cut to allow double sided welding of a nearby nodal joint and later closed by a single-sided welded plate. It is particularly important to be aware of misalignment between plates which can be a hot-spot for fatigue cracking.

When assessing an ageing structure for life extension, the assessment engineer should determine whether there are welds made from one side. Where these are in critical locations, inspection of these sites for fatigue cracking would normally be required. Experience in the North Sea has found fatigue cracks associated with single sided welds that have needed significant repairs.

4.12 Bolting Materials

4.12.1 Applications and Design Standards

Threaded fasteners used in offshore structures for energy production can be a key structural component. The fastener could be a threaded bolt with an integral head or a threaded stud without a head. The threaded bolt or stud can either screw into a nut or a threaded hole in the component. The term bolted connection will be used to refer to the general class of threaded fasteners in the following section.

Bolted connections are a well-established proven technology, particularly for onshore applications, and the parts are readily available and quick to install. For offshore applications the design of the connection has tended to follow normal onshore practice, but due consideration should have been made for the fatigue loading. Aside from fatigue, most of the differences from conventional design arise from the effects of the marine environment in relation to corrosion and hydrogen embrittlement.

While fixed tubular structures and mobile floating installations are generally of welded construction, bolted connections are used where parts may need to be disassembled or where access for welding is not possible. Bolted connections are used extensively within the topside structure, for example, the "tie-down" of topside modules and fastening of grating plates. Their subsea application is more restricted and is mainly for pipelines, as connections, flanges, and collars. Bolts are used in flanged joint connections, for example, in topside piping, process equipment, risers, blow out preventers and pipelines, although these applications are outside the scope of this book.

A bolted connection can be a cost-effective solution for the repair or modification of existing platforms with minimum impact on platform operations and avoidance of hot working. Bolted connections are used for the assembly of repair clamps for strengthening of damaged tubular members (Sharp and Ersdal 2021). They are sometimes used as temporary measures to allow time for a more permanent repair solution and may also be attractive if access is difficult for welding or grouting. In principle, bolted repair connections subsea are capable of being deployed by ROV or other diver-less methods thereby minimizing the need for divers.

Nowadays bolted connections for fixed steel offshore structures would be expected to meet the specification given in Sect. 15.2.8 of ISO Standard 19902 (Fixed Steel Offshore Structures). The standard sets out aspects of materials and manufacturing that should be considered together with strength and fatigue criteria and provides requirements and guidance for installation and inspection of the connection. The standard also provides a specification for the design of clamps to strengthen or repair tubular members and joints. When assessing bolted connections in an older or aged fixed steel structure for life extension or continued service it is useful to review the connections against this modern standard and determine the extent to which the connections comply.

Earlier guidance for the strengthening and repair of tubular steel offshore structures was provided in HSE report OTH 88 283 (HSE 1988). This considered the two types of bolting system that were used in repairs (high strength friction and long bolts) and noted that the typical bolting materials used offshore at that time were ASTM A193 B7 stud-bolts, Macalloy bars, nickel alloys such as Monel K-500 and high strength friction type short bolts. The report noted that while some failures had occurred in Macalloy stud-bolts that appeared to be due to stress corrosion cracking, the other materials were more resistant.

Bolted connections also play a key role in wind turbines. The sections of the tower, transition piece and monopile are typically joined using large internal bolted flanged connections with bolts in sizes up to M72. Bolted connections are also used to attach the nacelle bearing assembly to the top of the tower.

The main standards used for evaluating the design of bolted connections used in wind turbine support structures (towers and foundations) are the German DIN standard 18088-3 (DIN 2019) and the DNV Standard ST 0126 (DNV 2021). These documents specify suitable materials and construction and define the loads and actions that designers should consider. They provide methods that designers can use to size bolted flange connections to avoid plastic collapse and assess the fatigue life of the bolts and other components. In both standards, the method to assess the fatigue life of the bolts from fluctuating loads on the tower uses the simplified model according to Schmidt and Neuper (1997). With respect to fatigue design and plastic limit load, the Schmidt-Neuper model takes account of the effect of the prying bending moment from the tower on the flange and the bolts.

4.12.2 Materials Standards and Manufacturing

Bolting materials (bolts, studs and nuts) used on offshore installations and subsea are usually made of high strength carbon and heat treated chromium molybdenum alloy steels. The latter steels have a moderate carbon and manganese content and are strengthened by the addition of chromium and molybdenum and with appropriate heat treatment. Bolts threads are manufactured either by hot or cold rolling (forging) process or cut by turning from round bar. There are a range of possible thread forms. Some thread forms cannot be reused and retightened if they are undone after initial assembly.

Historically, the UK Department of Energy Guidance Notes (e.g. Department of Energy 1984) recommended that bolts and nuts for use in offshore structures should comply with relevant British Standards. The equivalent modern international standard covering the mechanical properties of pre-loaded fasteners is ISO 898-1 (ISO 2013).

The equivalent American standards for low alloy bolting are ASTM A193 for grade B7 (ASTM 2019) and A320 for grade L7 (ASTM 2022) based on AISI 4140 steel (SAE 2022) which has been liquid or air quenched. Grade B7 is recommended for medium to high temperature service and does not have a required impact test

Table 4.1 Bolt specifications with yield and ultimate tensile strength

Standard	Bolt category/ grade	Yield strength (min) 0.2% non-proportional elongation), MPa	UTS (min) MPa	References
BS 3692	8.8	640	800	British Standards (2014a)
BS 3692	10.9	900	1000	
BS 3692	12.9	1080	1200	
ISO 898-1	8,8	640	830	ISO (2013)
ISO 898-1	10.9	900	1000	
ISO 898-1	12.9	1080	1200	
ASTM A193/A320	B7/L7	720	860	ASTM (2019) and ASTM (2022)

value while the grade L7 is intended for low temperature service and is required to have a minimum Charpy impact value of 27 J at minus 10 °C.

Bolts used on offshore structures are typically in two or three strength grades. Higher strength bolts, typically with a minimum yield strength of more than 700 MPa, such as Grades 8.8, 10.9 and 12.9 or equivalent are heat treated to achieve the required mechanical properties. The yield strength and ultimate tensile strength for bolts specified in these standards are shown in the Table 4.1.

The specified chemical compositions according to these standards are shown in Table 4.2. The composition requirements are somewhat more complicated than shown in the simplified table below with respect to the addition of other elements and combinations of elements and the relevant standard should be consulted. The compositions for ASTM Grades B7 and L7 bolting material are the same and these bolts only differ in the latter requirement for being impact tested.

Bolts used offshore typically have higher yield strengths than bolts in normal industrial usage, and consequently can be more susceptible to hydrogen embrittlement and can have lower fatigue lives. The fatigue performance of bolts also depends on whether they are rolled or cut and in the case of rolled bolts, whether rolling takes place before or after heat treatment (see Sect. 6.7.16).

4.12.3 Bolt Assembly

Bolted connections are tightened to clamp parts together and transmit loads across a joint. The clamping force or pre-load applied to the bolt compresses the clamped parts, particularly around the bolt where a washer may be applied to distribute the clamping load over the surface. The different stiffnesses of the bolt and clamped parts

Table 4.2 Chemical composition of bolt materials

	C %	Mn %	P %	S %	Si %	Cr %	Ni %	Mo %
ISO 898-1 C steel Q&T	0.25–0.55		0.025 max	0.025 max				
ISO 898-1 Alloy steel Q&T	0.20–0.55		0.025 max	0.025 max		0.3 min	0.3 min	0.2 min
ISO 898-1 C–Mn steel Q&T	0.20–0.55	0.6–0.7	0.025 max	0.025 max				
ASTM A193/ A320 B7/L7	0.37–0.49	0.65–1.10	0.035 max	0.040 max	0.15–0.35	0.75–1.20	N/A	0.15–0.25

ensure that a correctly specified bolted assembly can carry load without a significant change in the tension of the bolt.

Bolts are generally tightened to a high but not excessive load. In offshore structures, the initial bolt pre-load typically ranges between 40 and 60% of the yield stress of the bolt, but 80% of the yield strength is achievable in the field with a bolt tensioner and up to 90% of the yield strength under laboratory conditions. The initial tension may relax after tightening by 10–12% over the design life of the structure. A high initial bolt tension offsets the effects of relaxation which can improve fatigue life and leak sealing.

To attain the required pre-load in a bolt, one of the four main tensioning methods are used:

- Torque control where control of pre-load is achieved from the torque applied to the nut by the tool.
- Torque and Turn of the nut control where control of pre-load is achieved from the angle of the turn of the nut.
- Tension control (stretch control) where pre-load is achieved from controlled stretching of the bolt.
- Pre-load control where the pre-load is achieved from direct measurement (strain gauge or similar) of the bolt tension.

These methods all have a degree of uncertainty in the pre-tension achieved ranging from 10 to 30% (Energy Institute 2021). Torque control typically has a 30% uncertainty while tension controlled (such as using a Hydratight bolt tensioner) typically has a 10% uncertainty. Such uncertainties particularly occurs if friction has not been considered and the bolt is tightened dry without lubrication. Whichever method is used, the application of the pre-tensioning tools to establish the required pretension is a challenge.

Like welding, the assembly of a bolted joint for a pressurised joint or other safety critical application is a process that should nowadays be undertaken by qualified personnel and subject to a written procedure. The relevant standards are EN 1591-4 (EN 2013) and ASME PCC-1 (ASME 2010). Historically there were no formal requirements to qualify personnel for assembling bolted connections (Energy Institute 2021). Hence, for bolted connections on older or aged structures, there may be even more uncertainty in the quality of the assembly and the subsequent performance of the joint.

If the bolt is not sufficiently tensioned, vibration may cause loosening of the assembly. The possible consequences could be fatigue of the bolt and leakage where the joint is used for sealing. Insufficient tensioning also lowers the fatigue life if there is not enough clamping force to reduce the fluctuation of loads in the bolt. In contrast, overloading of the bolts during tightening can lead to failure of the connection for various reasons, including permanent yielding of flanges and clamped parts, thread stripping and cracking of the bolt. More information on the fatigue performance of bolted connections is given in Sect. 6.7.16.

As a result, when assessing the integrity of an ageing installation for continued service or life extension, it is important to know how the joint was assembled and the bolts tightened and the intended pre-load and to determine the extent to which the bolt tension relaxation has taken place during life. If such information is not available, the replacement of the bolts should be considered for critical components.

4.12.4 Service Performance

A study of bolts for sub-sea installations was undertaken by DNV for PSA (DNV 2005). This study considered typical applications, the technological status and limitations and uncertainties associated with bolted connections. The report highlighted the need for continuous integrity management of bolted joints referencing the guidelines originally developed by the UK Offshore Operators Association and now published by the Energy Institute (2021). These guidelines identify seven essential elements of a management system for bolted joints:

- *Ownership*: Identification of responsibility for implementation, maintenance and communication throughout the organisation.
- *Awareness* with respect to the management system, its objectives, expectations and effect.
- *Tools*: Implementation of tools to meet expectations, such as risk assessment, competence management and control of the used practices.
- *Review*: System reviews, verifications and audits at agreed intervals.

- *Records and Data Management*: The recording of traceable data will encourage a best practice.
- *Learning* from incidents will improve performance.
- *Measurement:* Quantification of contribution being made by the management system. Evaluation of user satisfaction.

A study of incidents offshore involving bolted connections was published by the Petroleum Safety Authority (2019). A search of the incidents database found that 331 were bolt related. While it was difficult to find any trends relating to failure mechanisms and root causes, the report highlighted recommendations for the initial assembly and maintenance of bolted connections including the achieving of the correct bolt pre-tension. PSA recommended that new design procedures are validated based on optimized preload, fatigue, and fracture assessment.

The report comments that despite replacing many bolts and efforts to improve manufacturing, construction, and installation there have been continued failures of safety critical bolts used subsea. While mechanisms of failures are not fully understood, hydrogen embrittlement and stress corrosion cracking (SCC) are the most likely reasons for these failures.

The integrity of the connection between the transition piece and monopile, which is subject to large cyclic bending loads from the action of the wind on the blades through the tower and well as the action of waves and currents on the monopile, has prompted new research into the fatigue performance of large bolts and the effects of loss of bolt pre-tension and variations in the contact area between the flange surfaces, which is discussed more in Sect. 6.7.16.

4.12.5 Viking AD Platform Bolted Grouted Clamp Repair

An example of the use of a bolted connection to strengthen a tubular joint is the bolted grouted clamp fitted to the Viking AD platform. The Viking AD platform was installed in December 1970 in the Southern area of the North Sea in 24 m of water depth The clamp was fitted to the structure in 1982 below the mean water level where it was cathodically protected. The platform was decommissioned in 1993–1994 and the clamped joint removed and taken to the Health and Safety Laboratory at Buxton for examination.

The recovered joint (see Fig. 4.7) consisted of tubular members from the platform, the clamp steelwork, the threaded stud bolts and grout between the tubular and the clamp. Tests and examinations were subsequently carried out on each of these components after standing for a year during in the open. (HSE 2002a, b). At the time of the testing and examination the original steelwork was approximately 23 years old with the clamp less aged having been in sea water for only 12 years.

A visual examination found the nuts on the threaded stud-bolts to be in very good condition. The only components that appeared to have suffered any noticeable corrosion were parts of the threaded stud-bolts themselves, which had suffered a

Fig. 4.7 Bolted clamp used on the Viking AD platform after decommissioning (courtesy of John V. Sharp)

small, insignificant amount of corrosion, but could be undone after cleaning. There was minor general corrosive attack at locations where the bolts passed through the pockets on the main structure, but the thread form was still readily visible. This general corrosion could have taken place during the year in which the clamp was exposed to the air. Other areas of the threaded stud-bolts were in very good condition.

A metallographic sample from one of the stud-bolt ends revealed that the steel was in the hardened quenched and tempered condition. Subsequently chemical analysis was carried out. The certificate of analysis is given in Table 4.3 and indicated that the stud-bolts had been made from a medium carbon, low alloy steel, later confirmed as ASTM A320 Grade L7 in AISI 4140 steel, a grade frequently used for marine bolting applications.

A tensile test was then performed on a test piece manufactured from one of the stud-bolt ends. The tensile strength of the stud-bolt material was found to be 930 MPa, with 0.2% proof stress of 825 MPa, and thus the tensile properties are equivalent to those of a grade 8.8 bolt.

It is understood that the stud-bolts should have been tensioned in accordance with a note which stated, "Monel stud-bolts (12 No per clamp end) to be simultaneously tightened to 88 KIP (88,000 psi) including 10% preload using Hydratight 'Seaser-pent' bolt tensioners". The Hydratight bolt tensioning system enables a bolt to be tightened to high loadings without the need for large wrenches. HSL used a single

Table 4.3 Chemical composition of stud bold materials (ASTM 2019)

Element	C %	Mn %	P %	S %	Si %	Cr %	Ni %	Mo %	Co %	V %
	0.42	1.01	0.009	0.022	1.10	1.04	0.39	0.24	0.25	< 0.01

actuator version of this technique in order to dismantle one end of the clamp. It was found on testing that all the studs had a remaining pre-tension that was still above the design value.

References

API (1993) Planning, designing, and constructing fixed offshore platforms

ASME (2010) ASME PCC-1 guidelines for pressure boundary bolted flange joint assembly

ASTM (2019) ASTM A193 B7 Specification for high tensile bolts, screws, stud bolts, threaded rods, anchor bolts and other externally threaded fasteners, made from chromium-molybdenum, alloy steel

ASTM (2022) ASTM A320 Standard specification for alloy-steel and stainless steel bolting for low-temperature service

AWS (2020) AWS D1.1 structural welding code

Bailey N (1994) Weldability of ferritic steels. Abington Publishing, ISBN 1 85573 092 8

Bennett WT, Cadiau L, Caudreuse L (1992) Steels for jack-up legs. In: Boswell LF, D'Mello C (eds) Recent developments in Jack-up platforms

Bews R (2000) MMA welding gas from strength to strength. Welding and materials fabrication, July/August 2000

Billingham J, Blackman S, Norrish J (1998) Further assessment of high strength weld metals for use in offshore engineering applications. Cranfield Final Report, January 1998

British Standards (1982) BS 6235 code of practice for fixed offshore structures. British Standardisation Institution

British Standards (1984) BS 5135:1984 Specification for arc welding of carbon and carbon manganese steels. British Standardisation Institution

British Standards (1985) BS 4871:1985 Specification for approval testing of welders working to approved welding procedures—arc welding of tube to tube-plate joints in metallic materials. British Standardisation Institution

British Standards (1986) BS 4360 British Standard specification for weldable structural steels. British Standardisation Institution

British Standards (2014a) BS3692 ISO metric precision hexagon bolts, screws and nuts—specification

Cotton HC (1979), Welded steel for offshore construction, 15th John Player Lecture, Proc Inst Mech Eng 193(26)

Dawson GW, Judson P (1982) Procedural guidelines for the achievement of tough welded joints in structural steels for offshore applications, paper 2. In: Second international conference on offshore welded structures, London, November 1982

Department of Energy (1984) Third edition of the guidance notes for the design and construction of offshore structures. HMSO London, UK

DIN (2019) DIN standard 18088-3 Structures for wind turbines and platforms—Part 3 Streel structures

DNV (2005) Joining methods—technological summaries. DNV Report 2005-3394 DNV, Høvik, Norway

DNV (2020) Offshore standard OS B101—metallic materials

DNV GL Standard ST 0126 (DNV GL, 2018)

DNV (2021) DNV-ST-0126 Support structures for wind turbines

du Plessis J (2024) Welding of offshore structures. Published at Spesmet Technology Web-pages, Microsoft Word—Welding of Offshore structure 2012 (spesmet.co.za), visited August 2024

Energy Institute (2021) Guidelines for the management of integrity of bolted pipe joints

EN (2001) EN 1011-2 Welding. Recommendations for welding of metallic materials. Arc welding of ferritic steels. CEN

EN (2013) EN 1591-4 Flanges and their joints—Part 4: Qualification of personnel competency in the assembly of the bolted connections of critical service pressurized systems. CEN

Holdcroft P (1979) Welding process technology. Cambridge University Press

HSE (1988) Report OTH 88 283, grouted and mechanical strengthening and repair of tubular steel offshore structures—Part 2

HSE (2002a) Assessment of strengthening clamp from the Viking offshore platform: Phase I. Offshore technology report 2000/057. HMSO, London, UK

HSE (2002b) Assessment of strengthening clamp from the Viking offshore platform: Phase II. Offshore technology report 2000/058. HMSO, London, UK

HSE (2003) RR 105, Review of the performance of high strength steels used offshore. HMSO London, UK

ISO (2013) ISO 898-1:2013 Mechanical properties of fasteners made of carbon steel and alloy steel Part 1: Bolts, screws and studs with specified property classes—coarse thread and fine pitch thread. International Standardisation Organisation

ISO (2020) ISO19902:2020 Petroleum and natural gas industries—fixed steel offshore structures. International Standardisation Organisation

Karlsson L, Keehan E, Andren HO, Bhadeshia HKDH (2007) Development of high strength steel weld metals. Eurojoin 5, Vienna, May 2007

Kobelco Welding Today (2003) Welding of offshore structures Part 1. Kobelco Welding Today 6(1), Jan 2003, kwt2003-01.pdf (kobelcowelding.nl)

Krancioch JJ (1994) Fabrication of offshore structures and potential productivity increases. MTD 1994, The safe design and fabrication of offshore structures, Sept 1993, Publ. IBC

Moan T, Bekkvik P, Næsheim T (1981) Alexander L. Kielland-ulykken. Norges Offentlige Utredninger, NoU 1981:11, Universitetsforlaget, Oslo, Norway

MTD (1995) High strength steels in offshore engineering, MTD Publication 95/100, ISBN 1-870553-21-7

Pisarski 2023—private communication

Pisarski HG, Harrison JD (1978) Fracture resistance of materials used in offshore structures. Select seminar: European offshore steels research, Cambridge, November 1978. Welding Institute on behalf of the Department of Energy

Petroleum Safety Authority (2019) Bolted joints study on bolt incidents. Work performed by DNV-Gl for the Petroleum Safety Authority, DNV GL report number 2018–5333, bolted-joints-study-on-bolt-incidents.pdf (havtil.no)

Ramirez JE (2007) Characterization and the mechanical properties of High strength steel weld metals, In: Proceedings of the sixteenth international offshore and polar conference, Portugal

SAE (2022) Steel, bars, forgings, mechanical tubing, and forging stock 0.80Cr–1.8Ni–0.25Mo (0.38–0.43C) (SAE 4340) normalized and tempered AMS6484F, SAE International

Schmidt H, Neuper M (1997) Zum elastostatischen Tragverhalten exzentrisch gezogener L-Stöße mit vorgespannten Schrauben. Verlag Ernst & Sohn. Stahlbau 66, vol 3, pp 163–168

Sharp JV, Ersdal G (2021) Underwater inspection and repair for offshore structures. John Wiley & Sons Ltd, Chichester, UK

Steen Ø (1993) På dypt vann—Norwegian Contractors fra 1973–1993. Norwegian Contractors, Stavanger, Norway

TWI (2024a) Carbon equivalent formulae in relation to hydrogen cracking. TWI Webpage, . TWI (twi-global.com), visited August 2024

TWI (2024b) What is welding? Definition, processes and types of welds. TWI (twi-global.com), visited September 2024

UEG (1985) Underwater engineering group (UEG) design of tubular joints for offshore structures. UEG. ISBN 0860172317, 9780860172314 (Section 3.11)

Chapter 5
Steel Materials After Fabrication

5.1 Types of Fabrication Defects in Offshore Structures

Various defects which could compromise the structural integrity may be introduced to a steel structure during fabrication. These include:

- Welding defects and cracking (see Sects. 5.2 and 5.3),
- Misalignment of components during assembly which can lead to eccentricities and uneven stress distributions,
- Thermal expansion and contraction can cause distortion or warping of steel components which could affect their dimensional accuracy and fit-up,
- Surface irregularities such as scratches, gouges, or pits which can reduce the effectiveness of protective coatings and increase susceptibility to corrosion.

The normal procedure for fabrication of steel offshore structures is that all such defects are reported and, if found necessary, mitigated or repaired. However, experience has shown that these initial defects can lead to later deterioration at these locations.

To minimize these defects, the use of robust materials (see Sect. 5.5), proper welding procedures, quality control measures and adherence to industry standards and specifications (see Sect. 5.6) are essential during the fabrication process. Additionally, pre-fabrication inspection of materials, meticulous handling practices and post-weld heat treatment (see Sect. 5.4) can help mitigate the likelihood of defects.

© The Author(s), under exclusive license to Springer Nature Switzerland AG 2025

J. V. Sharp and J. Wintle, *Assessment of Materials in Ageing Offshore Structures*,
https://doi.org/10.1007/978-3-031-84489-8_5

5.2 Types of Fabrication Welding Defects in Offshore Structures

5.2.1 Introduction

When assessing an ageing structure for life extension the review must consider the prevalence of defects introduced during welding fabrication that were undetected by manufacturing inspections and remained in the structure when it entered service. The likelihood of fabrication defects remaining in a structure entering service depends on both the likelihood of their occurrence and the likelihood of detection by manufacturing inspection and subsequent removal and repair. A particular case study of defects of two offshore steel structures is given in Thurlbeck et al. (1996a), as discussed later. Such fabrication defects can act of initiators for fatigue cracks that can grow during service and threaten the integrity of the structure.

In this section we consider the various types of welding fabrication defects and their causes. Welding fabrication defects can be roughly divided into two types: those that are due to the process conditions of welding and those due to metallurgical factors, although the two sometimes interact. Examples of the former include lack of fusion, lack of penetration, undercut, porosity and inclusions, which the latter type includes hydrogen cracking, stress relief cracking and lamellar tearing.

Some types of defects, such as solidification cracks and stress relief cracks, occur mainly in the weld metal while other types occur in the parent material and heat affected zone including hydrogen cracks and lamellar tearing. Lack of fusion, lack of penetration and undercut are associated with the fusion boundary between the weld metal and the HAZ.

5.2.2 Solidification or Shrinkage Cracks (Hot Cracks)

Solidification cracking, otherwise known as shrinkage cracking, occurs in the weld metal when there is rapid cooling. These conditions often arise when large volumes of weld metal are deposited in a single pass with high heat input rates. Even when the cooling rate is controlled, cracking can occur where there are inclusions or other stress concentrators.

The cause of solidification cracking is the very high residual stresses generated when there are steep temperature gradients in the weld region. While generally associated with the deposited weld metal, excessive pre-heat or inter-pass temperatures can also cause cracking in weld metal that has already been deposited for the same reason. Solidification cracking can be controlled by maintaining the heat input rate of welding within the band defined in the weld procedure and avoiding higher heat inputs.

The likelihood of cracking increases when high production rates have been demanded. When reviewing the susceptibility of an ageing platform to this form of cracking reviewers should examine the welding procedure and welding records and search for any evidence of excessively high heat inputs and deposition rates. Although it is difficult to ascribe any period as being more vulnerable than any other, the welding of the early platforms with thick sections and platforms ordered under contracts with penalties for late delivery might be more at likely to contain solidification cracks.

Solidification cracks are generally surface breaking and wider than hydrogen cracks. They can be detected by surface inspection using for example MPI and dye penetrant methods. They tend to be transverse to the weld direction but are sometimes angled across the weld when they are known as chevron cracks.

5.2.3 Hydrogen Cracks (Cold Cracks)

Hydrogen cracking is one of the most common flaws associated with fabrication welding of offshore structures. It is also called cold cracking (because it usually forms when the joint has cooled sometimes after several hours), weld toe cracking or HAZ cracking (because cracks often (but not always) form in the heat affected zone). Although mainly associated with the HAZ, hydrogen cracking can also occur in the weld metal if the joint is highly constrained and or low carbon consumables are being welded without preheat. Cracks usually form at stress concentrations such as the weld toe or root regions or at voids or from other imperfections such as porosity or lack of fusion within the material.

The source of hydrogen can be from the flux coating or welding consumable or from paints grease or rust on the surface or from moisture on the surface or in the surrounding atmosphere. Hydrogen atoms created by the heat of welding and electric arc are absorbed into the molten weld metal. The hydrogen atoms migrate up stress gradients within the material towards sites of higher restraint and stress concentration where the density of hydrogen becomes sufficient to form metallic hydrides forcing the material to crack in coarse grained microstructures. The coarse-grained microstructure of the HAZ close to the fusion boundary is more susceptible to cracking since because it is normally harder and more brittle than the surrounding material.

The hardness of the HAZ is determined by the composition of the parent steel and the rate of cooling after welding. The critical hardness above which cracking can occur depends on the electrode hydrogen potential, the degree of restraint (for example thickness) of the site as well as the chemical composition of the parent steel. Early steels with high carbon equivalent values and later steels with lower carbon equivalents but with micro-alloying additions of niobium and vanadium were found to have more susceptibility to hydrogen cracking. As a rule of thumb, the offshore sector used a HAZ hardness of 350HV below which no cracking would be expected to occur.

Those assessing an ageing structure should examine the precautions taken to minimise the risk of HAZ hydrogen cracking during initial fabrication. In practice this could be done by using low hydrogen electrodes and/or by reducing the cooling rate by increasing the heat input into the joint and/or by using preheat. The preheat temperature specified should reflect the thickness of the weld, where thicker welds (which tend to cool more rapidly) should have a higher preheat temperature.

Pre-heat also has the advantage of driving away moisture and condensation away from the welding region. Therefore, the application and efficacy of suitable and sufficient preheat is a key factor when assessing the risk of hydrogen cracking in welds of an offshore structure. Structures welded in the open in windy and wet locations were more susceptible to hydrogen cracking particularly if the manner of applying preheat by means of heating pads was not effective in achieving the required temperature or sustaining it during welding.

The understanding of the causes hydrogen cracking and its avoidance improved markedly over the timeline for major construction of offshore platforms of the North Sea and those developments have been available to fabricators of more recent structures. There was awareness of the issue of hydrogen cracking before the construction of the Forties platforms but the real understanding took place in the following decade. Department of Energy Guidance Notes of 1974 (Department of Energy 1974) introduced requirements for pre-heat of thick section welds and these were developed in later editions. Recommended preheat temperatures are given in standards such as BS 5135 (British Standards 1984a).

Significant work was undertaken at the Welding Institute between 1980 and 2000 and led to an understanding of the factors underlying the susceptibility to hydrogen cracking and the development of a diagram relating these factors (Bailey et al. 1993). They discovered that there can be a lag time between the completion of welding and the appearance of a hydrogen crack. Structures where non-destructive testing was carried out immediately after welding could therefore be more at risk of containing undetected hydrogen cracks.

Hydrogen cracking acts as an initiating flaw for fatigue crack growth and reduces the incubation period before a growing fatigue crack forms. The failure by fatigue of the brace of the Alexander L. Kielland platform was initiated by a hydrogen crack associated with the welding of a hydrophone attachment to the surface of the brace after the structure had entered service (Moan et al. 1981).

5.2.4 Stress Relief Cracking

Stress relief cracking is associated with welds of higher strength steels and occurs at elevated temperatures during post weld heat treatment or at operational temperatures. Cracking in the HAZ is driven by residual stress created by welding, and higher restraint welds where plasticity and creep do not allow sufficient relaxation of residual stress are therefore more vulnerable. The problem can be managed by selecting steels with low phosphorus and sulphur content and minimising harmful trace element. A high heating rate of PWHT is also beneficial in reducing the likelihood of this type of cracking.

5.2.5 Lack of Sidewall Fusion and Other Types

Lack of sidewall fusion (see Sect. 4.4) occurs as a result of a failure of molten weld metal to fuse with the surface of the parent material leaving an unbonded area or discontinuity. This may result from incorrect electrode angle, insufficient welding current, incorrect welding technique or lack of adequate joint preparation. Other types of lack of fusion can occur where weld beads do not fuse between passes or at start stop positions when the welding rod is renewed.

Lack of fusions can be particularly difficult to detect and size by conventional NDT techniques because they tend to be buried within the weld and aligned along the fusion line/face. In multi-pass welds lack of fusion defects can be large if repeated pass-by-pass and can threaten the load carrying capacity of the joint. A review of the welding and inspection procedures would indicate whether the conditions for lack of fusions remaining in the structure are present.

5.2.6 Undercut and Lack of Penetration

Under-cut (see Sect. 4.4) is formed when parent metal is melted by the arc and falls away and is not replaced by weld metal leaving a notch. It can be caused by excessive travel speed, too much current (too high power) or incorrect angle of electrode. This type of defects is typically associated with structures that were fabricated with high rates of production welding.

Incomplete or lack of penetration occurs when the weld metal fails to reach, fill or penetrate the weld root. It leaves a void within the weld that weakens the weld and can be a site for hydrogen cracking. In MMA welding lack of penetration is usually caused by using a larger diameter welding rod than is necessary to fill the root. Double sided welds are usually back-gouged at the root to permit full penetration from the other although back gouging can also create lack of penetration if it is not done correctly.

5.2.7 Porosity and Inclusions

Porosity and inclusions occur within the volume of the weld metal and create a weakend zone. Porosity flaws are small bubbles of gas within the weld metal produced by the expulsion of dissolved gases from previous passes or from chemical reactions during welding when unsuitable consumables are used or when the welding surface is excessively rusty or scaled.

Inclusions are any non-metallic matter in the weld metal or parent steel. The main source of inclusions in MMA welds is due to slag from the electrode covering or flux where this has not been adequately removed by cleaning between successive runs.

Slag can be trapped in cavities or undercuts from where it is difficult to remove. Both porosity and slag can be readily detected by radiography and should be removed when above a density defined by welding standards.

5.2.8 Lamellar Tearing and Through Thickness Ductility

Lamellar type cracks may appear when the plate material is subjected to stresses in the thickness direction during welding. The lamellar pattern is caused by separation of weak planes of rolled inclusions (Almar-Næss 1985). A typical example of lamellar tearing is shown in Fig. 5.1.

The welding of thick plates for jackets and other structures brought about the potential problem of lamellar tearing of the parent material beneath where an attachment has been welded to the surface. Lamellar tearing is the delamination of parent material plate on a transverse plane parallel to the plane of welding a short distance away from the fusion face. It typically occurs at restrained joints, such as at nodes where brace members are welded to the surface of the main tubular chords.

Lamellar tearing is caused by the material having reduced fracture properties in the short transverse (through thickness) direction due to the presence of planar inclusions parallel to the surface. In aluminium-treated or vacuum degassed steels lamellar tearing is related to the presence of Type II manganese sulphide, manganese silicate and oxide inclusions, which were present in the early steels (Harrison 1981). It is not generally associated with hydrogen-induced cracking and can occur even when adequate precautions against hydrogen-induced HAZ cracking have been taken.

The inclusion content of an as-cast steel depends on numerous factors such as the type of steel. deoxidation practice, composition, position in the ingot, etc. (Farrar and Dolby 1972). The inclusions are usually formed as spheres, eutectic films, or small angular particles in the solidifying ingot. When the ingot is rolled to form steel plate the inclusions deform into plates or discs parallel to the plate surface. Different

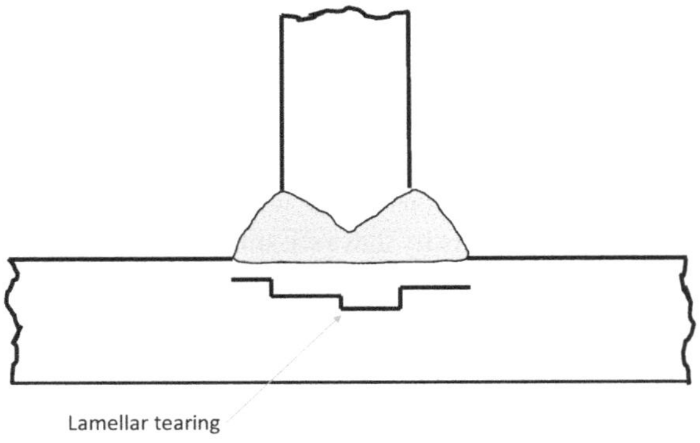

Fig. 5.1 An example of lamellar tearing associated with a welded tee with the distinct step characteristics

types of inclusions deform to different degrees relative to the steel matrix and some may fragment during rolling (Farrar and Dolby 1972).

Even though all steels contain deformed inclusions it is important to note that only a small percentage of steel plates are susceptible to lamellar tearing. Only certain plates have risk of tearing where concentrations of inclusions are coupled with unfavourable shape and type. Lamellar tearing does not so readily occur in modern steels where the chemistry is better controlled.

Three conditions must be satisfied for lamellar tearing to occur (Farrar and Dolby 1972):

1. Strains must develop in the short transverse direction of the plate perpendicular to the plate surface. These arise from weld metal shrinkage but can be greatly increased in the assembly of integrated structures where the weld shrinkage strain is unable to relax due to the restraint imposed by the neighbouring structure.
2. The weld orientation must be such that the fusion boundary of the weld with the parent plate is on or roughly parallel to the plate surface, so the strains act perpendicularly to the surface and through the plate thickness.
3. The material must have poor ductility in the short transverse (through thickness) direction and thus be susceptible to tearing.

Lamellar tearing is characterised by and distinguished from other forms of cracking by having the following features (Farrar and Dolby 1972).

1. The appearance in section is a series of steps with long flat delamination and small steps in height.
2. It lies in a plane parallel to the weld fusion boundary and the plate surface just outside the visible HAZ.
3. Being completely subsurface, it can be difficult to detect using non-destructive methods.
4. The surfaces of the tears are characteristic of low ductility fractures being fibrous and having the appearance of split wood.
5. The area of tearing may be very extensive extending all along weld in some cases. Where the plate ends it may be readily visible on the plate edges or on the surface at or near weld toe.
6. In some instances a loaded welded attachment may completely pull out from the plate.

Extensive studies (Farrar and Dolby 1972) showed that the solution to the problem of lamellar tearing was to specify a steel with a tight control of sulphur content and a prescribed minimum short transverse reduction in area (STRA) in the through thickness direction Guaranteed minimum values of STRA and were available in Z quality steels from the early 1970s. They were first used for the Forties platforms, although these steels were not specified for all structures of this era. In general, a minimum average STRA of 20% has been found to be adequate to avoid lamellar tearing even in highly restrained welding situations. Steels with below 10 to 15% STRA were only suitable for lightly restrained joints (TWI 2024a, b).

Providing the sulphur content is sufficiently low (< 0.005%) good STRA properties are virtually guaranteed (Harrison 1981) because Kanazawa (1974) had shown there is a reasonable correlation between sulphur content and STRA. Consequently, steelmakers reduced the number and size of MnS inclusions by keeping the level of sulphur low. Even better resistance to lamellar tearing could be obtained through modifying the sulphide shape. In some steel making practices, the addition of rare earth metals (REM) or calcium both reduced the inclusion content by acting as desulphurisers and deoxidisers and improved the shape of the remaining inclusions by making them more globular (Harrison 1981).

Thus, lamellar tearing was ultimately controlled by placing much tighter restrictions over the chemical composition of steels. This was achieved through the specification of a maximum value for carbon equivalent and through thickness fracture toughness properties to be achieved. In such cases, requirements for enhanced through-thickness ductility *additional* to the specified properties in, for example, BS 4360 (British Standards 1990) could be specified for the member onto whose surface the attachment is made. The actual level of STRA specified to avoid lamellar tearing was usually considered in the light of where the materials would be employed within the structure, level of constraint and through-thickness strains due to residual stress, with higher values specified in more restrained situations.

The UK's Department of Energy Guidance Notes (Department of Energy 1974) recommended specifying a minimum reduction of area of 20% or more in the through thickness direction for highly constrained joints and sections. Plate material that was required to have a quality level in respect of lamination and inclusions should have been examined in accordance with BS 5996 (British Standards 1993).

Detailed guidance for the avoidance of lamellar tearing was developed by the Welding Institute in 1972 (Farrar and Dolby 1972) and is still considered to be good practice today (TWI 2024a). A large part of the report is given over to welding configurations where it could occur and the measures available to avoid the problem. While no grade of steel can be singled out as being distinctly more susceptible than any other, there is some opinion and supporting evidence that aluminium treated steels can have higher susceptibilities than corresponding silicon-killed grades of comparable thickness unless special care is taken during manufacture.

Farrar and Dolby (1972) reported the results of a survey of reported instances of lamellar tearing. From the survey there appeared to be an increased risk of tearing with increasing strength level. Most reported occurrences of tearing were in plates greater than 25 mm in thickness, although thin materials (< 25 mm) could also have poor short transverse thickness properties.

When assessing an older steel material for life extension, it is therefore prudent to check the STRA properties and maximum sulphur content and carbon equivalent from the original material certificates, if these data are available. If such data are not available, then the necessary tests to determine the chemical composition should be performed. This should enable the assessor to determine if the original specification was adequate with respect to lamellar tearing. Even if there are no reports of it, some selective inspection (for example ultrasonic testing) for lamellar tearing at the most highly restrained joints would be prudent for an ageing structure about to enter into life extension.

5.3 Sizes and Frequencies of Welding Defects

5.3.1 Introduction

Welding of structures can produce fabrication defects in the weld and the heat affected zone. Defects can be grouped into surface breaking or embedded. Information about the type, likelihood and size distribution of fabrication defects for a welded structure may be obtained from an assessment of the welding and inspection quality records. The fabrication inspection may detect many welding defects some of which may be repaired, but other defects may remain in the structure when it enters service. Defects noted in fabrication inspection reports are either:

- Reportable defects. These defects are those which are above the specified reference level during inspections but are of a size or in a location that is deemed less critical to fatigue cracking.
- Repairable defects. Those above the specified reference level or of a type which are not permitted regardless of size or location (e.g. planar defects). Those welds containing defects which are outside the specification are sentenced for repair and subsequent inspection and NDT of the repair weld or allowed by engineering critical assessment (ECA). A repair may, however, introduce further defects.

Following any repairs, a structure leaving the fabrication yard could contain defects which are:

- Not detected or detectable by the fabrication inspection,
- Within the specification of allowable defects,
- Outside the specification but have been inaccurately sized,
- Outside the specification but assessed by ECA to be allowed to remain,
- Not fully repaired or removed,
- Created by the repair and not detected by the subsequent inspection.

Fabrication defects have an important effect on the probability of failure as a result of fatigue crack. Fabrication defects remaining in the structure can initiate fatigue cracks in-service and if these are not detected by in-service inspection and repaired, they can grow into larger fatigue cracks. Thus, information about the frequency and size of fabrication defects is an indication of the likelihood of fatigue cracking in-service.

The number of fatigue cracks detected and repaired or removed during fabrication and in-service affects the crack size distribution and frequency. Cracks detected during service are typically assessed as anomalies using, for example, engineering critical assessment procedures according to e.g. BS 7910 (British Standards 2019) and either are allowed to remain in-situ or repaired. Further information on fatigue crack development is given in Sect. 6.7.

Larger surface breaking or near surface fabrication defects remaining in welded joints of a structure exposed to high cyclic stresses (which are most susceptible to fatigue) are likely to produce observable fatigue cracks early in the structure's life. These tend to grow semi-elliptically along the line of stress concentration and in the thickness direction and eventually form a through-thickness crack, which may first become detectable by an otherwise dry member becoming flooded. In principle, these defects should be detected by in-service inspections early in life and repaired. These inspections are designed to detect fatigue cracks that initiate from surface breaking fabrication defects (for example, undercut or hydrogen cracks at weld toes) in joints at positions of high cyclic stresses. Smaller surface breaking fabrication defects may not develop into fatigue cracks if the cyclic stresses are low. Where cracks do develop then these will generally be on a much longer timescale.

Long embedded fabrication defects such as lack of fusion can also remain in butt welds between straight members and on the fusion faces of tubular joint welds. These defects can also initiate a fatigue crack which can grow over a longer crack front. However, these will not be detectable by surface inspections until later in life when they become surface breaking. A long embedded crack can grow rapidly through-thickness to a size that significantly increases the local stress intensity to a point where the joint is unable to carry further load (see Sect. 6.7.12). At this stage the crack can propagate and the joint fails. Experience from pressure vessels is that long cracks behaving in this way are of most concern in thicker sections where cracks can grow substantially in the longitudinal direction before they break through wall and propagate. Table 5.1 below summarises the behaviour in-service of different forms of fabrication defects and their relevance for life extension.

There have been several studies of size distributions and frequencies of fabrication defects associated with welds of offshore structures. These studies were generally based on observations of defects detected by fabrication inspections of particular structures. While the results of the studies were specific to the structures studied, a comparison of welding practices could be used to justify the transferability of the distributions and frequencies to other structures with similar welding processes and quality control. The size distributions and frequencies obtained enabled the quality of welding, inspection and repair to be evaluated and also to inform probabilistic fatigue and fracture analyses.

The main studies are listed in Table 5.2 Kirkemo (1988) reviewed the studies undertaken before 1988. The following sections provide a summary of these studies.

5.3.2 *Rodrigues, Wong and Rogerson (1980)*

An early study of weld defect size distributions in offshore structures was made by Rodrigues, Wong and Rogerson from Cranfield University and published at the 1980 Offshore Technology Conference (Rodrigues et al. 1980). They recognised the

Table 5.1 Summary of in-service behaviour of fabrication defects

Form of fabrication defects remaining in the structure on entry to service	Behaviour of fabrication defects in service	Relevance for life extension assessment
Large fabrication defects (Surface breaking or embedded)	Grow into fatigue cracks quickly and are likely to be detected by water ingress (e.g. FMD) when surface breaking and repaired early in life	Minimal. However, the repair may introduce new defects or residual stresses
Smaller fabrication defects, typically surface breaking, in the joints with the lowest fatigue resistance (e.g. tubular joints)	Initiate fatigue cracks that in time grow through wall into short stable cracks which are likely to be detected by water ingress with limited residual life if not repaired	Highly relevant as an increasing number of fatigue cracks should be expected in life extension (seen in real life particularly for floating structures)
Embedded fabrication defects in welds of moderate fatigue resistance (e.g. stub-to-brace butt welds in tubular sections)	Initiate long and large fatigue cracks that are often not detected in-service by visual or FMD inspection	Can seriously reduce the load carrying capacity of the weld later in life if they become extended before they become through wall

Table 5.2 Research work on types of defects in welds

Authors	Types of defects detected	Inspection method	Location
Rodriguez et al. (1980)	Embedded	UT—as welded	Node welds of tubular joints
Burdekin and Townend (1981)	Embedded	UT—corrected for NDT reliability	T butt welds at tubular joints
Bokalrud and Karlsen (1981, 1982)	Weld undercuts, Embedded defects	Surface replicas UT	Shipyard welds between plates
Rogerson and Wong (1982)	Embedded defects	UT corrects for NDT reliability and repairs	Node welds of tubular joints
Kirkemo (1988)	Reviews work by the authors above		
Baker and Kountouris (1988, 1989a, b)	Surface breaking Embedded	MPI UT	Welds of tension leg structures
Thurlbeck et al. (1996b)	Embedded	MPI, UT, RT	Butt and node welds
DNVGL-RP-C210 (2015)	Weld undercuts		Node welds of tubular joints

need for the determination of weld defect size distributions for predicting structural reliability and the probability of failure from fatigue and fracture.

The authors noted a study by Bokalrud and Korsgren (1979) who showed that surface defects at the weld toe such as under-cut or shape defects were the most important where fatigue crack growth is the failure mechanism. Where there was a possibility of a defect or crack propagating by fast fracture (of particular importance for low temperature operations), the authors recognised that embedded defects not revealed by surface inspections could also be an important contributor to the total probability of failure.

Bokalrud and Korsgren (1979) obtained their data on embedded defects from reports of ultrasonic NDT made on node welds of four different North Sea structures fabricated by four different European fabricators. A fifth set of data was obtained from a repeat inspection of one of the structures. Most of the welding was using the gas shielded metal arc process although some was gasless corded metal wire welding. The 100% ultrasonic testing of the node welds covered 250 m of welding and the defects detected were embedded.

The defect data was in the form of the type (planar and non-planar), size and location of each defect detected. The average number of defects per metre of weld varied between 0.5 and 1.53 over the four structures. This variation was attributed to differences in the sensitivity between the ultrasonic NDT used as opposed to actual differences in the quality of the welding.

The size dimension recorded was the length: there is no record of the depth or height of defect, which would have been more useful for fracture mechanics analysis. For the purposes of undertaking a reliability analysis the authors assumed a constant ratio of 4:1 between length and height but provided no justification from the data for this assumption. The authors fitted a Weibull probability density function to the defect lengths detected with appropriately chosen parameters related to the sensitivity of the NDT. This distribution fitted the variation of defect length for all four structures although the authors suggest that an exponential distribution could also have fitted the data to three of them.

This observation indicated that structures of similar type by different fabricators have a similar defect size probability density distribution once variations in NDT sensitivity had been allowed for. It suggested that it might be possible to derive a common probability density distribution function for structures of a particular type. This would enable estimates to be made of their individual failure probabilities taking structure specific variations of loads and material properties into account.

5.3.3 Burdekin and Townend (1981)

Burdekin and Townend of the University of Manchester Institute of Science and Technology (UMIST) published a paper in 1981 on the reliability aspects of fracture in stress concentration regions of offshore structures (Burdekin and Townend 1981).

They recognised that planar defects were the most severe as a cause of fracture and noted that the through wall height of lack of fusion and lack of penetration defects was the dimension of primary importance for determining the fracture reliability of tubular joints.

As the 1980 paper by Rodrigues et al. (1980) had only considered defect lengths, Burdekin and Townend derived a separate distribution of defects heights. This defect height distribution was based on good quality ultrasonic NDT records of embedded defects detected in the welds of a major offshore fabrication in the late 1970s. The data was obtained from records of 100% ultrasonic NDT from 1.9 km of manual metal arc T butt welds at tubular joint intersections in a series of jacket structures in BS 4360 Grade 50 steel (British Standards 1990). The authors commented that the not all defects would have been detected and reported and that the sizing would not have been 100% accurate, noting that small defects tended to be oversized and large ones undersized.

The as-reported data on weld defect heights were found to fit a Weibull probability density distribution. The estimated parameters gave a cumulative probability function $F_Z(Z)$ of the probability that a defect will have height equal to or less than z millimetres with the form:

$$F_Z(z) = 1 - \exp - \left(\frac{z}{7.38}\right)^{2.39}$$

This distribution did not allow for the reliability of the ultrasonic testing method, i.e. sizing errors and defects not reported. The authors then went on to make a correction for defects missed based on work undertaken as part of the PISC trials on nuclear components (OCED 1979) using the discrete approximation method. They assumed that the performance of the NDT operators taking part in the PISC trials was comparable to that for the offshore construction.

Using this approach, they gave the following probability that a defect in the tubular joint welds of the as-fabricated and inspected structure would have height z or less as:

$$F_Z(z) = 1 - \exp\left(-\frac{z}{2.55}\right)$$

Burdekin and Townend (1981) went on to combine this distribution with a distribution of critical defect heights for fracture considering the variation of fracture toughness. A separate study had been undertaken by the Welding Institute to determine the critical crack tip open displacement (CTOD) of manual metal arc welds and HAZ in BS 4360 50D steel in the as-welded and post weld heat treated (PWHT) condition at – 10 °C. From these results study [Data obtained by the Welding Institute for BP Ltd, privately communicated] the authors determined Weibull probability density distributions of CTOD.

From these distributions of defect height and CTOD the probability of a critical defect for fracture was determined using the PD 6493 design curve at different levels of applied strain. The probability of a critical defects was lower for the PWHT material. The authors concluded that to reduce the probability for fracture it would be better to improve the fracture toughness of the material rather than undertaking more exhaustive inspection.

5.3.4 Bokalrud and Karlsen (1981)

Bokalrud and Karlsen (1981) added to the work of Bokalrud and Korsgren (1979) and undertook a probabilistic fracture mechanics evaluation of fatigue failure from weld defects in butt welded joints. There results were presented in a conference paper on fitness for purpose of welded structures (Bokalrud and Karlsen 1981).

In their analysis fatigue cracks were assumed to develop either on the surface from undercuts at the weld toe or from long internal (lack of fusion) defects. The frequency distribution for the heights of undercuts defects was obtained from measurements of surface replicas taken at 827 randomly selected points on production welds at various shipyards of which 325 welds had undercuts. An exponential frequency function was fitted to the histogram of defect heights and had a mean value of 0.11 mm with 16 defects per metre.

$$f(z) \equiv 9e^{-9z}$$

The distribution for internal embedded defects was based on the results of ultrasonic inspections of about 3200 m of production butt weld selected randomly from a total weld length of about 40,000 m. About 6400 lengths of 0.5 m were inspected and 450 of the test lengths contained defects. An estimate of the defect heights was made by comparison with the signals from flat bottomed holes in a test piece.

The authors then took account of the probability of detection of the ultrasonic inspection to determine a corrected frequency distribution for internal defects including defects missed. This was fitted to a one-sided normal distribution function. This function was then further modified to take account of weld repairs of defects revealed by ultrasonic inspection to establish a one-sided normal frequency distribution of defects after inspection and repair with a standard deviation of 2.88 mm of the full normal distribution. The authors examined the sensitivity of the final distribution for different reliabilities of the ultrasonic inspection.

5.3.5 Rogerson and Wong (1982)

Further work was published by Rogerson and Wong (1982). In this paper they presented a detailed analysis of the distribution of the heights of unrepaired weld defects found in one (unnamed) offshore structure and derived an appropriate distribution function. The data was collected from reports of magnetic particle and ultrasonic inspections of over 1000m of weld from 12 vertical node joints and 6 horizontal weld joints, which were all located in the splash zone of the structure. All the reported defects were embedded and 90% were in the horizonal nodes, which were the most complicated connections.

The results reported and analysed found that a Weibull function provided the best description of defects height distribution because it closely fitted the observed distribution particularly at extreme values. The authors noted that planar (crack-like) defects had a different frequency distribution to non-planar defects (inclusions, porosity etc.). The average frequency of all defects was 7 defects in 10 m of weld.

The authors recognised that the observed height might not correspond to the actual heights because of the uncertainties and errors in the NDT measurement technique but noted that the errors in observed size tended to be proportionately smaller the larger the defect. The distribution of derived defect heights reputedly overestimated the defect height at extreme values. In terms of undetected defects, the authors estimated there was at least an 80% probability that the NDT had detected all defects greater than 5 mm height.

The parameters of the derived Weibull distribution of heights of all defects were $\beta = 0.8$, $\eta = 0.1$ and $\Upsilon = 1.12$. The authors considered that this distribution was appropriate for use to provide a conservative estimate of the structural reliability of fixed steel North Sea platforms constructed in the late 1970s and early 1980s. The authors noted, however, that weld repairs would significantly alter the distribution of defect heights in the tail of the distribution. If defects are removed by repair from the population this could increase the structural reliability, although this aspect was not considered within the study.

5.3.6 Kirkemo F. (1988)

In a paper on the applications of probabilistic fracture mechanics to offshore structures, Kirkemo argues that surface defects are more important than embedded defects because experience is that almost all fatigue cracks initiate from an initial surface defect. Consequently, Kirkemo reviews the studies on the sizes and frequencies of weld undercuts including the work by Signes et al. (1967), Watkinson et al. (1970), Eide (1976), Bokalrud and Karlsen (1981) and Recho (1984).

Kirkemo also reviews the work on embedded defects by Rogerson and Wong (1982) and Burdekin and Townend (1981) but does not produce any further data or insights. The paper considers the effect that inspection and repair can have on the size distribution of cracks in service and provides methods for modifying the initial distributions.

In an example, the author uses the exponential distribution of defect heights proposed by Bokalrud and Karlsen (1981) in an analysis to determine the probability of failure following fatigue crack growth. Based on current knowledge and accumulated experience on uncertainties associated with each design parameter, the author considered that the most important sources of uncertainty were the parameters in the fatigue crack growth analysis rather than the initial defect size distribution.

5.3.7 *Baker and Kountouris (1988, 1989a)*

In two Imperial College reports, Baker and Kountouris (1988, 1989a) presented a statistical analysis of both embedded and surface breaking flaws detected by the manufacturing inspections undertaken during the construction of the structure of the Conoco tension leg platform fabricated during the early 1980s. This structure comprised six cylindrical legs joined by buoyancy tanks and braces and was formed from tubular sections, tees and X-cruciform attachments using plate up to 50 mm thickness.

The authors considered 800 embedded defects that had been reported in 27 km of weld, making an average of 1 embedded defect per 30 m of weld. The largest weld defect reported was categorised as a lack of fusion 31 mm height, while the largest crack-like defect reported was 25 mm in height.

The authors investigated parameters like welded joint type, welding process, and defect type's influence on defect dimensions, drawing from a sample of around 8300 defects. Understanding defect occurrence rates and size distributions aids in non-destructive evaluation planning and safety assessments for structures vulnerable to fatigue and fracture. However, caution is advised when extrapolating findings from the TLP to other offshore construction forms due to differences in structural design.

5.3.8 *Thurlbeck, Stacey, Sharp and Nichols (1996a)*

The Offshore Safety Division of the UK Health and Safety Executive commissioned an extensive study of welding fabrication defects in two North Sea deep water structures (Thurlbeck et al. 1996a). These structures were constructed at the same UK fabrication yard between the mid 1970s and early 1980s spanning different eras of construction. Structure B was a 4 legged jacket structure, installed in 1986, while Structure A was an 8 legged jacket structure installed in 1991.

Both structures had been welded using MMA, SAW, and self-shielded flux cored arc welding. The defect data was collected from a detailed retrospective review of the records of inspections carried out using magnetic particle testing, ultrasonic testing and radiography. These included both the initial fabrication inspections and the post-repair inspections.

This study was based only on the reported inspection data. It did not allow for defects outside the specification that may have been wrongly reported or sized or missed. Typical defect types which were detected included profile defects, slag inclusions, porosity, lack of fusion, lack of penetration and cracks.

The joint types considered in this survey were the following and shown in Fig. 5.2:

- Type A: Tubular joint brace-to-chord weld, which is generally a single sided full penetration T-butt weld;
- Type B: Circumferential stub-to-brace closure weld, which is normally a single sided flat butt weld made from and inspected from one side only;

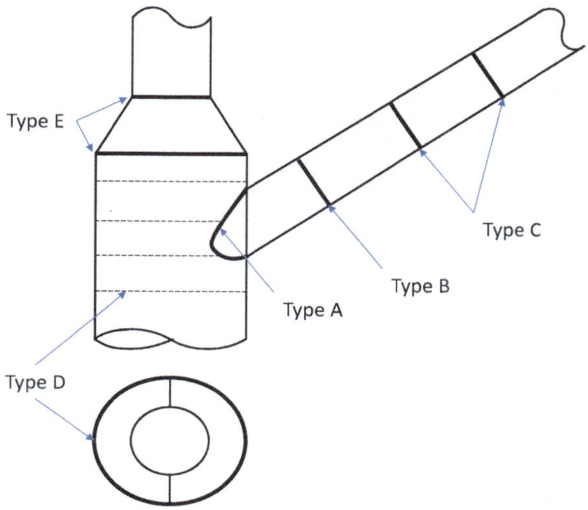

Fig. 5.2 Joint types A–E

- Type C: Circumferential brace-to-brace weld, which is usually a flat butt preparation which may be back gouged and where access is possible from both sides;
- Type D: Circumferential leg-to-cone weld. This is the connection between main leg sections and the weld is typically a flat butt preparation weld with access from both sides;
- Type E: Internal leg-node stiffener weld. This is the connections between the ring stiffener and the inside of the leg wall and is normally a full penetration T-butt weld with access from both sides of the weld.

The defect data collected was from ~ 9.35 km of welds in the two platforms. 3500 inspection records were examined which reported ~ 1400 defects. These were analysed for defect incidence rate, defect dimensions and distribution of defect dimensions. No surface breaking defects were found using MPI, which may have been due to either good welding practice or such defects when found being lightly dressed within the yard and not being subsequently reported.

The total number of defects reported in Structure A exceeded that for Structure B. While this was party because Structure A being the larger had a greater length of weld, the authors concluded that either the welding methods for structure B were better than those for Structure A or the inspection and QA procedures for structure A were better than those for structure B. It was noted that the defect sentencing criteria for structure A contained more detail than that for structure B which simply had a pass/fail criteria.

Joint types A and D showed the highest rate of incidence of embedded defects from ultrasonic inspections. This was mainly because these sections were thicker and the geometries difficult to weld. In addition, the welding process for these joint types was generally manual rather than being automated.

While both structures were subjected to radiographic inspection, no defects were found by radiographic inspection for structure B. This was because radiographic inspection for structure B was limited to joint types C and D only, whereas inspections for structure A included type B welds that were more likely to contain more defects by being a single sided flat butt weld. The length of defects detected by radiography of structure A was found to be greater than those found by ultrasonic inspections of the same weld, particularly for defects with lengths greater than 50 mm.

The maximum lengths of defects that were subsequently repaired were 2490 mm in structure A and 1430 mm in structure B. The maximum lengths of reported defects that did not meet the criteria for repair were 480 mm in structure A and 7068 mm in structure B. There defects were almost certainly lack of fusions due to incorrect set-up for automated welding.

The maximum defect heights reported were 4 mm for structure A and 3 mm for structure B. In addition, the maximum heights of defects introduced during repair were found to be ~ 1 mm smaller than those introduced during first welding.

For the embedded defects that were subsequently repaired, the report showed the frequency of a defect having a given length from both pre and post repair inspections. It was found that the mean length of repairable defects in structure A was less than that for structure B. It was claimed that the distributions of length followed Weibull and log-normal distributions. The authors concluded that the more reportable defects detected in the newer structure (A) were likely to be due to more stringent reporting criteria or that the QA procedures for Structure A were better than those for structure B. The remaining defects detected in both structures were of a size that could have led to in-service cracking but whether this would have been significant is not discussed in the paper.

5.3.9 Zhou and Stacey (2002)

In 2002 Zhou and Stacey reviewed the defect distributions then available for probabilistic structural integrity assessment of offshore structures. They noted that the subject is complex and that despite considerable research in the area and a significant body of data, the defect distribution that might exist in a specific structure at a point in time is subject to considerable variability and uncertainty. The paper reviewed the defect distribution data in the literature drawing data from structures from outside as well as from within the offshore sector and assessed their status and areas requiring further investigation.

The authors considered defect size distributions and frequency distributions for both surface breaking and embedded defects. For surface breaking and embedded defects, they reviewed the defect size distributions and frequency from the following types of structures:

- Nuclear reactor pressure vessels.
- Ship plates—weld undercuts.

- Pressure vessels.
- weld undercuts.
- Magnox reactor components.

They also considered models for probability of detection and defect sizing error.

The authors commented that different sets of defect data have been found to fit to different types of distribution (exponential, log normal, Weibull). As well as the uncertainty in the form of the defect distribution, the authors also commented on the variability and uncertainty in the frequency of defects or rate of defects per length of weld. These observations could be because the defect data in different studies were obtained using different inspection and sampling methods and form different populations.

Another uncertainty that the authors draw attention to is in the probability of defection curves or conversely, the probability of missing a defect. POD curves are used to correct the known defect occurrence rate for defects that have escaped detection. POD curves have been determined from round robin type tests, but the rates predicted should be adjusted for site conditions where more defects are likely to have escaped detection.

From their study of defect size distributions, the authors suggest that the population of fabrication defects in welds can be divided into three sizes: small; medium and large. These size regimes correspond roughly to the three types described at the beginning of this section.

- Small fabrication defects (e.g. weld undercuts) make up a large percentage of the initial defect size population, which under cyclic loading can initiate to form fatigue cracks which either grow and coalesce to form larger fatigue cracks or do not reach the threshold for growth.
- Medium size fabrication defects (e.g. longer embedded defects) that can remain undetected and could grow to form large cracks later in life.
- Large fabrication defects that are most likely to be detected in early inspections or detected early in service (and subsequently repaired). As a result, their initial size distribution may be truncated. However, the consequence of a rare large defect that escaped detection could be serious.

The authors suggest that the probability of fatigue cracking (as determined from a probabilistic analysis) is likely to be dominated by the frequency of larger cracks, whether these are at initial fabrication or have initiated from smaller defects and grown large over longer periods of service. The influence on the probability of fatigue cracking of the distribution of smaller defects was an area of uncertainty.

5.3.10 DNVGL RP C210 (2015)

The 2015 DNVGL recommended practice RP C210 provides guidance for the probabilistic planning of inspection of cruciform, X, K and T type tubular joints of offshore structures based on the probability of a detectable fatigue crack initiating at a weld toe (DNVGL 2015).

RP C210 section D6.2.2 (DNVGL 2015) is based on work by Lotsberg (2014), which in turn references the work of Bokalrud and Karlsen (1981) who observed that most fatigue cracks initiate from undercut-type surface breaking defects at the weld toe. Following Lotsberg (2014), RP C210 recommends an exponential distribution of initial defect heights with a mean size of 0.043 mm as an appropriate distribution for determining the probability that a crack will have grown at the weld toe of a specific joint to a size that would be detectable. This very small size of the mean value of defects may seem surprisingly low, but as stated in DNV-RP-C210 this provides a fatigue life in line with the normal S–N curves in standards.

There is no consideration of embedded lack of fusion type fabrication defects within RP C210. Large, embedded defects can be important contributors to the probability of failure, but their relative rarity compared with undercuts defects and the difficulty of detection before they become surface breaking would mean that the chances of finding one of them would be small particularly for subsea structures.

In conclusion, while this DNV recommended practice represents the view in 2015 on weld defect height distributions for probabilistic inspection planning, it may not be sufficient for estimating the probability of failure of a structure considering all possible types of defects. As probabilistic failure analysis is becoming an important feature of life extension cases, further work on defect distributions would be beneficial, particularly taking account of embedded defects.

5.4 PWHT and Residual Stresses

Historically, multi-pass welds and heat affected zones in parent material were prone to low Charpy energy and poor resistance to fracture at temperatures below – 10 °C. Work by Alan Wells (Woodley et al. 1964) succeeded in identifying strain age thermal embrittlement as the likely cause. While the use of fully killed steels could overcome strain age embrittlement in parent materials, for welds this could only be overcome by using post weld heat treatment. In his 1993 paper, Stacey (1993a) reviews the post heat treatment requirements for offshore structures in the UK sector of the North Sea and the development of the HSE guidance for this area (HSE 1994).

The application of PWHT generally improves the fracture resistance of as-welded joins for two reasons. Firstly, the thermal cycle reduces strain ageing effects and increases the fracture toughness of the material. Secondly, the stress relief lowers detrimental tensile residual stresses within the material, thus enabling structural integrity to be maintaining with a lower fracture toughness. Put together the effect of PWHT is to increase the size of defect that can be present before structure integrity is compromised. In addition to these benefits, PWHT of welded joints was found to lead to an improvement in fatigue performance of welded joints at low R-ratios thereby increasing the life of the joint before a critical defect size is reached.

In developing guidance for the application of PWHT, data (Burdekin 1967; Harrison and Pisarski 1978) indicated that poor fracture toughness and high tensile residual stresses became manifest where the section thickness was above around 40 mm. Below 40 mm thickness, cooling rates and lower constraint were sufficient to avoid the worst effects of strain age embrittlement and lower residual stress. Hence a requirement was introduced into the guidance (Department of Energy 1974) for structures in the UK sector that welds in plates greater than 40 mm should in general be post weld heat treated in the range of 580–620 °C.

The criteria and guidance for PWHT being required varied between jurisdictions. The regulations of the Norwegian Petroleum Directorate gave no guidance on PWHT but stated that fracture toughness tests were required for Class I and Class II materials for thicknesses above 50 mm, implying that a fitness for purpose approach would be acceptable. The DNV rules at the time required fracture toughness testing and fracture mechanics analysis for as-welded plates and sections with thicknesses greater than 60 mm at special locations and more generally 80 mm for primary structure. These thickness limits could be increased by 20 mm for joints that were given PWHT.

The UK requirement for PWHT for joints above 40 mm thickness was very unpopular in the industry. It was evident that the wide range of loading, stresses and material conditions that invariably exist across the joints of an offshore structure could with indiscriminate application of this requirement would lead to a considerable amount of unnecessary and costly PWHT in locations where the critical defect size in the as-welded condition was already large and the likelihood of fracture very low. Thus, the industry agreed that the requirement for PWHT of joints over 40 mm thickness could be dispensed with if a fracture mechanics analysis could demonstrate that the critical defect size was sufficiently large in the as-welded condition for the structure to be deemed fit for purpose.

The offshore industry has increasingly recognised the economic value of this approach. Despite the costs of undertaking facture toughness tests and undertaking the analysis, the industry placed greater reliance on fracture mechanics analysis and testing for the purposes of avoiding PWHT and the reducing the potential for fracture in-service. This drove the development of standards for fracture toughness testing and defect assessment to the point where these methods were universally recognised.

The principal application was to fixed steel structures where it was found that significant numbers of welds in topsides, jackets and foundations could be left in the as-welded condition. This was particularly so for butt welds and to a lesser extent to welds of nodal joints. The Shell Kittiwake fixed platform is an appropriate example of a fixed platform where with the exception of a few highly stressed nodes the entire structure was left in the as-welded condition in the basis of a fracture mechanics analysis (Inglis et al. 1989).

The PWHT requirements were originally justified on the basis of CTOD data from weldments with low toughness and rudimentary yet conservative defect assessment procedures (Burdekin and Dawes 1971 and British Standards 1980). Over the years steels with a lower carbon equivalent became available and with improved welding procedures and consumables led to better weld toughness. More accurate defect assessment methods complemented by improved sensitivity and accurate and reliable

NDE methods during fabrication resulted in a significant reduction in the amount of PWHT being carried out. As a result, there are many structures that were left in the as-welded condition.

While avoidance of PWHT could be and was justified for the original design life, this justification should be reassessed when life extension is envisaged for locations where significant fatigue cracks could have grown from welding defects that were either allowed for missed by the fabrication NDE. The reassessment should take account of new residual stress profiles and methods for fracture toughness testing and defect assessment that have evolved and consider whether the original methods and assumptions remain conservative compared with modern practice. Where this may not be the case, further inspection of the joint may be required to demonstrate that significant fatigue cracks do not exist.

The assessment of the critical crack size is very sensitive to the magnitude and distribution of residual stress in the weld and heat affected zone. Measurements of residual stress in the as-welded condition have indicated that the peak magnitude reaches the yield strength of the material at 0.2% strain in all types of joint. The defect assessment procedures then make an allowance for the effect of shake down under peak applied loading.

The distribution of residual stress in the as-welded condition is joint specific and depends on the specific welding procedure used. For butt welds the assumption of a pure uniform axial residual stress across the joint of yield magnitude is most conservative approach followed by the assumption of pure bending stress. Other approaches are to assume a cosine distribution or a distribution from the literature measurements on T butt joints (Stacey 1993b; HSE 1997) The question of how residual stress interacts with the applied stress is still a matter of research investigation.

5.5 Fracture Toughness of Steel Materials and Welds

5.5.1 Historical Background

The propensity for fast fracture of ferritic steel materials and welds and their behaviour and properties gained the attention of engineers during and just after World War 2 and has remained at the forefront of offshore engineering design ever since. During this period the welded hulls of several Liberty ships and T2 tankers fractured from the welds in a brittle manner without forewarning. In total, 3 Liberty ships broke in two and several T2 tankers experienced similar problems.

The loss of the Liberty ships, T2 tankers and other vessels highlighted that the brittle fracture properties and behaviour of welds were different to those of the parent plate. The mode of failure was by the initiation of a trans-granular crack in the weld or heat affected zone and then the propagation of the running crack into parent material and arrest many metres away from the initiation site. These failures proved that brittle fracture could occur at modest stress levels if other conditions were also present.

The understanding of fracture was at that time at an elementary level. Significant research in the UK, USA and elsewhere during the 1950s 1960s and 1970s advanced knowledge to a point where the resistance to brittle fracture could be quantified by standardised materials tests and analytical fracture mechanics methods. In his landmark paper, Cotton (1979) reviewed the control of brittle fracture in the USA and the UK and the development of methods for testing by Puzak and Pellini (1952) and Baker and Tipper (1956) respectively.

The US approach focused on quantifying the dynamic propagation and arrest of the running crack using drop weight testing and setting limits on the minimum toughness for crack arrest. By contrast, the UK approach was aimed at preventing fracture initiation at much lower strain rates by focussing on factors such as the temperature, size, shape and acuity of welding fabrication flaws, welding residual stress and its relief through the requirement for post weld heat treatment. Baker and Tipper (1956) introduced the notch tensile test to determine the temperature at which the fracture behaviour changes from being ductile to brittle.

The UK approach recognised that fractures were not caused by welding but as a result of the properties of the weld and associated welding flaws. Welding was a manual process involving the deposition of material by melting and fusion, so welds inherently contained fabrication flaws and defects and residual stresses. Welding fabrication flaws and defects could, on their own, be initiators of fracture when subject to peak loading and residual stress while at a sufficiently low temperature. They could also be sites for initiation of fatigue cracking that could trigger fracture when the combination of increased crack size, stress and fracture toughness at the local temperature became critical.

After the Sea Gem failure (Ministry of Power 1967), the industry recognised that once a crack had initiated by brittle fracture in a structure under high stress and containing large amounts of strain energy, no structural steel then available could arrest the running crack and prevent total failure. The focus was therefore on preventing the initiation of fracture. Attention then turned to the fracture properties of welds in the as-welded and post weld heat treated conditions.

The collapse of crude oil storage tanks at Fawley in 1955 led to the recommendation that structural steel greater than 12.5 mm thick should have a Charpy impact energy greater than 40 J at the minimum service temperature as a measure to provide appropriate resistance to fracture. For thicker strakes over 30 mm a requirement for 40 J at – 15 °C was recommended. In addition to thickness, strength was also recognised as a factor affecting the susceptibility to fracture in materials that could be expected to carry higher stresses with less plasticity. Thus, the Charpy energy requirement was raised to achieving over 48 J at the service temperature for these stronger steels.

The avoidance of fast brittle fracture at tubular joints was recognised as being of great important to the integrity of jacket structures despite their inherent redundancy to a single member failure. For structures in the North Sea, the relatively low local temperature combined with the thicker sections necessary for deep waters increased the likelihood of fracture over that of installations in the warmer conditions of the Gulf of Mexico and the Middle East. The main concern was that a running fracture initiating in a weld, or a heat affected zone could with sufficient driving force extend far into the parent material and cause an uncontrolled catastrophic structural collapse.

5.5.2 Structural Steels Fracture Toughness

It is important to understand that the initiation and propagation of a crack in metallic ferritic materials can be caused by both brittle and ductile mechanisms. Brittle and ductile fracture are separate processes and there is competition between them for which initiates first. Sometimes a crack can grow in a stable manner by ductile tearing and then revert to unstable brittle fracture, whereas in other situations a crack initiating in a weld by brittle fracture can divert into parent plate and arrest there or propagate by a ductile mechanism.

Brittle fracture can be distinguished by the crystalline appearance of the fracture surface with chevrons of texture pointing back to the initiation site. Brittle fracture occurs by a cleavage-type mechanism along weak or embrittled planes in the material and is triggered by stress concentration before there is significant plasticity in the material. Brittle fracture toughness reduces with decreasing temperature and increasing strain rate, thickness and constraint until a lower shelf threshold is reached. This makes brittle fracture the more likely mechanism in highly constrained thicker materials (e.g. those meeting conditions for plane strain) at low temperatures, or where there are shock or rapid dynamic loads.

Ductile fracture produces a fibrous fracture surface often with signs of plasticity such as shear lips. As the stress on a crack is increased, the crack grows forward by a small amount in a stable manner—a phase stable crack growth. If the stress is sustained and further increased, a point can be reached where the material does not have sufficient crack growth resistance and unstable ductile propagation occurs until the crack either arrests or fractures. Thinner materials with low yield strength and resistance to ductile fracture are more likely to fracture by a ductile mechanism at higher temperatures and under slowly applied and sustained loads. In the case of pipelines, running cracks can travel long distances.

The material property now used to characterise both brittle and ductile fracture is the fracture toughness. The variation with temperature of the toughness associated with each of these modes of fracture is shown in Fig. 5.3. Cleavage toughness rises with increasing temperature from a lower shelf in an upward curve at an increasing rate (solid line in Fig. 5.3). Ductile toughness varies with increasing temperature as a gently downward sloping series of straight lines (dashed lines in Fig. 5.3) associated with different amounts of stable crack growth and reflects decreasing yield strength. The region where the rising brittle fracture curve overlaps with the downward ductile line is called the transition zone where either brittle or ductile fracture can occur.

The load lines a-c in Fig. 5.3 represent the stress intensity factor of a crack at 3 different temperatures with increasing loads. At load line a) the failure mechanism is by cleavage fracture. At load line c) the failure is by a ductile mechanism, such

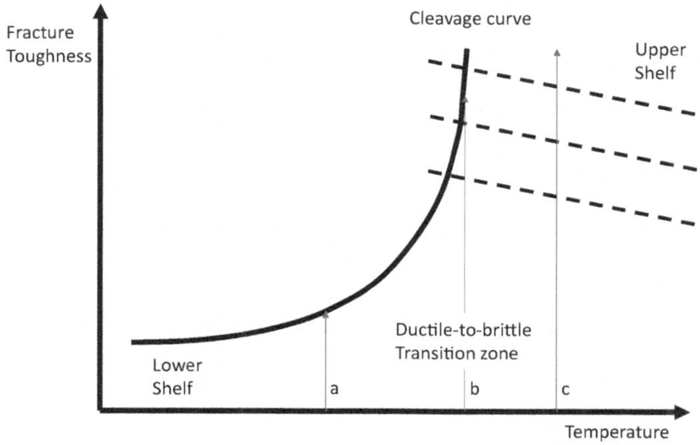

Fig. 5.3 Schematically plots fracture toughness versus temperature showing the brittle to ductile transition curve and lines of stable tearing

as ductile instability or plastic collapse. At load line b) the crack advances by stable ductile tearing until failure occurs by cleavage.

Burdekin and Stone (1966) proposed the concept of crack (tip) opening displacement (COD) for the characterisation of fracture resistance of materials in the brittle to ductile transition region and at upper shelf temperatures. This was accompanied by the development of testing methods for measuring COD. Meanwhile, the Charpy V notch impact test remained the most commonly used technique for determining ductility and fracture resistance of plates and welds, despite its poor correlation with COD. Relationships and correlations between Charpy V notch energy, COD and fracture toughness have been derived for practical purposes since often all the required test data is not available (BS 7910, British Standards 2019).

Fractures at lower temperatures could initiate in welds and then propagate into parent plate. Hence, efforts were made at several research institutes (e.g. TWI) to determine the toughness required to avoid fracture initiation in welds or heat affected zone material and to improve the toughness of welds by the development of welding consumables with improved composition. A requirement for post weld heat treatment had the benefits of improving toughness and reducing residual stresses close to welds that could reduce the external load at which initiation could occur.

The UKOSRP Phase 1 (Department of Energy 1988) fracture programme investigated a number of unresolved issues that existed in the 1970's over the most effective criteria for designing platforms to resist brittle failure. These included:

- Uncertainties over the maximum thickness of steel for which PWHT should be mandatory,
- Uncertainty as to whether it was the parent metal, HAZ region or the weld which provided the regions of lowest toughness,

- Concern as to whether both Charpy and CTOD minimum requirements were comparable.

Most of the tests were the common Charpy tests, together with CTOD testing. Some wide plate tests (thicknesses of 38 and 50 mm) were undertaken for comparison purposes. The tests were carried out on the weld metal, HAZ, both in the as welded and the PWHT condition. Several welding techniques were used with high and low inputs and some specimens were strain-aged to represent a structure which had been in service.

The materials were tested at temperatures below the minimum design and operating temperatures expected for North Sea structures. A minimum Charpy energy (27 J) was required from the tests for providing the necessary assurance of sufficient toughness.

For North Sea structures in the 1980s the design temperature was normally − 10 °C for components at or above the splash zone and typically + 4 °C for components below sea level (UEG 1985). Where structures could be subjected to appreciable stresses during fabrication in the open air some operators used a lower design temperature of − 10 °C for the whole structure (UEG 1985). For colder regions and particular arctic regions, even lower design temperatures need to be specified.

The fracture toughness of the weld and heat affected zone materials were usually limiting with parent plate having adequate toughness and ductility. Most testing effort has therefore been directed at the weld and HAZ materials with much more data being produced. As a result, the factors influencing fracture toughness of the weld and HAZ material such as the welding procedure and the selection of consumables became better understood.

5.5.3 Low and Medium Strength Offshore Steels Fracture Toughness

There was generally no routine testing of the fracture toughness of low and medium grade parent steels such as BS 4360 50D (British Standards 1990) when the early North Sea structures were fabricated. It had become known that despite considerable variations, the toughness of these parent steels was much better than the weld metal or heat affected zone material. Parent steel was selected based on meeting the relevant design code Charpy V impact test energy at the relevant test temperature.

The design codes used during the 1970s had different requirements for the Charpy V impact testing and the acceptance of material. It is therefore important to determine the actual design code that was used to specify the material when assessing an ageing structure for the purposes of life extension or evaluating a crack in service. The UK Department of Energy Guidance Notes (Department of Energy 1977) required a minimum impact energy of 27 J for lower strength C Mn steels with minimum yield strengths up to 288 MPa and 34 J for medium strength steels with minimum yield strengths up to 355 MPa at specified temperatures relating to the section thickness

Table 5.3 Parent steel Charpy requirements (Department of Energy 1977)

Thickness (mm)	Charpy specimen location	Charpy test temperature (°C)			
		As welded		PWHT	
		High stress region	Other	High stress region	Other
t ≤ 20	Sub-S	– 20	– 20	– 20	– 10
20 < t ≤ 100	Sub-S	– 40	– 30	– 30	– 20
40 < t ≤ 100	Mid-t	– 30	– 20	– 20	– 20
t > 100	To be considered on an individual basis				

and the intention to post weld heat-treat. The recommendations of the 1982 Guidance Notes are shown in the Table 5.3.

The equivalent DNV rules specified Charpy V impact energy and test temperature requirements based on the class of steel in terms of its function within the structure and its yield strength and thickness. Other standards bodies specifying fracture toughness requirements for offshore structural steels included British Standards, Lloyds Register of Shipping, the American Bureau of Shipping, The American Petroleum Institute, ASTM and the Norwegian Petroleum Directorate. In some cases, there were more onerous requirements for steels used to fabricate tubular joints.

Historically, there were significant variations between the fracture toughness requirements for a particular steel depending on its use and the design code specified at the time. It is therefore important to look closely at the original specification for the steel and the test results obtained. Structures fabricated today would normally be specified to meet the Charpy V impact energy and test temperature requirements of ISO 19902. Divergencies with modern standards should be considered but are not necessarily significant since each set of design rules should be treated as a whole package reflecting the knowledge and margins of safety that were considered appropriate at the time. However, this statement assumes that the present knowledge indicates that the combination of material and design standards at the time still provides sufficient safe structures.

5.5.4 High Strength Steels Fracture Toughness

Historically, high strength steels have tended to have lower toughness than conventional offshore structural steels, in the range 40–100 J at – 40 °C (Stacey et al. 1996). This may cause issues in the assessment of an existing older platforms containing high strength steels in major components. Particularly if the high strength steel is welded according to similarly old procedures, this may cause concern in the assessment on fracture toughness and hydrogen embrittlement.

As noted earlier, high strength steels in offshore structures are primarily found in jack-up legs, modern mooring lines and prestressing tendons in concrete structures. Considering that prestressing tendons in concrete structures are protected in grouted ducts and not very sensitive to fracture toughness issues and that only modern mooring lines use higher strength steel, the discussion here will primarily relate to high strength steels in jack-ups.

In contrast, the modern production of high strength steels has managed to achieve improved weldability, strength and toughness. This is due to their lower carbon levels, smaller amounts of alloying elements and a finer grain structure compared to older quenched and tempered steels (Stacey et al. 1996). The paper shows that modern developments in alloying and processing are able to achieve impact energies in excess of 200 J at – 40 °C (Stacey et al. 1996). However, the available information on the fracture properties of higher strength steels is somewhat limited.

In general, carefully controlled welding procedures are required for high strength steels to obtain the required mechanical and fracture properties. The use of consumables with very low levels of hydrogen and low inter-pass temperatures are required to avoid hydrogen cracking.

Research on high strength steels for offshore structures was carried out at Cranfield University (HSE 2003). The average Charpy impact energy at – 40 °C of the parent material was found to lie in the range 120–300 J for a wide range of higher strength steels tested in the Cranfield University programme.

In addition, a joint industry project on the fracture toughness properties of two Grade 690F jack-up steels, i.e. Nippon Steel Wel-ten 780 and Creusot Loire A 517 F, was completed by DNV (1995). Both the work at both Cranfield University and in the DNV joint industry project included Charpy impact energy, in addition to CTOD (crack tip opening displacement) tests on parent and welded material. The data suggest that high toughness can be achieved in modern high strength steel parent material, although the data for steels at the higher end of the yield strength range are limited. In these reports, it is also stated that the fracture toughness properties of the HAZ are generally poorer, with a wide spread of values. For some samples, fracture toughness energy values at a relatively higher transition temperature of – 10 °C were found. The CTOD testing in the DNV project was performed on both small-scale specimens and large-scale specimens manufactured from the full thickness of the material, i.e. between 180 and 250 mm.

The development of welding procedures for high strength steels is an important factor in the avoidance of hydrogen cracking and providing adequate toughness. As part of the assessment of an older existing structure, any information about the welding procedures used during fabrication is, as a result, of vital importance. These welding procedures need to be checked against the latest information regarding proper welding procedures and if found necessary, the typical issues that may be present in the older high strength steel and its welds.

5.5.5 Castings and Forging Fracture Toughness

Casting has been used to some extent for nodal joints in offshore structures following the development of casting steels with good fracture toughness and fatigue performance. In the 1980s Sheffield Forgemasters and Hoesch reported that a casting steel with minimum yield strength of 340 MPa gave a minimum CTOD value of 0.77 mm at 20 °C (UEG 1985). This value is comparable with that for BS 4360 Grade 50D plate at the same temperature. However, at − 10 °C the casting steel gave a maximum CTOD value of 0.35 mm, although the material was still sufficiently ductile for there to be ductile crack growth before cleavage failure.

Fracture toughness testing carried out on a prototype cast tubular joint obtained Charpy energies more than 34 J at − 40 °C on the thinner sections comprising the brace and chord walls (UEG 1985). Specimens taken from the thicker sections comprising the intersections between members showed considerable scatter with one value as low as 11 J. The CTOD tests showed a similar trend with values as low as 0.12 mm from specimens taken at the centre of a 200 mm thick intersection.

The conclusion from this work is that although the fracture toughness of cast steel is generally good, regions of low toughness can be present in thick sections where the casting and cooling conditions might be different. The minimum toughness will be specific to the design of the joint and the casting process variables, so it is necessary to cast test blocks to simulate the solidification and cooling rates in the centre of the casting if accurate values are required. Owners of ageing structures with cast joints are advised to be aware of the potential for low fracture toughness when considering the possibility of failure from fatigue cracking in extended life.

5.5.6 Grade a Ship Steel

As previously mentioned in Sect. 3.6, Grade A ship steel is a steel commonly used for parts of a ship structure that are assumed to be less critical by virtue of carrying lower stress or in a location where local failure could be tolerated. By its specification Grade A ship steel does not have guaranteed minimum Charpy Impact Energy or fracture toughness properties and there was generally no requirement according to traditional ship rules of the classification societies to undertake any systematic fracture toughness tests on batches of steel received. Many offshore floating production storage and offloading (FPSO) vessels, both new-built and converted conventional tankers, were constructed to these rules and used Grade A steel plates to a large extent. As noted previously, the proportion of steel in commercial ships (including oil tankers converted to FPSOs) that was Grade A ship plate was estimated to be between 80 and 95% (Randerson and Morrison 1989).

As more FPSOs constructed with Grade A steel came into service in the Northern North Sea, the offshore oil and gas industry and regulators became concerned about their susceptibility to brittle fracture. Although Grade A steel had successfully been used in traditional tankers for many years, FPSO vessels are in a fixed location for several years and must sustain the severe weather that a conventional tanker could avoid. Further, while most commercial tankers operated predominately at temperatures at or above 0 °C, the same could not be held for FPSOs operating in colder regions with temperatures low enough for brittle fracture to occur.

In addition, while conventional tankers could be brought regularly to a dock for inspection of the hull for cracking and damage, FPSOs had to be inspected on station (so-called wet class). Hence, an increased focus on the possibility of brittle fracture of these Grade A steels was therefore necessary. Like tankers, FPSO's contain large volumes of oil for long periods and the consequences of a crack or damage could be similarly environmentally adverse.

In 1995, the UK Health and Safety Executive published the report OTH 95 489 (HSE 1995) it had commissioned from British Steel reviewing the fracture properties of Grade A ship plate from different sources to better understand their range and distribution. The project gathered such data as was available from the literature covering Charpy impact, CTOD, initiation fracture toughness, nil ductility transition temperature and the fracture toughness for crack arrest.

The main conclusion of OTH 95 489 (HSE 1995) was that the Charpy impact properties of as-rolled Grade A plate were highly variable. Some material had very poor properties with low upper shelf energy, high transition temperatures and a pronounced orientation effect thereby making it vulnerable to ductile or brittle fracture where a defect or crack was present. From the test data reviewed the Charpy energy of Grade A ship steel at 0 °C was found to be in the range of 10–25 J with 80% surface crystallinity as a measure of the extent of trans-granular cleavage fracture. This places the steel in a regime where brittle fracture could occur at temperatures of around 0 °C or lower. The Charpy energies of HAZ regions of welds were surprisingly comparable to those of the parent plate, in contrast to those of other types of steel.

Another review of the properties of Grade A ship plate by Liston (1982) suggested that a Charpy Impact Energy of 27 J at + 10 °C could be assumed, which would just meet the minimum requirement for avoidance of brittle fracture at this temperature. However, this claim was never verified for all Grade A plates where the effect of the casting route has a significant effect on the resultant Charpy properties. Taking the different temperatures of the tests reviewed by Liston (1982) and HSE OTH 95 489 (HSE 1995) into account, the test data shows that the Charpy energy of Grade A steel can be highly sensitive to temperature in the range 0 to 10 °C. Such sensitivity could be important in the context of operations in colder regions of the world. Thus HSE OTH 95 489 (HSE 1995) recommends that operators test their material at different temperatures to establish the Charpy energy transition curve where there is uncertainty or lack of relevant data.

The HSE report (HSE 1995) found that the initiation fracture toughness values for Grade A ship plate based on CTOD testing were also very low compared with those of other steels. The report gave typical CTOD values at 0 °C at low strain rates in the range of 0.15 to 0.30 mm, although a value of 0.1 mm had also been reported for material taken from a failed barge (Wellman and Rolfe 1984). These values are not good, but neither are they out of line with the lowest toughness expected for welds and HAZ material of Grade A steel.

These CTOD values were not as low as would have been expected from standard correlations with the corresponding Charpy data. The authors of HSE OTH 95 489 thought that this may be due to the lower constraint of full thickness fracture specimens of ship plate typical 10-25 mm thick compared with higher constraint specimens upon which the correlations were based. It could also have been because of high strain rates associated with the Charpy impact test.

The HSE report (Page 26) notes that the effects of strain rate on fracture toughness. In some intermediate strain rate tests (Malik and Tomin 1991) a significant increase of + 50 °C over the CTOD transition temperature at low strain rates was reported, indicating decreasing toughness with increasing strain rate. The authors noted, however, that although the effect of strain rate on CTOD is significant, dynamic loading with high strain rates is generally less relevant for ship structures. They noted that the CTOD fracture toughness for crack initiation at more appropriate intermediate strain rates may still be adequate (> 0.1 mm at 0 °C) even if the Charpy impact energy was low. The probability of achieving a value of < 0.1 mm at − 25 °C was 16% according to a statistical analysis of British Steel internal data for HAZ of welded Grade A plate.

The HSE OTH 95 489 report (HSE 1995) assessed available data corresponding to fracture arrest. It found the nil ductility transition temperature (NDTT) values for Grade A plate from Pellini tests had a mean of − 5 °C (Hawthorne and Loss 1975). This was significantly higher than the standard requirement of the US rules for fracture control of primary load-carrying primary members in welded ship hulls for a NDTT of − 18 °C (Rolfe et al. 1974). The authors of HSE report (HSE 1995) commented that based on this data this requirement was "far from being consistently achievable with mild steel as-rolled ship plate". Hawthorne and Loss commented that the adoption of an NDTT of − 18 °C would result in a high rejection rate for non-heat-treated steels so they would cease to be used for ship applications. These steels have, in practice, continued to be used.

The crack arrest toughness data for Grade A plate were typically in the range of 30–60 MPa $\sqrt{m}$ at temperatures between − 10 and + 30 °C. The authors of HSE (1995) considered that these properties were barely adequate for hull plates and judged them inadequate for crack arrester plates. The variability of the crack arrest toughness data of Grade A plate was generally less than that of the toughness for initiation of brittle fracture and did not provide any confidence that a propagating crack would be arrested.

The HSE report concluded that without guaranteed minimum fracture properties, the control of fracture in Grade A ship plate depended on preventing crack initiation. This could be achieved from a combination of having low stress from adequate design, a reliable material supplier where the properties might be above the minimum, and the absence of defects from good fabrication and inspection. Where these conditions were met, the authors of the HSE report considered that the initiation toughness of as-rolled Grade A ship steels may often be sufficient for the set of operating conditions normally expected for a ship such as a FPSO.

Where circumstances such as fatigue cracking and crack growth create a distinct possibility of fracture initiation in these steels, the main concern is their ability to arrest an initiated and fast running crack. Bannister and Stacey (1999) suggested that the poor crack arrest toughness and low variability of the Grade A steels reported in the literature reviewed in HSE (1995) was probably inadequate in the event of a fracture initiating and that this factor may greatly contribute to the overall risk of structural failure. Later, Stacey and Sharp (2021) commented that the structural integrity of an FPSO vessel could not be guaranteed at a sufficient level of confidence in the event of initiation of a large fast running crack.

Following the publication of the review of existing data in HSE (1995), a further programme of work was undertaken to generate fracture property data from as-rolled ship plate material, nominally Lloyd's Grade 'A', representing the lower end of the Charpy property range. Seven plates, which were not necessarily certified Grade A plates were offered by stockists and originated from several steel makers and had different steel chemistries. The aim of the work was to examine the variability in properties which could be obtained from this steel type, both within and between plates.

The results are reported in HSE (2000). The fracture toughness of four of the plates was measured at a high loading rate. The results were compared with the criterion, proposed by Sumpter and Caudrey (1995) for the avoidance of dynamic brittle fracture in ships. This was that the toughness Kmat must exceed a value of 125 $MPa\sqrt{m}$ at a loading rate of 1×10^4 $MPa\sqrt{m}/s$ or higher at a test temperature reflecting operating conditions for ships.

Three plates with thickness up to 15 mm had adequate toughness such that K_{mat} exceeded the minimum toughness according to the criterion set by Sumpter above at a test temperature of -10 °C which is the minimum operating air temperature for the North Sea. The 20 mm thick plate in this study failed to meet the minimum toughness level. The authors concluded that it did not possess sufficient toughness for use as ship plate for the North Sea environment.

While in general, there was no minimum Charpy impact energy value specified for Grade A ship steel, amendments were made to the Lloyd's Register rules and regulations in 1997 for the Classification of new FPSOs and new conversions of existing conventional oil tankers into FPSOs. These amendments did not, however, apply to conversions existing before they were introduced. The amendments included a general requirement for the Charpy impact energy for Grade A plates of 27 J in the longitudinal direction and 20 J in the transverse direction at + 20 °C where this was an appropriate temperature. The Charpy impact energy for the plates in the test programme reported in HSE (2000) exceeded 27 J at + 20 °C in both orientations.

When a tanker was converted into a FPSO, it was considered that the FPSO would stay on the proposed production site and the service temperature would be the minimum temperature expected on the site according to a given return period. For a FPSO designed to operate in high latitude regions such as the North Sea, or where there was a change in use to such a site, it was necessary to assess that the steel grade was suitable for the service temperature. For example, a tanker structure built with Grade A steel would according to the Rules have the meet the material requirements of a Grade B steel if it was to operate in a region where temperatures could be as low as 0 °C. As it was obviously impossible to re-engineer and replace all Grade A plate with Grade B, a process of requalification was proposed (Cambos and Parmentier 2016).

Requalification required further tests to provide evidence that the Charpy impact energy of the material fulfilled the requirements of Grade B (27 J at 0 °C). This evidence had to be based on a minimum amount of testing and the use of random sampling and statistical models based on a Weibull distribution of Charpy energy. A probabilistic approach was therefore propsed for justifying that the fracture properties of the plate were adequate (Cambos and Parmentier 2016) when an existing tanker originally constructed of steel designated as Grade A was converted for use as a FPSO.

Sill et al. (2000) provide an example of a new conventional tanker being 'intercepted' during construction for service as a new build FPSO. Although the design originally specified Grade A for the tanker's side shell plates and Grade AH32 for the tanker's deck and bottom shell plates, the shipyard agreed to change the hull materials to Grade D and Grade DH32. Selection of these higher-grade materials was based on the improved toughness requirements for these grades. In addition, to the higher grade, the thickness of the deck and bottom shell plates was increased from 16.5 to 19.5 mm.

A comprehensive programme of materials testing was subsequently undertaken on the weld test panels to characterise the toughness of the weldments. The tests included CTOD fracture toughness tests at − 10 °C and Pellini drop-weight tests, to determine the nil-ductility transition temperatures (NDTT), in the three critical regions of the hull seam welds, i.e. weld metal, heat-affected zone (HAZ), and parent plate. In addition, an engineering critical assessment according to BS 7910 and weld toe improvement was undertaken. This example shows the lengths that an operator will go to prove that the material for a FPSO is suitable.

When the design life of a tanker converted FPSO has been reached and there is a proposal to extend service life particular attention to fracture toughness properties and tolerance to cracking is required because the likelihood of fatigue cracks existing is much greater than earlier in life. Under these circumstances, the probabilistic approach for justifying that the fracture properties of the plate are adequate by statistical means based on a Weibull distribution of Charpy impact energy may not be sufficient when set against the uncertainties and variabilities and other factors (e.g. increasing fatigue crack growth) making the probability of a fracture more likely. More tests on a greater range of plate may therefore be required combined with detailed inspection to build sufficient confidence that the probability of a fracture in life extension remains acceptably low.

5.5.7 Welds and Heat Affected Zone Fracture Toughness

Within an offshore structure the areas that are likely to have the lowest fracture toughness at a given temperature and thickness are the welds and the heat affected zones of the parent material. Knowledge of the fracture toughness of the weld metal and HAZ material is therefore very important for making the case for continued service and life extension of an ageing structures when cracks are more likely to have developed. Standards for fracture toughness testing of welds and HAZ have evolved over the period of construction of offshore structures. The current standard is ISO 15653 (ISO 2018).

The longitudinal and circumferential seams of tubular members are typically fabricated using submerged arc welds for their high deposition rate. Because the welding conditions of this process are rarely consistent, the fracture toughness of submerged arc welds is highly variable. CTOD values of 0.5 mm for as-welded material and 1.5mm for post weld heat treated material at − 10 °C have been reported (Garwood (1985), but these values were regarded as optimistic. Berkhout et al. (1978) found lower values, although these data could have been affected by an excessive number of MMA weld root runs (UEG 1985).

The fracture toughness of manual metal arc welds (MMA) used for joining tubulars to nodes and site closure welds has been investigated by various authors. The fracture toughness has been found to vary with the position sampled in the weldment with weld root material tending to have lower toughness than elsewhere. The position is complicated because of constraint effects of different thickness test specimens and the extent to which test specimens sample single or multiple microstructures.

The UKOSP 1 (Department of Energy 1988) investigated the fracture toughness at − 10 °C of double V bevelled MMA welds made using seven different consumables. Welds with nickel consumables had the best properties with as-welded COD values in the range of 0.15–0.20 mm. Post weld heat treatment resulted in a significant improvement in fracture toughness for most but not all consumables tested. UKOSP 1 (Department of Energy 1988) also included fracture toughness (COD) tests on heat affected zone material of typical offshore steels. Of 193 tests most had high COD values. Only two tests had values less than 0.2 mm and these were related to unusual compositions of the parent steel.

A separate study on HAZ material by Eide (1976) involving 230 COD tests found most gave values above 0.25 mm with a minimum of 0.15 mm. However, data from weld procedure tests on low carbon micro-alloyed steels containing niobium or a combination of Niobium and Vanadium showed that the HAZ toughness can be extremely low and did not improve after PWHT. Low HAZ toughness has also been found in other cases which cannot be attributed to Niobium or Vanadium segregation and could possibly have been due to overly high heat inputs. Therefore, when reviewing an ageing structure, the HAZ toughness should receive particular attention in the light of the material composition and the heat input specified by the welding procedure.

The requirements of design codes for the fracture toughness test temperature and the minimum Charpy energy for welds and HAZ were generally the same as for parent material discussed above, although with some differences in detail. For example, the UK Department of Energy Guidance Notes specified that weld metal and HAZ samples did not need to be tested at temperatures lower than − 25 °C for as-welded joints and − 20 °C for post weld heat treated joints with a minimum impact energy of 27 J at these particular temperatures. There were specific requirements relating to the number of and location and orientation at which test samples were to be taken. These have now been incorporated into modern standards for fracture toughness testing.

Recognising the limitations of this Charpy V impact energy requirements for areas subjected to high levels of plastic strain such as complex joints, the Guidance Notes provided the option to select welding consumables based on defect tolerance from fracture mechanics tests (e.g. CTOD, wide plate) in conjunction with Charpy V tests. Pisarski and Pargeter (1984) first put forward an approach for determining weld metal fracture toughness requirements using the CTOD design curve for a joint under severe storm loading. This approach has been modernised in line with developments in fracture mechanics and is now contained within the British Standard BS 7910 (British Standards 2019).

The implication for the assessment of an ageing structure is that the present-day standards in most cases will assume better ductility and toughness than what may be found in older structures. The assessment of older structures either needs to use the design standards contemporary to the structure (given that these standards are still regarded as acceptable), or preferably a fracture mechanics analysis, considering the assumed material properties (toughness and ductility).

5.5.8 Local Brittle Zones

In the weld metal and particularly the heat affected zone (HAZ), discrete local microstructures can form that exhibit significantly lower resistance to fracture initiation than the surrounding material (TWI 2024b). Fracture toughness testing of heat affected zone (HAZ) of offshore steel welds often showed high variability between the toughness of nominally the same material (Department of Energy 1988). Testing of specimens could result in complete specimen fracture or in an arrested brittle fracture (known as a pop-in) where the crack advanced slightly but then arrested and was then able to tolerate further increase in load before finally failing.

Microstructural examination revealed that the instances of significantly low fracture toughness were due to fracture initiation in grain coarsened HAZ material. This was a surprise because, in general, the fracture toughness of the weld metal was *lower* compared with the HAZ. Indeed, in some cases specimens nominally notched into the HAZ were shown to have initiated from the weld metal close to the fusion boundary.

The term local brittle zones (LBZs) was first used in 1983 to describe low toughness zones observed in the HAZs of steels used in the construction of the Odin platform. The fracture mechanics tests were conducted at -10C and the results were initially surprising given the generally favourable HAZ fracture toughness exhibited by similar steels tested in UKOSRP 1. The term LBZ was coined by Doug Fairchild of Exxon who in 1987 made an important contribution to the standardisation of fracture toughness testing of heat affected zones for the selection of steels for construction offshore platforms (API 2014). LBZs could result in fracture initiation under near linear elastic conditions, for example, in terms of crack tip opening displacement (CTOD), fracture toughness could be very low (of the order of 0.01 mm) in the LBZ while the adjacent material could have a toughness of greater than 0.1 mm at the same temperature (TWI 2024b).

The offshore industry was concerned by the low toughness results from local brittle zones and initiated research at TWI (Pisarski and Pargeter 1984) to determine their cause and whether they were of real significance. Post-test metallography revealed that the LBZs were associated with coarse-grained material adjacent to the weld fusion boundary and included the inter-critically reheated grain coarsened HAZ in multi pass welds. They were attributed to variations in heating and cooling cycles during multi-pass welding that could cause inter-critical heating resulting in grain coarsened material.

It is important to note Pisarski and Pargeter (1984) stressed the need to carry out post-test metallography when conducting fracture mechanics test on HAZs to confirm that the lowest toughness region has been tested successfully. As a result, BS 7448 Part 2 (British Standards (1997) and subsequently ISO 15653 (ISO 2018) and other codes require post-test metallography when testing HAZs. The occurrence and size of LBZs was sensitive to steel composition, the welding procedure including heat input and weld bead placement.

In addition to the welding procedure, particular combinations of micro-alloying elements of the steel were also found to increase the susceptibility of the HAZ to LBZs. For example, micro alloyed low carbon steels containing Nb + V in combination could show lower fracture toughness than steels where these elements were not in combination. Post weld heat treatment (PWHT) could improve the weld/HAZ fracture toughness in some steels by tempering the material but owing to precipitation effects could also reduce fracture toughness in other steels.

In general, Charpy impact testing is not considered to be a useful indicator of whether a HAZ is susceptible to low toughness from LBZs (TWI 2024b). These are not normally apparent during routine Charpy impact testing because the test specimens do not sample sufficient material along the notch front and the specimen contain a blunt notch compared with the sharp fatigue crack used in the fracture mechanics test. However, low Charpy energy can be due to presence of LBZs when they are favourably positioned within the specimen and the test temperature is appropriate (TWI 2024b).

LBZs that result in premature fracture or pop-in are most apparent in the conventional, deeply notched three-point bend specimens used in CTOD testing, see ISO 15653 (ISO 2018), ISO 12153 (ISO 2011), ASTM E1820 (ASTM 2023). Although

LBZs behaviour may be less apparent and fracture toughness values may be higher in other test specimen geometries. Example of such geometries include shallow notched bend specimens, single edge notch tension specimens and structurally representative tests such as wide plate tension tests (TWI 2024b). This is because of reduced constraint from crack tip stress triaxiality in these geometries. Nevertheless, it is important to note that fracture behaviour can be a fine balance between constraint and the inherent brittleness of microstructure present at the crack tip. This was vividly illustrated when a steel exhibiting LBZs at − 10 °C in a bend test resulted in complete fracture of a surface notched wide plate test. This welded plate was 900 mm wide and loaded in tension (Webster and Bateson 1993).

LBZ behaviour could be a warning of potentially low resistance to brittle fracture initiation in the structural component. However, LBZ behaviour did not necessarily mean that the structural component would failure by brittle fracture if crack arrest could be assured. The industry therefore adopted a position that where LBZ behaviour is observed in a test specimen, the results needed to be carefully considered along with several factors as described in more detail below.

Consequently, when assessing the structural significance of LBZs, consideration needed to be given to differences in crack tip constraint between the test specimen and the crack in the structural component (TWI 2024b). To assess these differences, it is necessary to either:

- determine fracture toughness using a lower constraint test geometry (but ensuring at the same time that the microstructure giving rise to the LBZ is tested); or,
- conduct a fracture mechanics assessment allowing for the reduced constraint in the structure (e.g. using a constraint modified failure assessment diagram FAD according to BS 7910).

Both the theoretical and experimental work showed that LBZ behaviour is influenced by the length of LBZ present along the crack tip front. Consequently, in addition to constraint effects, the structural significance of LBZs will depend on the likelihood of a crack tip intersecting a LBZ and the length of LBZ present. Although LBZ behaviour produces low initiation (usually cleavage) fracture toughness, the arrest toughness of other regions of the HAZ could be significantly higher. This is because the HAZ is a composite of different microstructures, some of which have high toughness.

Thus, brittle fracture initiation from a LBZ will be followed by crack advance along a front intersecting both LBZ and high toughness regions. The overall toughness of this ‘composite’ microstructure may be sufficiently high to arrest the brittle fracture. This has been observed in crack arrest tests conducted on HAZs that contain LBZs, and it is part of the explanation for the arrest of pop-ins observed in both conventional fracture mechanics tests (e.g. CTOD) and wide plate tests.

Finally, if the material adjacent to and ahead of the LBZ can be shown to have sufficient fracture toughness to arrest a brittle fracture initiating from the LBZ, the likelihood of brittle fracture in the structure will be significantly reduced. However, this necessitates determination of the arrest fracture toughness of the surrounding material by suitable crack arrest testing. If this argument is made, the significance of an arrested brittle fracture on other modes of failure, for example, fatigue crack growth, would need to be evaluated A procedure for assessing the structural integrity of steels containing LBZs is given the British Energy document R6 (in this document the term LBZ is not used, instead these are described as "enclosed brittle regions").

In an assessment for life extension of an offshore structure, both testing of materials and initial fabrication tests may have shown low fracture toughness in the heat affected zone (HAZ) due to these LBZs. However, earlier analysis and evaluations as described above may have shown that these were acceptable. It should be noted that LBZs occur during fabrication welding and no new such zones will develop unless repair welding is performed. It should also be noted that since the LBZ are dependent on the composition of the parent material (and not the weld material), repair welding procedures need to be carefully developed for the specific parent material to minimize the occurrence and effect of LBZs.

5.6 Fabrication and Testing Standards

5.6.1 Control of Material and Fabrication Quality

When assessing an ageing structure for life extension, it is important to know the specification of the original materials and fabrication. The materials and fabrication would normally have been specified to meet the standards at the time, and a review of how these older standards compare with their modern equivalents can be helpful in determining any gaps or uncertainties regarding the initial material and fabrication quality. Where standards have significantly changed (e.g., in the composition or weldability of steels) the implications from the use the original standards for the life extension of the ageing structure should be considered. Some guidance on these issues will be provided in later sections of this book.

The supply of steels for offshore structures supplied until the mid-1970s came mostly from European and American steelworks which generally had a proven history of supply and quality. The requirements for thick welded sections of new deep water North Sea structures were a challenge for the manufacturers to deliver steel with sufficient through thickness fracture toughness properties and a composition to minimise the possibility of lamellar tearing. As the specifications for steel became more demanding, manufacturers had to change their steel making practices to meet the enhanced requirements. Thus, when assessing a structure from this era for life extension, it is important to determine the extent to which the through thickness toughness properties and composition of the material had been appropriately specified, tested and the required properties achieved and verified. The relevant materials standards and types of steel used for offshore structures are reviewed in the next sections of this book.

By the end of the 1970s, sufficient experience in the construction and service of offshore structures for the North Sea had been gained for the development of standards specifying the design, materials and fabrication techniques and tolerances that would represent generally accepted good practice. While the American API 2A Recommended Practice for Planning, Designing, and Constructing Fixed Offshore Platforms already existed, and some early North Sea structures had been constructed to this practice, it was primarily intended for structures in the Gulf of Mexico where fatigue is not such an issue. Accordingly European companies, government bodies and standards organisations developed standardised rules for materials and fabrication quality (e.g. DNV Rules, BS 6325, UK Department of Energy Guidance Notes) that took fatigue and the evolving understanding of lamellar tearing and fracture more fully into account.

The extent to which there can be confidence that the specification was achieved in practice is through the implementation of an appropriate quality control plan during materials procurement and fabrication. According to modern practice (e.g. ISO 19902), general quality control is typically performed by the construction or installation contractor, while inspection and testing is normally conducted by the owner of the structure or by a 3rd party. A QC Plan, usually approved by the owner, should reference the procedures and standards for all materials testing, inspection and NDT techniques employed at all stages of the construction to ensure that materials and workmanship meet the specified standards.

ISO 19902 lists the main contents of a quality control plan for the fabrication of the structure. These include the requirements for checks on:

- Materials certification and testing during procurement and verification of receipt of the correct material by crosschecking with appropriate original mill certificates and heat stamps or with other appropriate documentation.
- Inspection personnel and their qualifications appropriate to their roles and tasks, particularly considering the qualifications for any diving inspectors.
- Storage of materials and welding consumables on site; tolerances, fit-up and alignment; welding including welding personnel and weld inspection; and installation aids and appurtenances.
- Welding and inspection procedure qualifications.
- Inspections and testing made during the load-out, sea fastening and transportation phases and during structure installation covering launch and up-ending, piling, and grouting where applicable.

The resulting reports and documentation should contain the evidence that the necessary checks have been made and of compliance with the specification. A substantial permanent record of documentation would be produced and retained, which would include the documents shown in Table 5.4. In ISO 19902, the requirements for reporting documentation are linked to the exposure level for the structure, which is related to considerations of life safety and of environmental and economic consequences in the event of failure of the structure.

Structures constructed for operation in UK waters would until the mid-1990s would have needed to have complied with the Department of Energy Guidance Notes (Department of Energy 1984). These specified the construction standards that were applicable at the time including controls on materials and fabrication. The

Table 5.4 Reporting requirements for fabrication in ISO 19902 (ISO 2020)

Document description	Exposure level		
	L1	L2	L3
Quality assurance manual	X	X	X
Material tracing records	X	X	–
Shop drawings	X	–	–
Dimensional control/Survey records	X	–	–
Mill certificates	X	X	–
Welding procedure matrix	X	X	X
Welding procedure specification	X	X	X
Welding procedure (PQR) test results	X	X	X
Site welding instruction sheets (Short WPS)	X	X	X
Welder qualifications	X	X	X
NDT test results	X	X	–
NDT personnel qualifications	X	X	–
Inspection records	X	X	–
Pressure/Leak test reports	X	X	–
Weight reports	X	–	–
Foundation verification report (pile driving records etc.)	X	X	X
As built drawings	X	X	–
Weld numbering system	X	X	–

"X" indicates the minimum requirements for fabrication,
"–" indicate this is not required unless specified by owner

Guidance Notes did not, however, specifically cover the need for a quality plan and retention of documentation and reports.

Thus, when assessing an ageing structure for life extension, the extent to which the original documentation is still available needs to be determined. In many cases documentation will be incomplete or lost altogether. In these cases, the assessment for life extension is impeded and has to be made based mainly on the performance of the structure in service and any retrospective tests and inspections that may be carried out at the end of design life.

5.6.2 Development of Standards

The first standard specifying detailed requirements for the materials, fabrication, welding, inspection and testing of offshore structures for the North Sea was the 1977 DNV Rules for the design, construction and inspection of offshore structures which was reprinted with revisions in 1980. The original 1977 DNV Rules were updated periodically and later replaced in 2001 by the offshore standard DNV-OS-C401 Fabrication and Testing of Offshore Structures. Further editions of DNV-OS-C401 were published in 2004, 2014, 2015 and 2019 leading to the latest edition of DNV-OS-C401 in 2021 (DNV 2021).

The first edition of the UK Department of Energy Guidance Notes was published in 1974 as regulatory guidance for designers and constructors (Department of Energy 1974). In addition, the UK Department of Energy appointed five certification authorities who also developed their own requirements. In 1982 British Standards Institution standardised the requirements by publishing BS 6325 Code of Practice for Fixed Offshore Structures (British Standards 1982) with a second and final edition in 1984 (British Standards 1984b). BS 6325 complemented the UK Department of Energy Guidance Notes over its later editions which remained in general use until the late 1990's. Since then, the international standard ISO 19902 has evolved covering a wider scope and is considered suitable for the design, construction and operation of most fixed structures in the North Sea and elsewhere in the world particularly where fatigue is a major issue.

Assessing an ageing structure for life extension requires a knowledge of the standards to which it was originally constructed. As noted above these standards have evolved over time. It is important to have an awareness of limitations in the original standards compared with modern practice and to make allowances or mitigations as appropriate. The latest standards take account of a large amount of research and new knowledge and experience from operating practice.

5.6.3 1977 DNV Rules and DNV GL Offshore Standard C401

The DNV GL Offshore Standard C401 contains requirements for fabrication and testing of offshore units as described in the DNV GL rules for classification of offshore units. The specific offshore units to which this standard applies are floating structures including offshore drilling and support units, floating production storage and off-loading units for oil and natural gas, liquid natural gas liquid petroleum gas, and self-elevating units (jack-ups).

The objectives of this standard are to provide an internationally acceptable standard to ensure the minimum quality of: (1) all welding operations used in offshore fabrication, through identifying appropriate welding procedures, welder qualifications and test methods and (2) surface preparation, coating application and fabrication and installation of sacrificial anodes and impressed current systems. The standard specifies other requirements for offshore units and installations subject to DNV GL certification. The standard was written for general world-wide application, although it notes that local jurisdiction regulations may include requirements more than the provisions of this standard depending on the size, type, location and intended service of an offshore unit.

The standard contains technical provisions covering:

- Welding focusing on welding procedure qualification and the qualification of welders,
- Fabrication and tolerances including material identification, cutting and forming, heat treatment, assembly and repairs,
- Non-destructive testing covering the qualification of inspection personnel, inspection scope and methods, and other tests, for example, tests on leak-tightness of tanks,
- Corrosion protection systems including coatings and cathodic protection systems,
- Mechanical fastening including bolts.

There are also provisions relating to requirement for certification and classification, including records and documentation. The thoroughness and detail of this standard reflects modern practice and is a far cry from the original 1977 DNV Rules.

5.6.4 BS 6235 Code of Practice for Fixed Offshore Structure (1982)

Basis on the experience gained from the early UK offshore developments, it was decided that some standardisation for fixed offshore structures was needed. This need resulted also from known gaps in knowledge and results of relevant research undertaken. As a result, a draft code of practice for development (DD55) was published in 1978. Interested organisations were invited to submit comments for improvement to this draft stand and the responses reflected the considerable interest at the time in offshore engineering. Comments were received from designers, builders and from the six "Certification Authorities". All comments were considered in preparing the first version of BS 6235 which was published in 1982. As a result of further comments on this published standard a revised edition was issued in 1984. The standard includes aspects dealing with the design, construction, operations and maintenance of both steel and concrete fixed offshore structures.

Regarding materials, it was noted that selection of steels for fixed offshore structures is governed mainly by the need for adequate resistance to brittle fracture, corrosion fatigue and infrequent high loadings caused by storm waves. Other factors which affected selection include the ability of the steel to be formed by hot or cold working and its weldability, including resistance to lamellar tearing. On this basis the standard stated that the range of structural steels which comply with the full chemical composition and mechanical property requirements of BS 4360 were suitable. It also added that other specifications including quenched and tempered steels, forgings and castings are not precluded provided that the properties are demonstrated to be adequate to the satisfaction of the parties concerned.

Due to the serious consequences of failure from brittle fracture it is necessary to use a steel which exhibits adequate toughness to guard against the initiation of fast fracture. BS 6235 (British Standards 1982a) provided information linking the material thickness to the design minimum temperature for a minimum Charpy V notch energy of 27 J. The standard also added that, for the relevant locations of the North Sea, the design minimum temperature should be − 10 °C for exposed parts of the structure.

In respect of avoiding lamellar tearing in nodal joints BS 6235 (British Standards 1982a) proposed an additional specification to that of BS 4360 (British Standards 1990). The additional specification was for a guaranteed minimum value for the reduction of cross-section area in tensile tests of short transverse test specimens (ductility requirement). It stated that the value of reduction of area value required to avoid lamellar tearing was dependent on the severity of the through thickness stresses, but a 25% reduction in area would be adequate to avoid lamellar tearing.

Regarding welding, BS 6235 lists general requirements for welding procedures including the need for procedure trials. The fracture toughness of weldments (both HAZ and weld metal) need to be demonstrated by carrying out Charpy V notch tests or COD type tests on specimens extracted from the HAZ and weld metal of the welding procedure test plates. The required minimum average impact value of a set of three specimens taken from the procedure test plates was 35 J with no individual value below 25 J. The Charpy V notch test temperature depended on the region of structure for which approval is sought (4 °C underwater, − 10 °C exposed—splash zone and above).

BS6235 states that post weld heat treatment is required for all nodal regions when the thickness of the plate being joined by the weld exceeds 40 mm. There is an exception for the treatment of welds joining plates with thicknesses greater than 40 mm, if satisfactory performance of the as-welded condition can be demonstrated by fracture mechanics. Procedures for PWHT are given indicating that the temperature range should be within the range 580–620 °C.

Regarding acceptance of defects BS 6235 stated that no defects larger than specified in BS 5500:1982 (British Standards 1982b) should be accepted. However larger defects can be assessed using procedures given in PD 6493 (British Standards 1991).

With respect to fatigue, the S–N design procedure for tubular joints should be based on the Q curve using the applied stress range and the number of cycles to failure (see also Sect. 6.7.3). It is noted that there is little relevant experimental data on the effect of seawater on fatigue. It is stated that the effect of CP systems on fatigue performance is thought to be beneficial although over-protection can be damaging. The published S–N curve (Q) for tubular joints has no cut-off at high numbers of cycles but further investigative work is required which may lead to an effective cut-off. Comments are also made about fatigue life improvement techniques such as grinding.

Monitoring: BS6235 (British Standards 1984b) states that consideration should be given to incorporating permanent monotiling equipment into the structure. This could include strain gauges in locations where it is considered that critical strains might be induced by extreme conditions of load or temperature. Addition of inclinometers or accelerometers could be made to measure the attitude of the structure and its response to applied loads.

References

Almar-Næss (1985) In: Almar-Næss A (ed) Fatigue handbook. Tapir, Trondheim

API (2014) API RP 2A recommended practice for planning, designing, and constructing fixed offshore platforms

ASTM (2023) ASTM E1820 standard test method for measurement of fracture toughness

Baker MJ, Kountouris IS (1988) Defect assessment: analysis of the dimensions of defects detected by magnetic particle inspection in an offshore structure. Imperial College Report Ceslic OR6, 1988

Baker MJ, Kountouris IS (1989a) Reliability of non-destructive examination of welded joints. Imperial College Report Ceslic OR7, 1989

Baker MJ, Kountouris IS (1989b) Defect assessment: analysis of the dimensions of defects detected by ultrasonic inspection in an offshore structure. Imperial College Report Ceslic OR8, 1989

Baker JF, Tipper CF (1956) The value of the notch tensile test. Proc Inst Mech Eng 170(1):65–93

Bailey N, Coe FR, Gooch TG, Hart PHM, Jenkins N, Pargeter RJ (1993) Welding steels without hydrogen cracking. Woodhead Publishing Series in Welding and Other Joining Technologies

Bannister, Stacey A, Proceedings of OMAE99, 18th international conference on offshore mechanics and Arctic engineering July 11–16, 1999, St. Johns, Newfoundland, Canada OMAE99/MAT-2071

Berkhout CF, van den Brink SH, Vaessen GHG (1978) Controlling weld metal toughness of manual metal arc and submerged arc welds in carbon-manganese steel. In: Proceedings European offshore steels research seminar. The Welding Institute, Cambridge 1978, Paper 1

Bokalrud T, Karlsen A (1981) Probabilistic fracture mechanics evaluation of fatigue failure from weld defects in butt welded joints. Paper 28, Proceedings conference on fitness for purpose validation of welded constructions. Institution of Mechanical Engineers, London

Bokalrud T, Karlsen A (1982) Control of fatigue failure in ship hulls by ultrasonic inspection

Bokalrud T, Korsgren P (1979) Some aspects of the application of probabilistic fracture mechanics for design purposes. Colloquium on the practical applications of fracture mechanics, Bratislava, July 1979

British Standard (1980) BS PD 6493:1980 Guidance on some methods for the derivation of acceptance levels for defects in fusion welded joints. British Standards Institution

British Standards (1982a) BS 6235 1982 Code of practice for fixed offshore structures. British Standards Institution, London

British Standards (1982b) BS 5500:1982 Specification for unfired fusion welded pressure vessels

British Standards (1984a) BS 5135 Process of arc welding of carbon and carbon manganese steels

British Standards (1984b) BS 6325 Code of practice for fixed offshore structure

British Standards (1990) BS 4360 British Standard specification for weldable structural steels. British Standardisation Institution

British Standards (1991), PD 6493 Guidance on methods for assessing the acceptability of flaws in fusion welded structures. British Standardisation Institution

British Standards (1993) BS5996 Acceptance levels for internal imperfections in steel plate, strip and wide flats, based on ultrasonic testing

British Standards (1997) BS 7448 Part 2, Fracture mechanics toughness tests, method for determining KIc, critical COD and critical J values of welds in metallic materials

British Standards (2019) BS7910 Guide to assessing the acceptability of flaws in metallic structures, 2019

Burdekin FM (1967) Initiation of brittle fracture practice in structural steels. Brit Weld J 14:1967

Burdekin FM, Dawes MG (1971) Practical use of linear elastic and yielding fracture mechanics with particular reference to pressure vessels. In: Proceedings conference IMechE, p 28, May 1971

Burdekin FM, Stone DEW (1966) The crack opening displacement approach to fracture mechanics in yielding materials. J Strain Analysis 1(2):1966

Burdekin FM, Townend PH (1981). Reliability aspects of fracture on stress concentration regions in offshore structures. In: 2nd International symposium of integrity of offshore structures, Glasgow
Cambos P, Parmentier G (2016) OMAE 2016 paper 54325, Statistical method to requalify steel grades during conversion of tankers to FPSO
Cotton H (1979) Welded steel for offshore construction, John Player Lecture, Inst Mech Eng, ISSN 0020 3483
Department of Energy (1974) Offshore installations: guidance on design, construction and certification, 1st edn. HMSO, London
Department of Energy (1977) Offshore installations: guidance on design, construction and certification, 2nd edn. HMSO, London
Department of Energy (1984) Offshore installations: guidance on design, construction and certification, 3rd edn. HMSO, London
Department of Energy (1988) United Kingdom offshore steels research project—Phase 1 Final report (UKOSRP I), OTH88 282, HMSO London, UK
DNV (1995) Toughness acceptance criteria for rack materials and weldments in Jack-ups, Report No. 95-3056, Det Norske Veritas, Oslo
DNV (2021) DNV-OS-C401 Fabrication and testing of offshore structures. Det Norske Veritas, Oslo, p 1995
DNVGL RP C210 (2015) Recommended practice: probabilistic methods for planning of inspection for fatigue cracks in offshore structures
Eide OI (1976) Fatigue crack initiation and propagation at the weld toe of cruciform joints. M.Sc. Thesis, Div. of Ship Structure, NTH, Trondheim (in Norwegian)
Farrar, Dolby (1972), Lamellar tearing in welded steel fabrication. Welding Institute report
Garwood (1985) The measurement of critical values of crack tip opening displacement (CTOD) and J on parent steels and weldments for use in fracture assessments. The Welding Institute, Abington, Cambridge, CBl 6AL, United Kingdom
Harrison JD (1981) Material selection considerations for offshore steel structures. Welding Institute Research Report 150/1981, July 1981
Harrison JD, Pisarski H (1978) Fracture resistance of materials used in offshore fabrications. European offshore steels research seminar, Paper 1 (TWI)
Hawthorne JR, Loss FJ (1975) Fracture toughness characterisation of shipbuilding steels, US Coastguard Ship structure Committee, SSC 248, p36
HSE (1994), Guidance notes for the design and construction of offshore structures, Fourth edition, Health and Safety Executive
HSE (1995) OTH 95 489, Literature review of the fracture properties of grade A ship plate, prepared by British Steel plc, HMSO London, UK
HSE (1997) OTH 482 A review of residual stress distributions in welded joints for the defect assessment of offshore structures. HSE Books
HSE (2000) Fracture properties of Grade 'A' ship plate, offshore technology report OTO 2000/001, 2000, HMSO London, UK
HSE (2003) RR 105, Review of the performance of high strength steels used offshore, prepared by Cranfield University, HMSO London, UK
Inglis RB, Mathieson RJB, Sneddon NW, Staton VA (1989) Shell Expro Northern operations structural integrity project. Offshore Europe, Paper SPE 19271-MS, 1989
ISO (2011) ISO 12153 Welding consumables—tubular cored electrodes for gas shielded and non-gas shielded metal arc welding of nickel and nickel alloys—classification
ISO (2018) ISO 15653:2018 Metallic materials—method of test for the determination of quasistatic fracture toughness of welds
ISO (2020) ISO19902:2020 Petroleum and natural gas industries—fixed steel offshore structures. International Standardisation Organisation
Kirkemo F (1988) Applications of probabilistic fracture mechanics to offshore structures. Appl Mech Rev 41(2), Feb 1988

Kanazawa S (1974), Lamellar tearing resisting steels and the directions for using them, IIW Doc IX-873-74

Liston R (1982) Modern ship steels—a brief review—July 1982, pp 8–11

Lotsberg I (2014) Assessment of the size effect in fatigue analysis of butt welds and cruciform joints. In: Proceedings 33rd conference OMAE, OMAE2014—23187, June 2014

Malik L, Tomin M (1991) Evaluation of toughness of conventional ship steels at intermediate loading rate and its implications, DSS File No 06SQ-23440-7-9227-CANMET 1991

Ministry of Power (1967) Report of the inquiry into the causes of the accident to the drilling rig Sea Gem, HMSO Comnd 3409, October 1967

Moan T, Bekkvik P, Næsheim T (1981) Alexander L. Kielland-ulykken. Norges Offentlige Utredninger, NoU 1981:11, Universitetsforlaget, Oslo, Norway

OCED (1979) On the ultrasonic examination of three test plates using the PISC procedure based on ASME XI code, Nuclear Energy Agency Committee on the Safety of Nuclear Installations, OECD November 1979

Pisarski H, Pargeter R (1984) Fracture toughness of weld heat affected zones (hazs) in steels used in constructing offshore platforms

Puzak PP, Pellini WS (1952) Standard method for NRL drop weight tests. Naval Research Laboratory, Washington DC, US

Randerson, Morrison (1989) Materials for ship—British steel technical, Swindon Laboratory, report no PP-R-S1196-37-89-D, March 1989

Recho N (1984) Potential use of fracture mechanics in the fatigue design of fillet welded joints. In: Proceedings 5th European conference on fracture, Lisbon, vol 1, pp 159–170

Rolfe ST, Rhea DM, Kuzmanovic BO (1974) Fracture-control guidelines for welded steel ship hulls. Performed by the University of Kansas under Department of the Navy Naval Ship Engineering Center Contract No. NOO024-72-C-531

Rodrigues P, Wong WK, Rogerson JH (1980) Weld defect distributions in offshore platforms and their relevance to reliability studies, quality control and in-service inspection. Paper OTC 3693, 12th offshore technology conference

Rogerson A, Highmore PJ (1983) UK developments in structural integrity aspects of NDT. Nucl Eng and Design 76(3):1983

Rogerson A, Wong WK (1982) Weld defect distributions in offshore on structural reliability offshore technology conference Houston OTC paper 4237

Signes EG, Baker RG, Harrison JD, Burdekin FM (1967) Factors affecting the fatigue strength of welded high strength steels. Br Weld J March:108–116.

Sill JR, Speck JB, Razjmoo JR (2000) The naval architect. Int J R Inst Naval Architects:28–36

Stacey A (1993a) Review of the post-weld heat treatment requirements for offshore structures. In: International conference on offshore mechanics and Arctic engineering (OMAE-13), Houston, TX (United States)

Stacey A (1993b) The significance of residual stresses in defect assessment of offshore structures. In: Proceedings 12th OMAE conference, ASME, vol III, Part B, pp 729–739

Stacey A, Sharp JV (2021) Proceedings of the ASME 2021 international conference on asset integrity management ageing and life extension AIM-ALE2021 November 30–December 2, 2021, Virtual, Online AIM-ALE2021-76756

Stacey A, Sharp JV, King RN (1996) High strength steels used in offshore installations. In: Proceedings of the 15th international conference offshore mechanics and Arctic engineering, Florence, Italy, June 1996

Sumpter JDG, Caudrey AJ (1995) Recommended fracture toughness for ship hull steel and weld. Marine Struct 8:345–357

Thurlbeck SD, Stacey A, Sharp JV, Nichols NW (1996a) Welding fabrication defects in two offshore steel jacket structures. In: International conference on offshore mechanics arctic engineering (OMAE), Florence (Italy)

Thurlbeck SD, Mohammed A, Burdekin FM (1996b) Fabrication defects in offshore jacket structures. Report prepared for HSE Offshore Safety Division, EQE Report No 45-12-R-01 Issue 1 April 1996

TWI (2024a) Defects—lamellar tearing. https://www.twi-global.com/technical-knowledge/job-knowledge/defects-lamellar-tearing-047, visited August 2024

TWI (2024b) What are local brittle zones (LBZs)? TWI (twi-global.com), visited August 2024

UEG (1985), Design of tubular joints (Section E2.3). UEG, ISBN, 0860172317, 9780860172314

Watkinson F, Bodger PH, Harrison JD (1970) The fatigue strength of welded joints in high strength steels and methods for its improvement. In: Welding institute conference on fatigue of welded structures, vol I, pp 97–113

Webster SE, Bateson PH (1993) Significance of local brittle zones to structural behaviour. Mater Sci Technol 9(1):83–92. https://doi.org/10.1179/mst.1993.9.1.83

Wellman GW, Rolfe ST (1984) Engineering aspects of CTOD fracture toughness testing. Weld Res Counc Bull 299:1984

Woodley CC, Burdekin FM, Wells AA (1964) Mild steels for pressure equipment at sub zero temperatures. Brit Weld J 7:1964

Zhou W, Stacey A (2002), Review of defect distributions for probabilistic structural integrity assessment. In: Proceedings 21st conference OMAE, OMAE2002-28616, June 2002

Chapter 6
Steel Materials Degradation and Remediation

6.1 Introduction

Degradation and deterioration have a common meaning but in this book we will primarily use degradation as the process of something being damaged or made worse. The degradation mechanism can result in metal loss (as uniform or localised attacks), cracking or changes to the material properties (e.g. increased brittleness). In addition, damages from impact, high loads from temperature expansion/contraction and quick pressure changes may influence the material to be more exposed to degradation mechanisms (e.g. dents from impact damage may form crack initiators that may lead to a fatigue crack) and can also lead to changes in geometry of the structure or structural parts.

An absolute definition of which type of degradation the various deterioration mechanisms cause is not possible. A general idea is given in Table 6.1.

Materials selection at the design stage of offshore installations requires that several factors are taken into account, such as the ability to meet strength, temperature, fatigue and durability requirements of the structure. The properties of most materials used in offshore structures are now well understood as a result of operating performance and research. However, the fatigue properties of higher strength steels used, for example in the legs of jack-ups with yield strengths in excess of 600 MPa, are still less well known (HSE 2003).

When older structures were designed there were significant gaps in knowledge such as:

- through-thickness and fracture properties for large steel sections;
- the susceptibility to hydrogen embrittlement, particularly for high and higher strength steels;
- the fatigue performance of large tubular joints in seawater.

There are a number of deterioration mechanisms due to ageing, the principally are corrosion, fatigue, wear and erosion, which are covered in in this section.

© The Author(s), under exclusive license to Springer Nature Switzerland AG 2025

J. V. Sharp and J. Wintle, *Assessment of Materials in Ageing Offshore Structures*,
https://doi.org/10.1007/978-3-031-84489-8_6

Table 6.1 Degradation mechanisms in materials versus effect

Effect	Changes to form/surface changes			Changes to material properties
	Metal loss/wall thinning	Cracking	Geometric changes	Intrinsic/inherent changes
Degradation mechanism	Corrosion (chemical) – General – Pitting – Crevice – Corrosion Under Insulation – Galvanic – Bacterial Flow induced metal loss (mechanical) – Erosion from solids Wear and tear	Fatigue Hydrogen related cracking – Blistering – Stress Corrosion Cracking Creep	Dents from impacts Bowing (out of alignment) Permanent plastic deformations Patch and asymmetric corrosion, wear and tear	Hydrogen embrittlement Hardening – Accumulated plastic deformation Environmental deterioration – Exposure period – Temperature – Bacterial Microcrack development Heating

6.2 Hydrogen Embrittlement and Hydrogen Induced Cracking

6.2.1 Introduction

Hydrogen embrittlement (HE) is the most important intrinsic mechanism causing deterioration of the mechanical properties of steel in offshore structures, particularly in applications using high strength steel such as mooring chains, the legs of jack-up platforms and pipelines. It occurs when atomic hydrogen present at the metal surface is absorbed into a material with a susceptible microstructure. It can then diffuse, migrate and accumulate in the metal lattice resulting in a severe change in the local properties of steel. Several processes associated with offshore structures (e.g. cathodic protection, corrosion, welding) can generate atomic hydrogen, as will be discussed later.

The most important effects of hydrogen embrittlement are lowering the ductility and reducing fracture toughness resulting from a transition from ductile to brittle fracture behaviour. This reduces the stress at which failure occurs and hence the load carrying capacity. Figure 6.1 shows how the extension (% strain at failure), the stress at failure (strength) and fracture toughness reduce as the concentration of hydrogen increases in a high strength steel. Beyond a certain level of hydrogen concentration (e.g. 2 ppm), the % strain at failure is very low implying brittle behaviour with a corresponding reduction of the strength at failure.

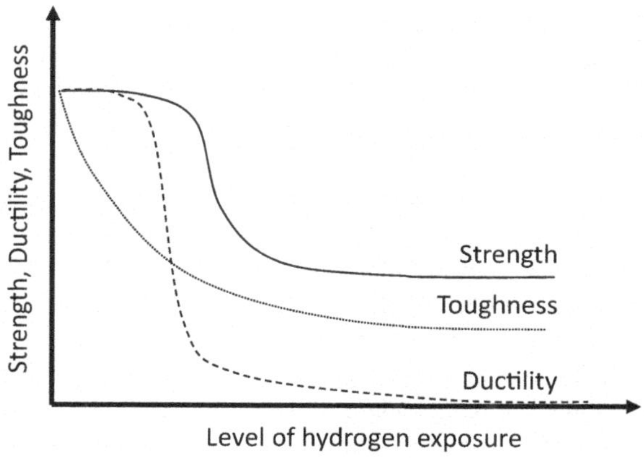

Fig. 6.1 Effect of hydrogen embrittlement (HE) on strength, ductility and toughness

Where there is loss of ductility and toughness due to hydrogen embrittlement, as shown in Fig. 6.1, cracking can result when the stresses are sufficiently high. Stresses causing cracking can be present because of external loads or as residual stresses from processes such as welding, while the presence of hydrogen in voids in the metal can also generate stress from within the material.

It should be noted that hydrogen embrittlement and cracking interact with each other. Because of the high local stress near a crack tip, the crystal lattice will expand, which allows the local hydrogen concentration to increase (Force 2023). This causes enhanced further embrittlement of the zone around the crack tip, which, combined with the high local stress, results in the formation of microcracks ahead of the initial crack tip. Over time, the microcracks coalesce with the initial crack resulting in crack growth and the crack will then continue to grow by repeated microcrack formation and coalescence, see Sect. 6.2.2. Hydrogen uptake under cyclic stress also has an effect on fatigue crack growth, see Sect. 6.7.11.

6.2.2 Hydrogen Cracking

As noted above, the embrittlement of steel from the ingress of hydrogen is a cause for the steel under stress to crack and for cracks to grow and propagate. Hydrogen cracks can occur during welding fabrication, which are known as Hydrogen Assisted Cold Cracking (HACC). Fabrication related hydrogen cracking is further described in Sect. 5.2.3. Hydrogen cracks can also occur while the structure is in service from the corrosion process and cathodic protection, which is known as Hydrogen Induced Stress Cracking (HISC). It should be noted that where a hydrogen crack (or other weld defect) has occurred during fabrication and not been removed, then it can further grow in-service by HISC.

Hydrogen induced stress cracking (HISC) is due to the combination of local hydrogen embrittlement caused by having a sufficient hydrogen concentration and a sufficient level of stress. Generally, a stress above the yield stress is required, such as residual stress of weld without post weld heat treatment. However, nominally lower stresses suffice if local stress concentrations are involved, i.e. if the steel structure already contains crack-like flaws, for example, in the form of HACC or other weld defects.

When a crack has formed or there are pre-existing crack-like defects, both hydrogen and stress are concentrated at the crack tip and the crack propagates. The HISC susceptibility increases with the strength and hardness of the steel. For it to take place, HISC requires a sub-critical threshold value of the linear elastic stress intensity factor K_{TH} to be reached at the crack tip. The crack propagation rate may initially rise with increasing stress intensity (K_I) but will stabilise at a more or less constant rate independent of the K_I value, as shown in Fig. 6.2.

Cracking will continue only if the concentration of absorbed hydrogen is maintained ahead of the crack tip and the K_I value remains above K_{TH}. HISC requires the continued generation and supply of hydrogen to the crack tip during the cracking process. In marine jack-up structures the source of hydrogen is mainly the cathodic reaction taking place on bare (metallic clean) steel surfaces, i.e. uncoated surfaces. Cracking may cease if the source of hydrogen falters, or if the crack enters a stress field where the K_I value falls below K_{TH}.

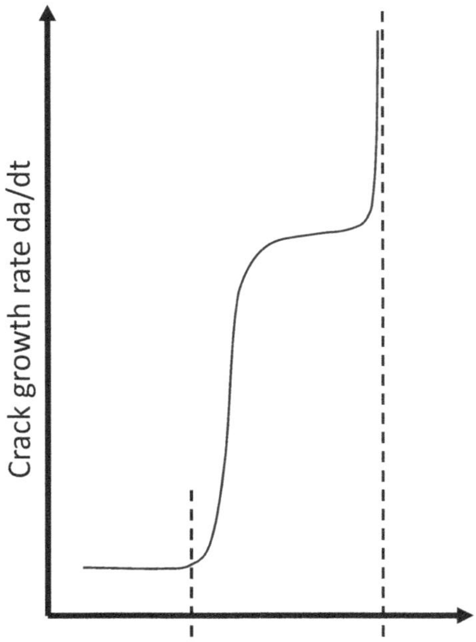

Fig. 6.2 Rate of crack growth rate versus stress intensity (based on Force 2023)

Where loading is displacement controlled in a redundant structure, the load carried by a cracked member may be redistributed through other loading paths as the crack grows. In this case the member with the crack may no longer carry sufficient stress for the crack to grow at the same rate and the rate of growth may slow and eventually cease. In load-controlled conditions, the stress acting on a cracked member increases as the crack grows and the rate of crack growth will be primarily dependent on the time for hydrogen to diffuse through the material to the crack tip.

The application of a surface coating may prevent or slow the ingress of surface hydrogen into the material. Experience (Force 2023) has identified the main cause of hydrogen related cracking of coated marine structures to be:

1. Fabrication cracks or flaws (HACC and traditional weld defects) present prior to coating
2. Fracturing of the coating and exposing of bare metal when these fabrication cracks and flaws are loaded during operation,
3. The growth of cracks in-service due to HISC.

A strongly negative cathodic potential may also be a source of dis-bonding and subsequent degradation of the coating (MTD 1990).

It should be noted that while hydrogen cracking is a result of local embrittlement around the crack tip, surrounding material and other parts of the structure may remain ductile. Consequently, the failure of a structure from a hydrogen crack can often be from a combination of brittle cracking in the local high hardness zone followed by ductile propagation and plastic collapse.

6.2.3 Industry Experience and Resulting Research

Hydrogen embrittlement was found to be the cause of the cracking that was discovered in the leg chords and spud cans of jack-up drilling rigs operating on the UKCS in the late 1980s (HSE 1991). More recently, a review of failures of mooring chains (DNV 2022) found that hydrogen embrittlement was implicated in 70% of the failures. In addition, cracks have recently been observed in jack-ups on the Norwegian continental shelf (Force 2023).

This discovery of cracking in jack-up drilling rig leg chords and spud cans caused considerable concern with the UK regulator and offshore industry. It led to a significant HSE research programme to understand the causes and effects of hydrogen embrittlement and hydrogen related cracking in high strength steels. The programme included a review of crack tip electrochemistry in relation to corrosion fatigue (HSE 1992a), a review of the role and effect of sulphate reducing bacteria (HSE 1998a) and a review of the effects of microstructure on hydrogen embrittlement (HSE 1999a).

A review of hydrogen assisted cracking in high strength steels (Barker 1996) examined the work by Capcis, UMIST and Cranfield University, in addition to work by DNV on anchor chains and other work on low alloy steels by TWI and the University of Glasgow. The conclusions highlighted the impact that cathodic protection level and sulphate reducing bacteria could play on hydrogen embrittlement and cracking. The HSE programme concluded with a comprehensive review of the performance of high strength steels used offshore (HSE 2003).

While high strength steels were considered to be generally the more susceptible to hydrogen embrittlement, lower strength steels could also be susceptible under certain metallurgical conditions. Several other factors were also found to be important and it was difficult to generalise. The HSE guidance (HSE 1994) contained in Sect. 33 of the HSE Guidance Notes (HSE 1995) was therefore that all structural steels and weldments intended for use offshore should be assessed for their vulnerability to hydrogen embrittlement and cracking. This led to the development of new test methods.

High strength steels have been and are being used increasingly in the construction of offshore platforms as they offer a significant weight saving compared to lower strength steels. Steels with yield strengths in the range 550–1200 MPa are used for a variety of marine applications, particularly in mooring chains, tendons and the legs of jack-ups. The advantages of using higher strength steels need to be weighed against their increased susceptibility to hydrogen embrittlement and cracking, particularly when the applied tensile stresses are high. Under these circumstances, appropriate testing and measures to control damage should be undertaken.

It has been found to be possible to use high strength steels grades in the range 500–690 MPa when appropriate codes and standards have been properly applied (FORCE 2023). However, the possibility of HACC followed by HISC in-service remains high where there have been time-constrained construction phases often combined with challenging wet climatic conditions, deviations in welding procedure, or unauthorised weld repairs.

6.2.4 Mechanisms of Hydrogen Embrittlement and Cracking

The deleterious effects of hydrogen on the physical properties of steel have now been realized for over a century. However, despite extensive research, there are still significant questions remaining for ferritic steels due to specific challenges associated with these steels. The primary challenge is that ferritic steels have high diffusivities with respect to hydrogen often leading to difficulties with measuring the hydrogen content, and, historically, to test specimens that exhibited in-air properties (i.e. limited hydrogen embrittlement) shortly after removal from the hydrogen environment.

The scientific community agrees that HE is a complex material and environment dependent process for which no single mechanism applies exclusively, and which may be caused by several different or separate mechanisms. Underlying all the mechanisms, the essential prerequisites for HE to occur are:

1. A source of atomic hydrogen for absorption and diffusion into the material,
2. A material microstructure that is susceptible for hydrogen to amass near internal stress raisers resulting in the material being in an embrittled state.

For offshore structures, the source of the hydrogen is normally either atomic hydrogen generated by electrochemical reactions or welding. The hydrogen then diffuses either through the metal lattice or through grain boundaries and amasses near dislocations, crack tips, microstructural defects and other traps in the metal matrix, such as voids, grain boundaries and precipitates (Martin et al. 2020). It is observed that decreasing grain size in steels provides an increased resistance to hydrogen embrittlement (HSE 1999a).

A great variety of mechanisms have been proposed and investigated as to the cause of the embrittlement once atomic hydrogen has been absorbed and diffused into the metal, but the exact physical mechanisms of hydrogen embrittlement in ferritic steel are not yet agreed upon by the scientific community. Mechanisms for hydrogen embrittlement that have been reviewed by Martin et al. (2020) and include the following:

- Internal pressure,
- Phase transformations,
- Hydrogen enhanced localised plasticity (HELP),
- Hydrogen enhanced decohesion (HEDE),
- Hydrogen decreased dislocation creation,
- Hydrogen enhanced strain induced vacancies (HESIV).

6.2.5 Sources and Uptake of Hydrogen

A primary requirement for HE is a source of atomic hydrogen at the metal surface. Hydrogen atoms can be generated from welding processes or from over-protective cathodic protection systems on station. If hydrogen atoms are not combined as hydrogen molecules to form gaseous hydrogen, they can be absorbed and diffuse into the metal causing embrittlement and cracking.

Welding can cause high levels of hydrogen absorption resulting in hydrogen cracking in the weld metal and HAZ if sufficient care is not taken during welding to avoid moisture, for example, by drying consumables and applying preheat and inter-pass heating before and during welding. In-service, excessively negative cathodic protection potentials, as can happen offshore, can also be a source of atomic hydrogen due to an electrochemical reaction with water at the steel surface and have been the cause of hydrogen embrittlement and cracking in high strength structural steels (HSE 1999a).

The effects of hydrogen charging from the CP system should therefore be fully considered. Over cathodic protection can be mitigated using potential limiting diodes or using sacrificial anodes with a less protective potential, although, in practice, both methods have disadvantages. The Department of Energy (now the Health and Safety Executive) guidance on limiting cathodic protection potentials for high strength steels used offshore in vulnerable situations recommended that the CP levels should be limited to a maximum negative value of − 850 mV Ag/AgCl, although this requirement was found to be very difficult to achieve in practice.

6.2.6 Effect of Microstructure and Strength on Susceptibility

The susceptibility to HE has been thought to increase with the strength of the steel and it has been common practice to assess the likelihood of it occurring in a particular grade of steel based on its strength or hardness. However, research has found this rule to be simplistic and that HE susceptibility is more sensitive to the specific nature of the microstructure than to the strength level of the material. Quenched and tempered (Q&T) steels and controlled rolled (CR) steels can be produced with the same strength and hardness, yet they have very different microstructures and different susceptibilities to HE.

While for a particular grade of steel, strength or hardness can give a useful first indication of its likely HE susceptibility, it has been shown (HSE 2003) that when steels of different grades are considered together, the strength or hardness is often poorly correlated with HE susceptibility. Even within steels of the same strength range, produced by the same manufacturing method, some alloys exhibit better resistance to HE than others. Some modern high strength offshore steels have been produced which have HE susceptibilities that are as low as those of BS4360 Grade 50D steels (British Standards 1990) and significantly lower than would be expected based on their higher strength alone. It is probable that small changes in the chemical composition and differences in microstructure are responsible for this behaviour.

In work for the UK HSE (HSE 1999a), Robinson and Kilgallon explained why some steels in the same strength range seem to be inherently more resistant to embrittlement than others. They considered the HE of offshore constructional steels with yield strengths of 450 MPa and showed that the embrittlement behaviour of these steels is broadly comparable to that of other steels within the same strength range. This report had the following conclusions:

(1) Microstructure has a controlling influence on HE susceptibility and is a more significant factor than the alloy composition. The susceptibilities of different microstructures to HE can be ranked in the order from least to greatest (HSE 2003):

- Tempered martensite (Least susceptibility) $<$
- Tempered bainite $<$
- Spheroidised ferrite and pearlite $<$
- Coarse ferrite and pearlite $<$
- Untempered martensite (Greatest susceptibility)

Untempered martensite, that may be present in the heat affected zone (HAZ) of welds, is generally regarded as having the highest susceptibility, whereas a tempered martensite microstructure has the lowest susceptibility. The reasons are thought to be a combination of the effects of grain size and hydrogen traps.

Refining the grain size has a beneficial effect in reducing the extent of hydrogen embrittlement. It is thought that this effect results from the increase in area of grain boundaries per unit volume at which hydrogen can accumulate innocuously (HSE

1999a). It is thought that the austenite grains before transformation through heat treatment are the most important in acting as effective boundaries to crack growth.

(2) The way in which hydrogen is trapped by features in the microstructure is particularly important. Typical traps for hydrogen include vacancies, voids, interfaces, grain boundaries and dislocations. Hydrogen traps should be sufficiently numerous and dispersed so that the hydrogen is held innocuously and not allowed to locally concentrate.

The optimum resistance to hydrogen embrittlement can be achieved with a structure containing fine carbide particles, uniformly distributed throughout the material, which can trap large quantities of hydrogen and prevent local concentrations from reaching the level required to cause decohesion and crack propagation. However, other individual features that are most effective hydrogen traps are also the most undesirable in terms of contributing to hydrogen embrittlement.

(3) Tempered martensite has been shown to be the microstructure with least susceptibility to HE. The steel should have a fine austenite grain size and sufficient hardenability prior to tempering to produce 100% martensite through the material thickness. Microalloying and appropriate heat treatments can be used to produce the required size and distribution of precipitates to act as effective hydrogen traps. Tempering at high temperatures is beneficial in reducing the density of dislocations that are reversible hydrogen traps and increase the solubility of hydrogen in the steel. Good tempering resistance is required to avoid the growth of the carbide particles at high temperatures.

In other cases, there are differences in the behaviour of new and old steels of the same grade. These might be explained by microstructural changes associated with the advances that have occurred in steel making practice. Consideration of the strength level is a good starting point but other factors including the microstructure, manufacturing method, heat treatment and chemical composition can all be important.

As noted above, the strength or hardness of a particular grade of steel may be a useful first indication of its likely HE susceptibility. It is, however, recommended that each steel or weldment should be considered individually and should be subjected to thorough testing in the relevant environment before being accepted for use.

The increased HE susceptibility in the HAZ of welds of medium carbon steels is due to microstructural changes that occur during the welding thermal cycle, particularly the formation of martensite, which may not be adequately tempered by subsequent welding passes. Twinned martensite has a particularly high susceptibility to HE. The related welding difficulty of cold cracking, which is a form of HE caused by hydrogen absorbed during welding, results from the extreme susceptibility of twinned martensite in the HAZ of medium carbon steels. Low carbon weldable steels have been developed to avoid this problem.

6.2.7 Effect of Sulphate Reducing Bacteria

In a wide-ranging review (HSE 1998a, b, c), Robinson and Kilgallon showed that the generation and uptake of hydrogen by steel in a marine environment is strongly influenced by the combined effects of the surface electric potential from CP and the presence of sulphate reducing bacteria (SRBs). SRBs are common micro-organisms present in marine bio films on the metal surface or within sediments on the seabed and can be found in oil-based lubricants and drill cuttings used or produced during past operations until oil-based products were phased out.

Biogenic sulphides produced by SRBs have the effect of promoting the absorption of atomic hydrogen on the steel surface produced by cathodic protection. These sulphides 'poison' or hinder the recombination of atomic hydrogen to form hydrogen molecules (and eventually hydrogen gas), which effectively keeps atomic hydrogen on the metal surface for longer resulting in a smaller fraction being absorbed by the metal. With the presence of SRBs, the hydrogen uptake can increase by as much as an order of magnitude compared to that in sterile conditions at the same CP potential.

6.2.8 Testing Methods

Testing is particularly important for steel used in critical locations and where there is a possibility that hydrogen charging from CP over protection may occur. This is especially important in the case of higher strength steels used in the construction of fixed jack-up platforms and for mooring chains, which are intended to remain in position for the life of the field and where inspection is more difficult to carry out.

The susceptibility of steels to HE and their behaviour can be determined using three types of test (EFC 1995):

1. The use of smooth specimens under constant ~~static~~ uni-axial tensile loads to generate time-to-failure and threshold stress data.
2. The use of conventional smooth specimens in a tensile test where the strain of the specimen is controlled and slowly increased at a constant rate until the specimen fractures (slow strain rate testing (SSRT)).
3. The use of pre-cracked specimens in conjunction with fracture mechanics concepts to determine threshold stress intensity values and to generate curves of the time-based crack growth rate, da/dt, against stress intensity, K.

These tests are not equivalent, and each has a specific role. For example, the use of smooth uniaxial tensile specimens under static loads will indicate the fitness for service of a specific material at a specific stress level in the given environment in real time. It can be used as a method for monitoring the condition of a steel in parallel with the operation of the structure.

The slow strain rate test will provide a result in a much shorter time and can be used to rank the performance of different steels under different environmental conditions. The HE susceptibility of steels in seawater is frequently assessed using slow strain rate testing (SSRT) as it is a relatively rapid, comparative method. A smooth plane specimen is immersed in a tank containing the relevant environment and the strain (and stress) slowly increased until the specimen breaks. The time and strain or elongation at failure are key indicators of the state of embrittlement.

SSRT is a severe test (to failure) and does not accurately represent the conditions found offshore and it is important to appreciate the limitations of the technique as erroneous results can occur. For example, when welded specimens are tested most of the strain sometimes occurs preferentially in the softest part of the microstructure. This leads to preferential failure in that region, whereas in practice HE usually occurs in the hardest region of the weld.

Fracture mechanics based tests are applicable for a fracture mechanics design approach based on a limiting defect size. Common testing configurations are four-point single edge notch bend tests, C-ring tests and double cantilever beam (DCB) tests. These fracture mechanics specimens will be fatigue pre-cracked and more accurately reflects the behaviour of the embrittled steel in the presence of the crack.

Example test results

The results of SSRT of two Q&T high strength offshore steels, SE500 (500 MPa yield strength) and SE700 (700 MPa yield strength), are shown in terms of the embrittlement index in Fig. 6.3. They are compared with results for typical E36 steel (Coudreuse et al. 1997), described by the authors as BS4360:50D (British Standards 1990) equivalent steel but having a somewhat unusual composition. The tests were conducted at two different strain rates (5×10^{-7}/sec and 20×10^{-7}/sec) in two different solutions (NaCl and NACE saturated with H_2S).

The hydrogen embrittlement index I_{HE} was defined by Loidl et al. (2011) in relation to the elongation of a specimens in a tensile test as follows

$$I_{HE} = \frac{\sigma_o - \sigma_H}{\sigma_o} \times 100\%$$

where σ_o is the elongation of specimens in air and σ_H is the elongation of specimens with hydrogen charging. Hydrogen charged specimens show a reduced elongation compared with specimens in air. Therefore, IHE can vary from 0% with no ductility loss, to 100% with total loss in ductility.

The effect of strain rate is seen with a higher embrittlement index (EI) measured in tests conducted at the lower strain rate. When tested in seawater (NaCl) at the same strain rate the Q&T steels performed slightly better than the BS4360:50D (British Standards 1990) equivalent steel demonstrating that microstructure has more influence than strength. Even in a saturated H2S solution, the 700 MPa steel had a similar EI to the BS4360:50D equivalent steel. It would be concluded from these results that the SE700 would be resistant to HE in seawater by virtue of the similarity in behaviour to the BS4360:50D equivalent steel.

Figure 6.4 gives more SSRT results for the same steels, showing the performance in seawater containing very low levels of H_2S (Coudreuse et al. 1997). The EIs

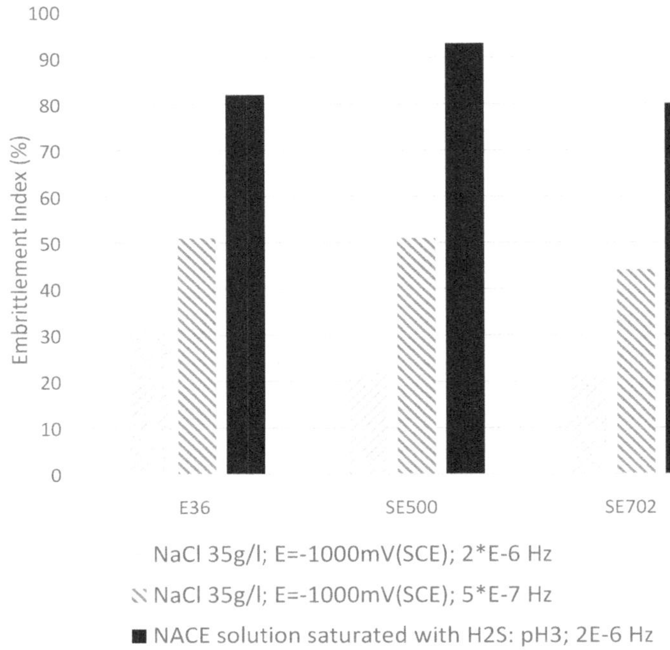

Fig. 6.3 EI for SSRT on E36 (BS4360:50D equivalent) and two high strength Q & T steels showing effects of strain rate and sulphide (based on Coudreuse et al. 1997)

were lower than those in saturated H_2S but in this case the 700 MPa steel was appreciably more susceptible than either of the two lower strength steels. As low levels of H_2S could arise from microbial activity in the marine environment, this could be an important difference and the 700 MPa steel could be at risk of cracking in marine applications. As low values as 100 ppm have been suggested by Robinson and Kilgallon in HSE (1998a). This example illustrates the essential requirement for embrittlement testing of high strength steels to be carried out in environments that closely resemble the expected service conditions.

Results reported by Pircher et al. (1990) also demonstrate that some Q&T steels can be more resistant to embrittlement than normalised 50D steel. These results shows reductions in area from SSR tests for two normalised and two Q&T steels tested in natural seawater and different levels of CP. The Q&T 690 MPa yield strength steel had similar HE performance to the normalised 355 MPa yield strength. The Q&T 500 MPa steel was more resistant to HE than the other steels. These results indicate that superior embrittlement resistance that can be achieved with a Q&T microstructure compared to that of a typical normalised steel.

Cole et al. (1988) and Cole and Brook (1993) compared the HE of two high strength steels (1%CrMoNb and 31/2%NiCrMoV) and one medium strength steel (21/4%CrMo) that could be considered for marine applications. Three point bend

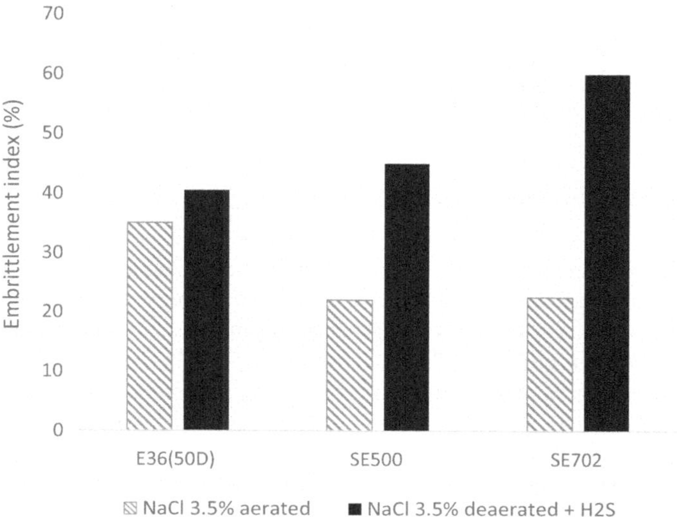

Fig. 6.4 EI for SSRT tests showing the effect of trace quantities of H_2S at a potential of − 1000 mV (SCE), based on Coudreuse et al. 1997

tests were carried out in natural seawater with the addition of approximately 250 ppm H2S to represent the effects of microbial activity. The specimens were in the as-received condition or heat treated to simulate the microstructures that would result from welding.

The low threshold stress intensity factor, Kth, values found for the HAZ microstructures indicated that they would be particularly susceptible to embrittlement due to the presence of untempered martensite. Cole concluded that in the Q&T condition the two high strength steels performed as well as the medium strength steel and increasing the yield strength does not necessarily increase susceptibility to embrittlement. Robinson and Kilgallon (1994) measured Kth values for two Q&T offshore steels, SE500 (500 MPa yield).

In terms of assessing ageing structures, particularly those with high strength steels, their history with respect to cathodic over-protection needs to be considered to assess the possible accumulated embrittlement and cracking from HE over the sustained period of operation. The type and microstructure of the steel and its known susceptibility to HE are important factors in assessing the likelihood of embrittlement due to hydrogen. Where there is doubt, testing should be undertaken.

6.2.9 Recent Research into Hydrogen Embrittlement

With the advent of the hydrogen economy requiring the storage and transportation of high-pressure hydrogen, the topic remains one of significant practical and research interest. Recent work includes that by Djukic et al. (2014), Troutmann et al. (2020) on hydrogen uptake, Wang et al. (2022) on hydrogen diffusion and a comprehensive review of hydrogen embrittlement in ferritic steels by Martin et al. (2020).

6.3 Sulphide Stress Cracking Due to H_2S

Sulphide stress cracking (SSC) is a form of environmental cracking due to the presence of H_2S arising from either bacterial activity or the presence of drill cuttings is primarily a problem for carbon steel production pipelines, storage tanks and other parts exposed to sour products. The susceptibility for SSC is dependent on a number of factors including the partial pressure of the H_2S, the total tensile stress, chloride ion concentration, and presence of another oxidant etc. SSC is not expected to occur below a critical partial pressure of H_2S. However, for partial pressures above this limit the environmental condition is termed as sour and there is an increasing likelihood of SSC occurring.

Cracking is the resulting failure mode and can be abrupt in nature. SSC is controlled by specification of the material properties (particularly hardness) and the manufacturing process. For susceptible materials, environmental cracking is more likely to occur during the initial phase of production. However older installations may experience a souring of the wells (the produced amount of H_2S increases) and the production environment turns from sweet to sour. This can lead to a higher likelihood of environmental cracking which is dependent on the material properties and the changed service conditions.

6.4 Hardening and Plastic Strain

Significant plastic strain can occur in a material when the loading and stresses exceed the elastic regime. Unlike elastic strains, these plastic strains will not be reversed when the loading is removed, making the material harder because the elastic yield strength has been increased. This hardening will increase the strength and reduce the ductility of the material. These effects are due to changes in the internal structure of the material and occur due to dislocation movements and dislocation generation within the crystal structure of the material. The exact extent of ductility reduction depends on factors such as the degree of plastic deformation, grain size and the specific alloy composition of the steel. The hardening of materials with subsequent ductility reduction could lead to formation and nucleation of microcracks, their growth and premature fracture.

Cyclic plastic strains occurs when a material undergoing plastic deformation in one direction and then is unloaded and reloaded in the opposite direction. This will result in a so-called hysteresis type stress–strain curve, in contrast to the linear elastic stress–strain curve which repeatedly follows the same "straight" path. During such cyclic plastic loadings, the material can experience several effects, including low-cycle fatigue, the Bauchinger effect and ratcheting.

The Bauchinger effect manifests as a reduction in the yield strength of the material in the reverse loading direction compared to its yield strength in the original loading direction. Hence, the material requires less stress to begin yielding when loaded in the opposite direction after experiencing plastic deformation. The Bauchinger effect is attributed to the previously mentioned rearrangement of dislocations and other microstructural changes that occur during plastic deformation. When the material is unloaded and reloaded in the opposite direction, these microstructural changes influence the mechanical behaviour of the material, leading to an observed reduction in yield strength. The Bauchinger effect can affect the fatigue resistance, ductility and overall mechanical performance of the material for materials when subjected to cyclic stresses.

Ratcheting is the gradual accumulation of plastic strain in the material that occurs when a material undergoes cyclic plastic strain over an extended period. During each loading cycle the material undergoes plastic deformation and with successive loading cycle the steel accumulates additional plastic strain. This occurs because the material does not completely recover its original shape during unloading due to the presence of residual stresses and microstructural changes induced by the previous loading cycles. Hence, the accumulated plastic strain over time leads to a gradual increase in the overall deformation of the steel. This progressive deformation manifests as an increase in the strain or displacement under cyclic loading, even when the applied stress remains constant or fluctuates within a certain range. The accumulated plastic strain can cause the material to fail from ductility exhaustion and increases the likelihood of fatigue crack initiation and propagation, leading to premature failure under cyclic loading conditions. Hence, ratcheting is closely related to low cycle fatigue as discussed in the next section.

Plastic cyclic strains can further promote the initiation and propagation of cracks in the steel. Microscopic cracks can nucleate at stress concentrations or defects within the material and propagate under cyclic loading conditions. These cracks can eventually lead to macroscopic cracks and possible failure of the material. Plastic cyclic strains can also lead to a reduction in the fatigue strength of the steel material.

6.5 Low Cycle Fatigue

Extreme loading of offshore structures exposed to the marine environment in severe storms can result in high amplitude load cycles, beyond the yield stress into the tensile and compressive plastic regimes. As mentioned above, such plastic deformation under cyclic loading can cause the nucleation and initiation of fatigue cracks after a relatively low number of cycles leading to crack growth. Low cycle fatigue (LCF) is an important degradation mechanism whose treatment has been reviewed in publications Standard Norge (2015) and HSE (2004).

The low cycle fatigue (LCF) mechanism includes the accumulation of dislocations increasing with each loading cycle, contributing to the progressive deformation of the material. The microstructural changes that occur during low cycle fatigue in steel are complex and depend on factors such as alloy composition, loading conditions and heat treatment history. Understanding the low cycle fatigue behaviour and performance of steel components is essential for predicting the life of the structure. Until the late 1990s this understanding was limited because of a lack of experimental data in the high stress region.

As a precaution, the HSE "Guidance Notes" (HSE 1994) placed restrictions on the applicability of S–N curves in the high stress—low cycle region. For welded plates and tubular joints in air or seawater (with or without CP) the Guidance Notes considered that values of the stress range greater than twice the yield stress needed special consideration. The Guidance Notes noted that some tubular joints could experience stresses beyond this limit during severe conditions, although this mostly occurred in older platforms constructed from low yield stress steel, or when the joint geometric SCF exceeded 5, or both.

The Guidance Notes did not identify what fatigue damage might be expected when these limits were exceeded. There was concern that the fatigue process might change from a stress controlled high cycle fatigue mechanism to strain controlled low cycle fatigue. Re-analysis of screened data suggested that there was no need to impose such a restriction (HSE 2004).

HSE (2004) reviewed earlier work (TWI 1991) on the low cycle fatigue resistance of plate details and tubular connections as well as more recent data identified by a literature search. This data indicated that for lower cycles the standard S–N design curve was unconservative. The report found data that showed a fatigue life *less* than that predicted by extrapolation of standard design curve was possible in the high-stress low-cycles region, but *only* under laboratory conditions as explained below. There was no evidence of components in structures on station failing under low cycle fatigue.

In addition, HSE (2004) found that the very high levels of cyclic stress required to promote low cycle fatigue behaviour would not prove acceptable in design code static strength checks. They would therefore *not* occur in actual offshore installations providing they had complied with the design standards at the time. Thus, when the extended life of an older structure is being assessed it is important to reassess the static strength taking changes from the initial design state into account to abnegate the possibility of low cycle fatigue. However, modern standards such as NORSOK N006 (Standard Norge 2015) provide methods for checking the storm cyclic resistance of steel frame structures to accumulated strain and low cycle fatigue of joints.

NORSOK N006 (Standard Norge 2015) provides a procedure in Annex A to develop a design fatigue curve to facilitate the low cycle fatigue check of joints. This is based on the recognition that laboratory data indicated a low cycle fatigue life lower than would be expected from an extrapolation of the standard fatigue design curves, as noted by the authors of HSE Research Report 207 (HSE 2004). This curve applies when the number of cycles falls less than 10^5 and has a lower negative inverse slope than above 10^5 cycles. For example, for a hot spot stress of 1000 MPa, the reduction

Table 6.2 Parameters for SN fatigue curves (Standard Norge 2015)

Environment	Low cycle fatigue	High cycle fatigue	
Air	for $N \leq 10^5$ $m = 7.486$ $\log \bar{a} = 23{,}655$	for $10^5 < N \leq 10^7$ $m = 3.0$ $\log \bar{a} = 12{,}476$	for $N > 10^7$ $m = 5.0$ $\log \bar{a} = 16{,}127$
Seawater with cathodic protection	for $N \leq 10^5$ $m = 6.508$ $\log \bar{a} = 20{,}566$	for $10^5 < N \leq 1{,}77 \times 10^6$ $m = 3.0$ $\log \bar{a} = 12{,}175$	for $N > 1.77 \times 10^6$ $m = 5.0$ $\log \bar{a} = 16{,}127$

Where in the usual S–N equation:
$\bar{a}$ = intercept of the design S–N curve with the log N axis
m = negative inverse slope of the S–N curve

in life is a factor of approximately 100 whereas the effect is much reduced for lower stresses.

The fatigue design curve for the whole cyclic range is thus in three parts, as given by NORSOK N-006 (Standard Norge 2015) (Table 6.2).

The low cycle fatigue check can be performed like an ordinary fatigue analysis as presented in DNV-RP-C203 (DNV 2021) with the use of linear structural analysis. An alternative but more complicated way of analysing low cycle fatigue would be with fracture mechanics using procedures such as BS 7910 (British Standards 2019) assuming an initial size of crack.

6.6 Corrosion

6.6.1 Corrosion

Corrosion is a surface phenomenon and the result of a chemical or electrochemical reaction between a metal and its environment, that leads to deterioration of the material on the surface. The electrochemical reaction (Fig. 6.5) involves both an anodic reaction (where metal is consumed) and a corresponding cathodic reaction. At anodic sites metal atoms dissociate to produce two ferrous ions leaving two electrons in the metal, as shown in the following formula.

$$\underset{\text{Steel}}{2Fe} + \underset{\text{Dissolved oxygen}}{O_2} + \underset{\text{water}}{2H_2O} \rightarrow \underset{\text{Ferrous ions}}{2Fe^{++}} + \underset{\text{Hydroxyl ions}}{4OH^-}$$

These electrons migrate to the cathodic site where they react with the available oxygen and water to form hydroxyl ions (OH^-). Finally, the negatively charged hydroxyl ions migrate to the positively charged ferrous ions to create a corrosion product (e.g. ferric hydroxide), see Table 6.4.

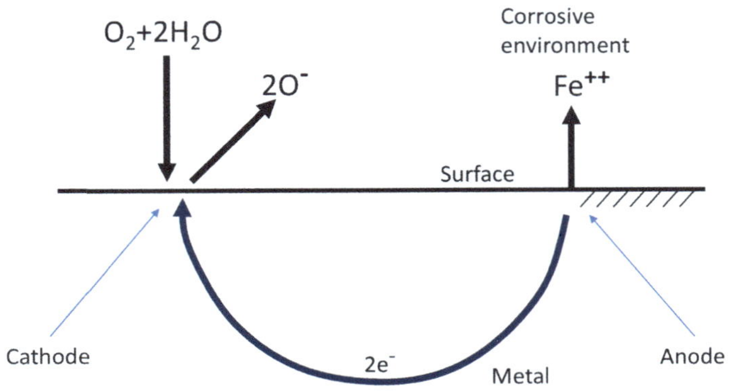

Fig. 6.5 Cathode anode reaction during corrosion

The following basic conditions must be fulfilled for corrosion to occur:

- A metal surface is exposed to a potentially corrosive environment (e.g. bare steel in physical contact with the local environment)
- The presence of a suitable electrolyte able to conduct an electrical current (e.g. seawater containing ions)
- An oxidizing medium able to cause corrosion (e.g. oxygen, carbon dioxide).

However, no corrosion will occur if one of these conditions is not present.

Corrosion of steel can result in significant metal loss over time, unless the steel is adequately protected. The removal of material can result in pitting, degraded surfaces and more general material loss and wall thinning.

Peak corrosion rates are found immediately below and in the splash zone at a rate between 0.25 and 0.75 mm per annum (MTD 1990) for unprotected fully immersed steel. However, if corrosion products accumulate, the reaction may become limited locally and the corrosion rate may decrease. Hence, a value of 0.13 mm per annum can be taken as a reasonable estimate for fully immersed steel for most parts of the world (MTD 1990).

If the anodic reaction occurs randomly on the surface, the effects of metal loss can leave the surface of the corroded steel with larger (uniform) areas of reduced thickness. If the cathodic reaction is limited to smaller areas of the surface, the effect will be localized (patches) and local pits. The uniform corrosion will primarily lead to a reduction of the general wall thickness, while local corrosion patches may also lead to unsymmetric structural elements. Local pits and defects in the general areas of corrosion can cause stress concentrations and defects for crack initiation.

The rate of corrosion is dependent on the level of oxygen and the temperature of the seawater. In very deep water, the amount of available oxygen may be insufficient for corrosion to occur. However, in the North Sea the seawater is normally saturated with oxygen at approximately 6 ml per litre, due to more or less continuous circulation of the water layers (Ersdal 2001). In other offshore areas the oxygen level in deep water may be much less, and corrosion would be inhibited. Corrosion is particularly a problem in more temperate areas of the world, such as Gulf of Mexico and the Persian Gulf. In colder regions, such as the North Sea and the Norwegian Sea, corrosion is less but still a significant issue.

Structures submerged in seawater could be protected from corrosion by cathodic protection using sacrificial anodes or impressed current. The typical anode materials are those with more negative electrochemical potential than that of steel, such as zinc and aluminium. In some cases, the use of external corrosion coatings (see Sect. 9.4) is employed.

The splash zone is a particular concern, as the structure in this region has high levels of available oxygen and cathodic protection is ineffective, because there is no continuous immersion in seawater, which is needed for the electrolytic cell to be active (Trethewey and Chamberlain 1995). This problem was initially addressed using thicker sections of steel (allowance for corrosion) in splash zone regions or the use of corrosion resistant paints and protective coatings (see Sect. 9.4).

6.6.2 Effects of Corrosion on Material and Structural Strength

Wall thinning alters the geometric dimensions of a structural member and hence its deformation behaviour and load carrying capacity. Where there is differential corrosion on opposite sides of a single member, this can give rise to asymmetry in the structural performance. This can increase the likelihood of the member buckling under compressive loading. In severe cases when corrosion has caused severance or near severance of a member the load will be redistributed to other members of the structure which are then more likely to subsequently fail.

In addition, corrosion pits act as local stress raisers and can be initiators of fatigue and other types of cracking. These local stress concentration factor at pits will also influence the material in tension and compression. As a result, the load carrying properties and structural behaviour of pitted material is often reduced compared with the original pit-free product.

The surfaces of corroded material are often degraded with connected areas of local thinning. Assessment methods are available for evaluating the strength of locally thinned areas. ISO 19902 (ISO 2007) and NORSOK N-006 (Standard Norge 2015) have methods for assessing the strength of member with locally thinned areas. The ASME post construction standard for refinery equipment also has methods for assessing local thinned areas primarily for pressure containing structures.

An important product of corrosion reactions is the production of atomic hydrogen at the surface. This can result from either corrosion of the structural member or from corrosion of an adjacent cathodic protection anode when the electrical potential is too negative. When atomic hydrogen is absorbed into steel the metal can become embrittled and generate hydrogen cracks. These topics are dealt with in detail in

Sect. 6.2. For underwater structures the state of the CP system and rate of depletion of the anode is an important indicator of the corrosion status of the structure.

6.6.3 Cathodic Protection

External corrosion is usually mitigated by the use of a cathodic protection (CP) system or an impressed current cathodic protection systems (ICCP). As noted earlier, sacrificial anodes are the most generally used system. Anode materials are mainly alloys of either zinc or aluminium, both of which are more electronegative and active than carbon steel. As a result, these corrode preferentially when they are electrically connected to the steel in an electrolyte (e.g. seawater). Typical anode materials are shown in Table 6.3.

The addition of alloying elements (mainly zinc) to aluminium leads to breakdown of the usual oxide film formed when aluminium corrodes in seawater, which is necessary to make it useful as an anode material. Proprietary anode materials may add additional small amounts of elements to increase performance. The anodes are consumed with time during operation and may need replacement at some stage.

A typical CP system consisting of a zinc anode is shown in Fig. 6.6. The zinc ions preferably go into solution rather than iron ions as zinc is more negative than iron. At the steel surface oxygen and water interact with electrons to produce hydroxyl ions, which create an alkaline environment.

The design of the CP system is dependent on the design life of the equipment and the type and quality of the external coating system in question. There are recommended levels of cathodic protection (e.g. – 800 to – 950 mV vs. Ag/AgCl) provided

Table 6.3 Typical sacrificial anode materials (MTD 1990)

Alloy	Closed circuit electrochemical potential (Ag/AgCl) negative volts
Al–Zn–Hg	1–1.05
Al–Zn–In	1.0–1.10
Al–Zn–Sn	1.0–1.05
Zinc	0.95–1.03

Table 6.4 Main corrosion mechanisms, based on DNV (2006)

Corrosion mechanisms in an oil and gas environment	
Corrosion mechanism	Chemical reaction
O_2 corrosion	$2Fe + H_2O + 3/2\ O_2 = 2FeO(OH)$ (rust)
CO_2 corrosion	$Fe + H_2O + CO_2 = FeCO_3 + H_2$
Environmental cracking due to H_2S	$Fe + H_2S = FeS + H_2$
Microbiologically induced corrosion (MIC)	Fe + (bacteria related oxidant) $\rightarrow Fe^{2+}$

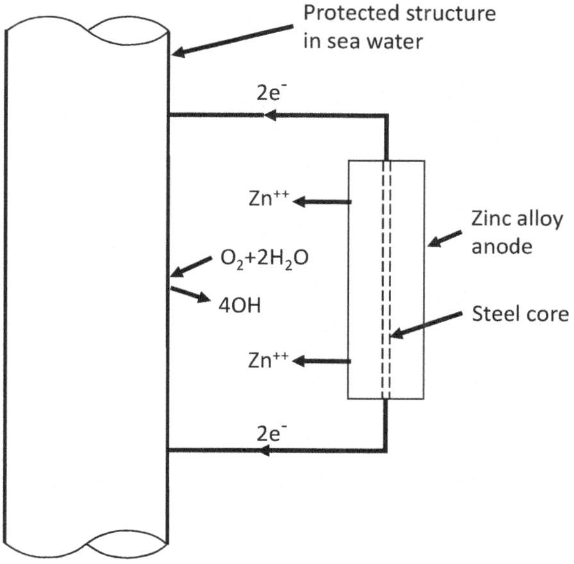

Fig. 6.6 Representation of cathodic protection using a zinc alloy anode electrically connected to a steel structure in seawater (based on MTD 1990)

in several classification document. On some installations there are more negative levels of protection and this over-protection can lead to the production of hydrogen with adverse effects on the steelwork (see Sect. 6.2). High strength steel (with yield strength greater than 500 MPa) are more susceptible to this and more stringent requirements are recommended for the level of cathodic protection (HSE 2003), see Sect. 3.5. Shielding can limit the effectiveness of the CP system e.g. in areas where anode location is difficult. However, assuming the CP protection is effective there should be limited loss of steel due to external corrosion. However, as already noted the CP system in not effective in the splash zone and alternative means of protection are required (e.g. coatings, plus a corrosion allowance). A number of coating systems have been used offshore with epoxy-based systems more typical (see Sect. 9.4).

Some components (e.g. chains) are not usually provided with a CP system due to the complexity of the chain unit and as a result a corrosion allowance is normally provided in design. This allowance will depend on the expected corrosion conditions and the expected life of the chain. In some areas where there may be a limited supply of oxygen (e.g. contained areas) the corrosion rate may be lower and a smaller corrosion allowance could be acceptable. For wire rope moorings, corrosion is managed either by adding zinc wires into the rope during manufacture or using a special blocking compound to resist seawater entry. Another alternative is to use a plastic sheath, but this is vulnerable to mechanical damage.

Where there is evidence that critical parts of the structure are inadequately protected or unprotected, some action is necessary, particularly if life extension is being considered. For structure at or above sea level a damaged or degraded coating should be replaced with a modern equivalent, see Sect. 9.4 on coating. Sacrificial anodes protecting underwater structures can also be replaced or supplemented (Sharp and Ersdal 2021).

6.6.4 Various Forms of Corrosion

Prospective corrosion mechanisms are summarized in Table 6.4 for subsea oil and gas production equipment. More details of corrosion processes are given in DNV (2006).

As already mentioned, O_2 corrosion is the most common type of corrosion leading to typical rust products. As the name suggests, O_2 corrosion occurs in oxygenated environments. In addition, environmental cracking due to H_2S is considered as a cracking mechanism and in Sect. 6.3 described in more detail. Further, carbon steel can be subjected to CO_2-corrosion. The corrosion rate is dependent on the partial pressure of CO_2, the temperature, the flow regime and the water in-situ pH. The corrosion is a time dependent degradation mechanism and generally is localised in the form of pitting. CO_2-corrosion can be managed by the use of corrosion inhibitors and/or by pH-stabilisation of the process fluid, which is primarily applicable for pipelines.

Microbiologically Induced Corrosion (MIC) is a form of degradation that can occur as a result of the metabolic activities of bacteria in the environment. The bacteria that cause MIC can accelerate the corrosion process because the conditions that apply already have elements of a corrosion cell. Sulphate Reducing Bacteria (SRB) are the most aggressive micro-organisms that enhance the corrosion of steel. SRB bacteria live in oxygen-free environments, making use of sulphate ions in the seawater as a source of oxygen. H_2S is produced as a waste product from the SRB, producing a local corrosive environment in connection with the bacteria. MIC has been observed on steel (e.g. anchor chains on the seabed) buried within seabed sediments. The risk of MIC occurring is difficult to predict as it depends on the availability of nutrients, water temperature, and local flow conditions. Further information on MIC is given in Energy Institute (2017).

6.6.5 Special Issues Related to Corrosion in Hulls and Ballast Tanks

The integrity of floating installations depends on intact hulls and ballast tanks where corrosion is the main threat. Ballast tanks are particularly vulnerable as seawater is used for ballast purposes. Corrosion protection systems are usually provided to limit the extent of corrosion, either by coatings or cathodic protection or by a combination of both. Surveys required for example for maintenance of Class require thickness measurements of critical parts of the structure and monitoring of the protection system.

DNVGL-RU-OU-0300 (DNVGL 2018) lists the relevant survey requirements and has a special section on "Special provisions for Ageing Units" which is mainly concerned with fatigue and the introduction of the "fatigue utilisation index" (FUI). When the FUI exceeds one then special measures are required including additional surveys, in addition systematic thickness measurements need to be performed at renewal surveys and inspection of the corrosion protection system to establish its effectiveness. For critical areas, a detection system is recommended to establish any water ingress as a result of corrosion or cracking. DNVGL-RP-B101 (DNVGL 2015) for corrosion protection for floating production and storage units lists both design and survey requirements for corrosion protection systems. This document points out that it is a challenge to provide more than 10 years' service life for the corrosion protection of an FPSO.

Whilst more traditional vessels dock every five years for detailed inspection and repair an FPSO will be in continuous operation for its service life. The DNVGL (2015) document points out that it is therefore necessary to develop an improved specification for the corrosion protection of an FPSO with the service life of 10 years or longer. This should be based on experience for the corrosion protection of fixed offshore structures with design lives exceeding 25 years. Clearly, a case for life extension would need evidence of thickness measurements of critical areas, as well as evidence of the continuing performance of relevant CP systems and coatings protecting against corrosion.

Ballast tanks are significantly exposed to corrosion, especially in areas where cathodic protection anodes are not effective. Oil tanks are also susceptible to corrosion, especially if the oil has a low pH value. This has a tendency to form pitting corrosion in the bottom of a tank. Hence, corrosion protection in form of coating and anodes are important in such areas.

6.6.6 Corrosion of Bolted Connections and Their Protection

Corrosion of bolted connections takes many forms but for bolts used underwater on a cathodically protected structure, the most serious aspect is likely to be hydrogen embrittlement (HE), see MSL (2004). The susceptibility to HE of steels tends to increase with strength and the level of negative cathodic potential (see Sect. 6.2). Even levels of cathodic protection that are acceptable for protecting the structure against corrosion can cause hydrogen induced failure of higher strength bolts. The optimum level of cathodic protection for a bolted structure is complex depending on the bolt alloy, its manufacture, including heat treatment, and the loading environment, applied stress and preload (MSL 2004).

It is not possible to give general guidance on the susceptibility of bolts to HE. When assessing an older or aged structure for life extension or continued service MSL (2004) recommends that it is advisable to seek specialist advice. The use of relatively low-strength bolts could avoid this problem but could render the bolts more susceptible to fatigue as the pre-tension is necessarily limited.

It should be noted that cathodic (over)-protection is not the only way of introducing hydrogen into the material. Electroplated coatings, such as cadmium plating used to protect the bolt, can in fact embrittle it because of hydrogen evolution during plating. To avoid hydrogen evolution, the plating needs to have been done properly or steps taken to remove absorbed hydrogen by a low-temperature heat treatment.

Corrosion may also take the form of general attack and waste the exposed parts of the bolt or the first few threads (this is especially serious as these take most of the load). Crevice corrosion at the thread root can be a problem as can galvanic effects caused by using the wrong combination of materials. Protective coatings such as PTFE improve the resistance to galling and corrosion and minimize thread friction during tightening MSL (2004).

Coated bolts are more suitable for bolts exposed to marine atmosphere. Splash zone bolts are normally galvanised (DNV 2019). Carbon steel bolting materials (bolts, studs, nuts, washers) used offshore in the splash zone are normally coated with Xylan (or similar approved coating) for corrosion protection (DNV 2006). Electrodeposited Zinc Plating is also an approved coating. The chosen coating is acceptable only if electrical continuity to the cathodically protected structure is confirmed with a test after assembly.

A study of incidents involving bolts was carried out by DNV for PSA and published in 2019 (PSA 2019). Bare carbon steel bolts have suffered from excessive corrosion and stress corrosion cracking of stainless-steel bolts has resulted in the release of contaminants. The application of conventional painting and greasing to steel fasteners has proven to be ineffective in minimizing corrosion and hot dipped galvanized bolts also fail to provide adequate protection. Premature failure of steel fasteners due to the harsh marine atmospheric environment in the Arabian Gulf has led to premature failures resulting in increased maintenance and inspection activities (Muzghi 2024).

Fluoropolymer coatings provide the best resistance to rusting, but their inherent softness has rendered them susceptible to damage. Ceramic composite coatings containing fluoropolymer has been found to improve the hardness of fluoropolymer coatings and minimize damage to the coating during and after fastener assembly. For permanent installation a composite coated fastener is the preferred choice, whereas a frequently unscrewed fastener requires a system that provides superior corrosion protection while still retaining adequate mechanical properties and durability (Muzghi 2024).

In order to prevent stress corrosion cracking, bolts, studs and nuts manufactured from carbon or alloy steel used in subsea service should satisfy all of the requirements of ASTM A 193/A 193M Gr B7, B7M, L7, or Grade 660 SS with a hardness not exceeding 35 Rockwell C, as specified in API SPEC 17D:2000.

6.7 Fatigue

6.7.1 Introduction

Fatigue cracking is characterised by progressive damage of the material due to decohesion of the material caused by cyclic stresses during the life of the structure. Fatigue cracks typically initiate from areas with high cyclic stresses and at the location where defects are present. However, fatigue cracks can also occur in materials without defects, but the initiation period of cracks will be significantly longer. The presence of both defects and geometric discontinuities are typically found at welded joints which will increase the likelihood of fatigue cracking at these locations.

Fatigue cracking is a time-dependent and accumulative degradation mechanism that has been known since Wöhler's studies on failures of railway axles in the 1850s. However, the understanding and quantification of fatigue cracking for offshore structures has only been fully realised since the mid 1970s with the advent of more experience and major programmes of research.

Soon after the oil and gas exploration and production began in the North Sea in the 1960s, it became apparent that the steel structure design developed for offshore structures in the Gulf of Mexico was not adequate when transferred to the more rigorous North Sea environment. In particular, fatigue cracks evolved as a result of wave action during the severe winter storms (Almar-Næss 1985; Department of Energy 1988). This was highlighted by the occurrence of several instances of cracking and the catastrophic accident of the Alexander L. Kielland semi-submersible (Moan et al. 1981).

The offshore industry recognised that the fatigue of large welded joints with both simple and complex geometries (typically found in offshore structures) was not sufficiently understood as a result of lack of relevant test data. As a result, extensive programmes of research were initiated to provide new data and a better understanding of the fatigue process for offshore structures. The limitations of designs from inadequate or failure to perform a fatigue assessment required in many cases expensive repairs and in a few cases resulted in serious accidents. In particular, the importance of welding defects remaining from the fabrication process were not fully recognised. An example of this is the above-mentioned catastrophic accident due to a fatigue crack that formed from a small defect from the welding of an attachment to the structure of the Alexander L. Kielland flotel and eventually caused the platform to capsize in 1980 (Moan et al. 1981).

The environment to which the material is exposed also has a significant effect on the fatigue life. The test data on fatigue that formed the earliest S–N curves used in design resulted from testing in air. A key task of the research programmes for offshore structures in the 1970s and 1980s was to study the detrimental effect of the seawater environment on fatigue performance.

The UKOSP programme fatigue tested large cracked tubular joints in seawater under free corrosion and applied cathodic protection of − 850 mV and compared the results with equivalent tests in air. The tests showed a significant reduction of the fatigue life in the seawater environment by a factor of two compared to tests in air. The HSE fatigue Review Panel (HSE 1992b) reviewing the data from the UKOSRP programme decided to recommend a reduction factor of 2 for tubular joints with lives < 10^6 cycles. However, it was considered that an environmental penalty was not appropriate for joints with longer lives > 10^7 cycles. For welded plates an environmental reduction factor of 2.5 was considered appropriate for the higher stress range. For free corrosion the data indicated that an environmental reduction factor of 3 was appropriate for both tubulars and welded plates.

Cathodic protection (CP) or induced current (IC) protection were able to slow the rate of fatigue damage and at lower stress ranges, the fatigue life was found to be very close to the life found in dry air. This understanding now forms the current guidance for the fatigue design of welded connections (ISO 2007).

This can be seen in Fig. 6.7 where a closed form fatigue analysis with Weibull distributed stress ranges are performed for "in Air", "Cathodic Protected" and "Free Corrosion" T' S–N curves, see Sect. 6.7.5. It can be seen that for low Weibull max stresses the "Cathodic protected" gives accumulated fatigue lives similar to those of "in Air". For increasing Weibull max stresses the "Cathodic protected" curve converges towards the free corrosion accumulated damage curve. The large effect of increasing damage due to the presence of seawater is clear both for cathodically protected (CP) and particularly for free corrosion (FC).

Another important effect on fatigue life that was studied in these research programmes was the so-called thickness effect. Increasing the thickness of a given type of fatigue specimen while maintaining all other parameters will in general cause a decrease in fatigue life. This is partly because in thicker material there is a greater

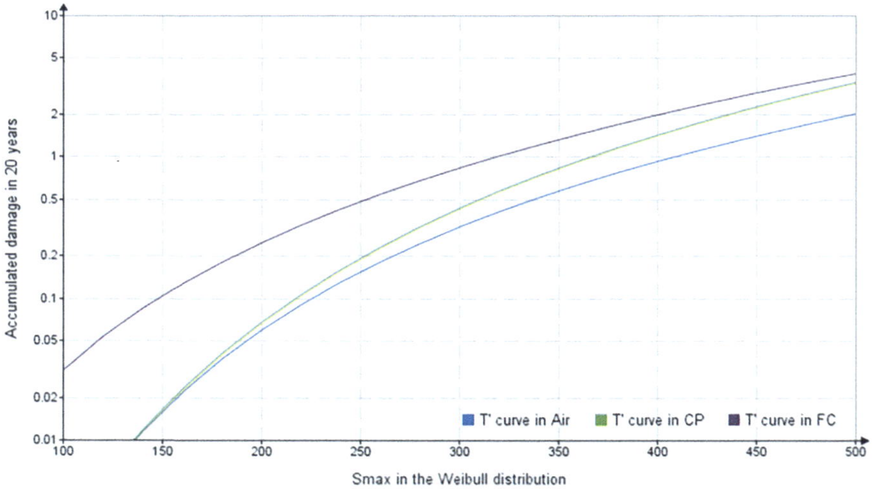

Fig. 6.7 Closed form fatigue damage based on Weibull distribution of stress ranges

probability of a larger microstructural defect existing (Haagensen 2013). The thickness effect has been quantified from testing of large specimens and is now included in the design codes as a modification to the stress input.

6.7.2 Fatigue Research Programmes

As mentioned above, the offshore industry recognised the need for improved data and knowledge and conducted extensive programmes of research to provide new data and models to enable the development of more appropriate fatigue design guidance. A considerable amount of work on the factors upon which fatigue cracking depends was undertaken in research programmes. These included:

- the development of the Q, T and T' S–N curves where the latter two included the effect of increasing thickness and different environments on the fatigue life (UKOSRP I and II, ECSC and the Norwegian FOSS programme),
- the development of stress concentration factors for tubular joints, including the work of Marshall (1974), Visser (1975), Kuang et al. (1975), Gibstein (1978), Wordsworth and Smedley (1978), Efthymiou (1988) and Smedley and Fisher (1991),
- the effect of joint flexibility on the fatigue life (Buitrago et al. 1993)

As the environments in the North Sea and Norwegian Sea are the main areas in which structures are most exposed to fatigue loading, much of the fatigue research was conducted in UK and Norway. The most relevant background references from this work includes Gurney (1979), Maddox (1991), Almar-Næss (1985), Lotsberg (2016) and the reports from the UK Offshore Steels Programme (Department of Energy 1987, 1988).

The most notable research programme was the United Kingdom Offshore Steels Research programme (Department of Energy 1987, 1988) in two phases which are described in further detail below. The first phase commenced in 1975 and led to significant amendments on fatigue design in the Department of Energy Guidance Notes (Department of Energy 1990). A second phase of United Kingdom Offshore Steels Research programme (UKOSRP) into the fatigue of steels in a seawater environment, the effect of thickness on fatigue performance and other factors (Department of Energy 1987). An example of a welded tubular joint tested as a part of the UKOSRP programmes is shown in Fig. 6.8.

The results from this programme led to a new edition of the Department of Energy (to become the Health and Safety Executive) Guidance Notes (HSE 1995). This was recognised as the most advanced design guidance on fatigue and was adopted by the Norwegians in NORSOK N-004 (Standard Norge 1998) and later by ISO in their standard on design and operation of offshore fixed structures, ISO 19902 (ISO 2007).

An important aspect of the UKOSRP programme was careful attention to details of the materials being tested in terms of chemical and physical properties. Steels to BS 4360 Grade 50D (British Standards 1990) had been specially developed for the

Fig. 6.8 Large fatigue crack in welded tubular joint (photo provided to J.V. Sharp during the UKOSRP programme)

first generation of northern North Sea platforms. Most of the tests in UKOSRP-1 were undertaken using this steel, although a node quality steel with very low levels of sulphur and phosphorus was also used which had through thickness properties providing resistance to lamellar tearing. In addition, some tests were also carried out on Grade 40E steel that was typical of some of the earlier gas platforms.

The main elements of the programme were;

- Advanced methods of stress analysis, including finite element methods, strain gauging of acrylic models, together with 3D photo-elastic stress.
- Main fatigue test programme involved tests on 175 welded tubular steel joints tested in air with diameters ranging from 168 to 1830 mm. Wall thicknesses of the main members ranged from 6 to 76 mm. Welding was to offshore standards. The tests included T, K and KT joints with and without overlaps, tested in in-plane and out-of-plane bending together with some axial tests.
- Influence of factors such as sea water, cathodic protection and loading spectrum were investigated by testing pre-cracked plates and flat plate welded specimens. This was aimed at providing information on the effect of these factors on larger welded tubular specimens without undue expense. These showed an effect of

thickness with thicker specimens having a shorter fatigue life. An effect of environment was also observed with tests showing half the life in air, under feely corroding conditions in seawater. Cathodic protection at a level of – 0.85 V (Ag–AgCl) tended to restore lives to those in air.

The main results were as follows:

- Tests on tubular joints showed that the hot spot stress correlated reasonably with the fatigue life found in the tests. The hot spot stress was determined by linear extrapolation of the strain gauge readings to a stress close to the weld toe.
- Fatigue strength was found to decrease as the chord dimensions increased.
- PWHT treatment was found to be generally beneficial to increase fatigue life under reversed loading and was advisable or even necessary for thicknesses over 38 mm when 1970s weld consumables were used.
- In most cases weld toe grinding or similar weld improvement methods increased fatigue life.
- A correlation was established between the Charpy impact energy and CTOD providing greater resolution of the ductile to brittle transition temperature.
- Limited wide plate tests suggested that the design procedure for determining acceptable fabrication quality based on CTOD contained sufficient margins of safety.

In general, a greater understanding of the relative fatigue performance of different types of welds and the importance of welding defects was being gained from these extensive test programmes on tubular joints as well as observations from inspection of structures on-station. This resulted in more optimised welding processes and procedures together with improvements to manufacturing quality standards linked to design life. This led to additional requirements on weld profile, permissible undercut and throat thickness that might not always have been achieved with earlier fabrications.

A Norwegian funded 5-year project was initiated in 1977 for the intensified research on fatigue of offshore steel structures. The aim of the project was to establish more reliable methods for fatigue analysis of steel structures and developing understanding of mechanisms and conditions governing the fatigue life of steel structures. The methods and analysis of relevant types of offshore steel structures were included in the project, including semi-submersibles, tension leg platforms and jack-ups in addition to jacket structures. The resulting Fatigue Handbook (Almar-Næss 1985) formed the basis for fatigue design of offshore structures in Norwegian waters (and possibly elsewhere) for decades.

An overview of the key achievements from these research programmes are indicated in Fig. 6.9.

Most of the early fatigue research work for offshore structures was undertaken on medium strength steels, such as BS 4360 50D (British Standards 1990). However, the fatigue performance of higher strength steels was also investigated and found to be similar in air but reduced with high cathodic protection potential in seawater.

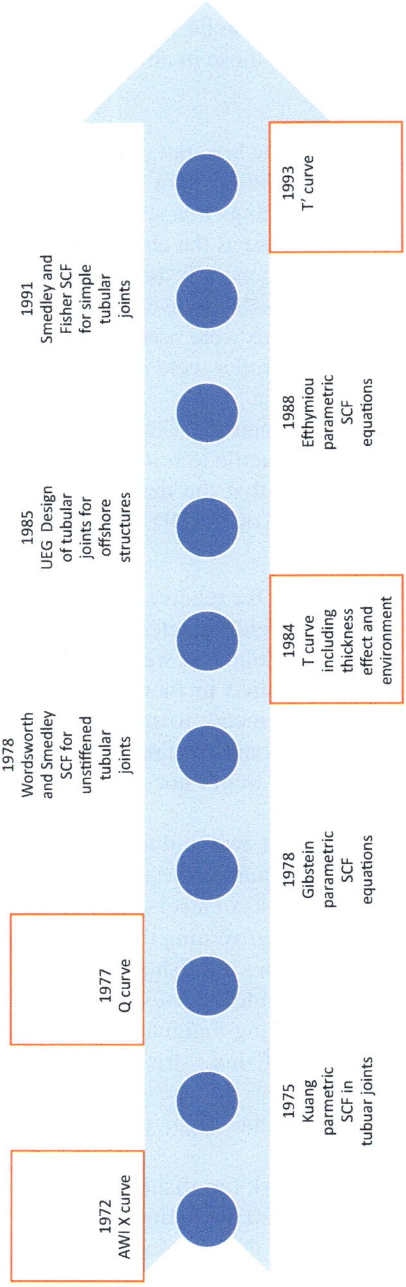

Fig. 6.9 Key achievements from fatigue research (references quoted in the text above)

Later, the FPSO Fatigue Capacity joint industry project (2001–2003) addressed fatigue capacity of FPSOs. Numerical investigations and fatigue testing were performed in order to improve the accuracy, robustness and efficiency of finite element modelling and hot spot stress evaluation for typical FPSO details (Lotsberg 2004).

6.7.3 Fatigue Testing and the Q and T S–N Curves

Stress-Life (S–N) curves are developed from fatigue tests where for tubular joints the hot-spot stress range is plotted on the ordinate axis against the number of cycles to failure on the abscissa. The construction of an S–N curve requires a large amount of experimental data from fatigue tests on representative geometries and materials (which is expensive and time consuming as each test generates only a single data point).

Different definitions of the cycles to fatigue failure exist, including:

- N1: First detected surface cracking as noted by any available method (e.g. drop in strain gauge monitoring of 15% where crack initiated). This stage is considered to have passed if the initial surface length is greater than 20 mm.
- N2: Intermediate surface cracking detected by visual examination (without the use of crack enhancement fluids or optical aids). If NDT techniques indicate a surface length of greater than 30 mm then this stage is considered to have passed.
- N3: First through wall cracking as detected either visually or more accurately by loss of internal pressure or by monitoring of strain gauges positioned adjacent to the deepest part of the crack.
- N4: End of test as occasioned by complete severance of the brace member, extensive cracking leading to loss of load symmetry or exhaustion of actuator stroke.

For the construction of S–N curves for tubular joints, the N3 definition has generally been used.

The hot-spot stress range is typically defined as the stress range which includes the geometric stresses due to the joint detail, but not the stresses due to the weld, i.e. as stated in BS6235:1982 the stress ranges "*which are as near as possible to the connection without being influenced by the weld profile*" (British Standards 1982). In the experiments used to define S–N curves this hot-spot stress (HSS) has normally been determined by linear extrapolation to the weld toe from the readings from a series of strain gauges on the detail tested.

The fatigue design of tubular joints in early platforms was based on S–N curves derived primarily from the performance of welded plates tested in air (Reynolds and Sharp 1991). The first published S–N curve for welded tubular joints, the X curve (American Welding Society, Structural Welding Code AWS 1972), was based on the results of small-scale fatigue tests on plates in air of the typical strength of steel used at the time (Reynolds and Sharp 1991).

Later, the UK Department of Energy published the Q curve in their 1977 Guidance (Department of Energy 1977), which was based on the same database of results as AWS X curve (American Welding Society 1972). The Q-curve was a basic curve without any distinctions of the environment or thickness. The Q curve was written as (British Standards 1982):

$$\log(S) = 2.571 - 0.242 \cdot (\log N - 4)$$

where S is the stress range and N is the number of cycles.

The Q curve can be rewritten to the form that is more commonly used at present as:

$$\log(N) = 14.623 - 4.132 \cdot \log(S)$$

In early 1980s the UK Department of Energy set up a panel to draft new fatigue guidance (Reynolds and Sharp 1991) as there were concerns that the Q-curve may be unconservative under certain conditions (seawater with cathodic protection and with free corrosion in addition to large thicknesses). This work was based on data mainly from the UKOSRP and ECSC research programmes. The resulting T curve was introduced in 1984 and was the first that was derived from tubular joint tests specifically. The curve was funded on statistical analysis of appropriate data and the design curve was defined as the mean curve minus two standard deviations of $\log(N)$ (thus corresponding to 97.6% probability of survival). The T curve also included a correction factor for joint thickness, which was determined from the fatigue performance of welded plates of different thicknesses. In addition, the effect of the environment was included which also was based on testing of welded plates in seawater both with and without cathodic protection (Reynolds and Sharp 1991).

The T curve for in air conditions was formulated in NS3472:1984 (Standard Norge 1984) as:

$$\log(N) = \begin{cases} 12.16 - \frac{3}{4}\log\left(\frac{t}{32}\right) - 3 \cdot \log(S) & \textit{if } N \leq 10^7 \\ 15.62 - \frac{5}{4}\log\left(\frac{t}{32}\right) - 5 \cdot \log(S) & \textit{if } N > 10^7 \end{cases}$$

where t is the wall thickness of the joint.

The Q and T curves are compared in Fig. 6.10 for different thicknesses of the joint.

As can be seen, the Q curve is unconservative for cycles above 5×10^5 for the base case T curve (t = 32 mm) and more unconservative for larger thicknesses.

The T curve for seawater condition with cathodic protection was formulated by the Department of Energy (1984) and adopted by several other standards such as NS3472:1984 (Standard Norge 1984):

$$\log(N) = 12.16 - \frac{3}{4}\log\left(\frac{t}{32}\right) - 3 \cdot \log(S)$$

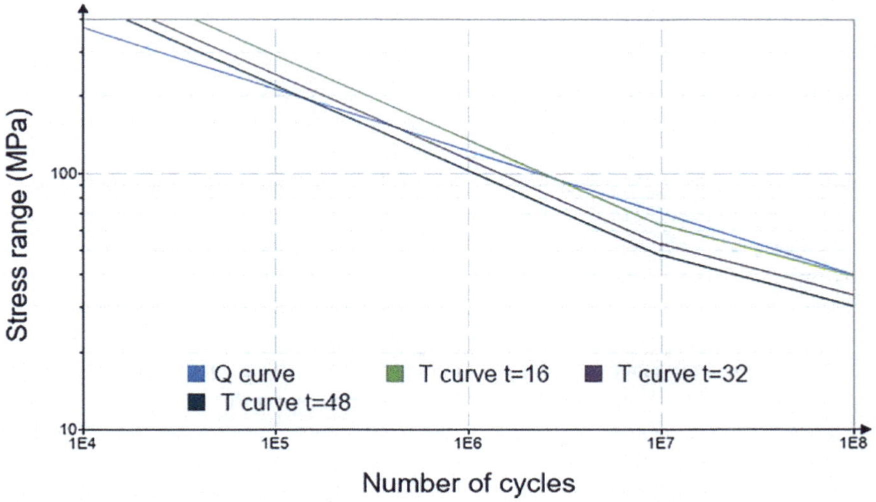

Fig. 6.10 Comparison of Q curve and T curves for different thicknesses

As can be seen from the formula for the T-curve for seawater condition with cathodic protection, the curve is merely a continuation of the upper part of the "in-air" T curve with no change in slope for lower stresses, as shown in Fig. 6.11.

Figure 6.11 shows that at low stress range levels (less than ~ 100 MPa) the Q curve is unconservative and the effect of seawater with the T curve is evident for stress ranges less than 50 MPa.

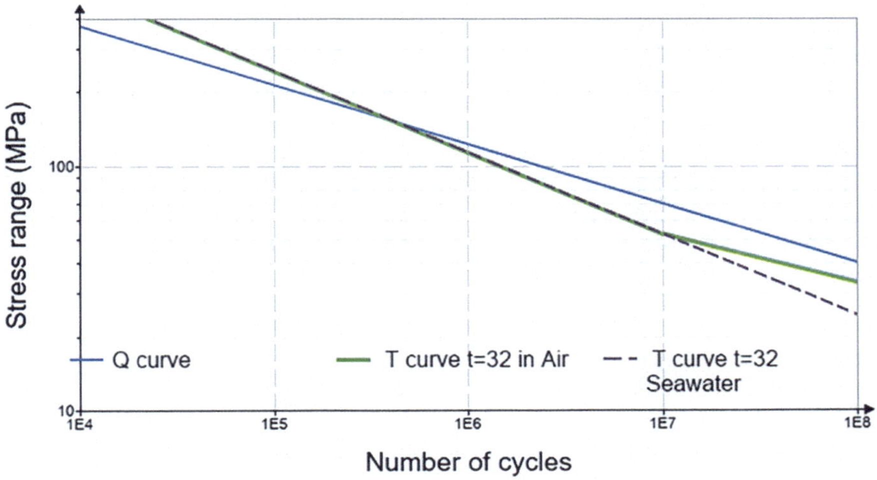

Fig. 6.11 Comparison of Q curve and T curves for different environments

6.7.4 The T' S–N Curves

To enable improved design against fatigue, S–N curves were further developed by testing the fatigue performance of model offshore components and joints in the research programme UKOSRP II (Department of Energy 1987). As noted earlier, components tested included welded plates and a large number of welded tubular joints of varying scale, some of which were close to full-scale. This testing included loading the components in tension and both in-plane and out-of-plane bending. The resulting T' SN-curve in-air was found to be as presented in Table 6.5.

In order to represent subsea and splash zone conditions, tests were carried out in seawater, both with and without cathodic protection. In a corrosive environment such as sea water, it was generally accepted that fatigue was both time and temperature dependant. For example, tests had shown that crack growth rates are doubled if the temperature is increased from 5 to 20 °C (Department of Energy 1987). For this reason, seawater fatigue tests were normally carried out at temperatures in the range 5 to 8 °C, representing North Sea conditions. The seawater used in testing was either natural seawater from a suitable marine site or artificial seawater made from a standardised formula of mineral salts representative of those found in the sea. The use of artificial seawater had the major advantage of being reproducible.

Many of the tests in air were performed at loading cycle frequencies of typically a few hertz. This enabled fatigue lives of up to 10^7 cycles to be evaluated which were relatively independent of frequency. For tests in seawater, it was found that the fatigue crack growth rate could be strongly affected by the loading cycle frequency (Department of Energy 1988). Tests by Scott and Silvester (1975) showed that in the range 1 to 10 Hz the cycle frequency only had a small effect on the fatigue crack growth rate compared with tests in air, whereas crack growth rates could be up to four times faster than those in air when testing at 0.1 Hz. Such low frequencies allow more time for corrosion effects to increase the crack growth rate, thereby reducing fatigue life in seawater. As a result, all further testing in seawater, in the UKOSRP and similar programmes, was carried out at frequencies close to 0.17 Hz. As a result, the duration of these seawater tests was up to 20 times longer than corresponding tests in air. Thus, a year's testing in seawater (which is quite lengthy) only results in lives in the range of 5 million cycles, whereas a typical offshore platform after 25 years exposure will have experienced ~ 10^8 cycles. This is particularly a problem for the testing at lower stress amplitudes ($\Delta S \leq 100\,\text{MPa}$), which requires rather many cycles (>10^6) for the specimen to fail.

Table 6.5 T' S–N curves for in-air conditions

$\log(N) = 12.476 - 3 \cdot \log\left(S\left(\frac{t}{t_{ref}}\right)^k\right)$	For N < 10^7
$\log(N) = 16.127 - 5 \cdot \log\left(S\left(\frac{t}{t_{ref}}\right)^k\right)$	For N > 10^7

where t_{ref} is a reference thickness of 16 mm and $k = 0.25$ for tubular joints

This results in important uncertainties in the S–N curve for steel in seawater at low cyclic stress levels, which are particularly relevant to ageing structures. This lack of data at high cycles is shown for the T' curve in Fig. 6.12, for tubular welded joints. However, the change of slope is based on crack growth data, where at low values of ΔK the crack growth data for tests in air and in seawater with cathodic protection (CP) converge (King et al. 1996). The T' S–N curve for cathodic protection in seawater was determined as shown in Table 6.6.

For free corrosion where there is no mitigating effect of cathodic protection the evidence is that the fatigue lives are shorter (see Fig. 6.12). Crack growth data indicate that at low values of ΔK the rates for free corrosion exceed those for either air or seawater with CP. On this basis it is assumed that the design curve continues to follow the line based on the test data for shorter lives and that a life of 10^8 cycles would correspond to a cyclic stress as low as 20 MPa.

The test data and the T' curves for air and seawater are shown in Fig. 6.12.

Most fatigue tests were carried out with constant amplitude loading. However, a number of tests used a load spectrum simulating North Sea wave conditions. A block length of 10^5 cycles was selected in UKOSRP-I, representing approximately one week of testing in real-time. The results compared to constant loading in UKOSRP-I showed that the fatigue life under variable amplitude loading was a factor of up

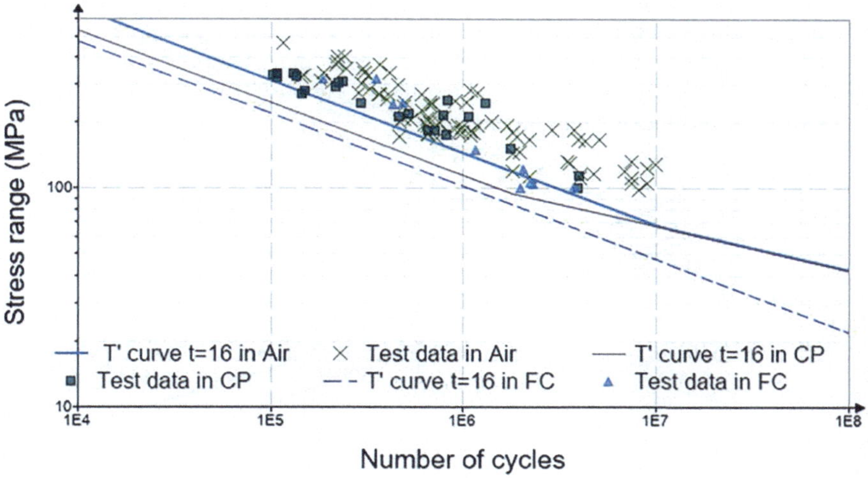

Fig. 6.12 S–N curve for tubular joints in air and seawater (HSE 1992a, b)

Table 6.6 T' S–N curves for seawater conditions with cathodic protection

$\log(N) = 12.18 - 3 \cdot \log\left(S\left(\frac{t}{t_{ref}}\right)^k\right)$	$N \leq 1.8 \times 10^6$
$\log(N) = 16.13 - 5 \cdot \log\left(S\left(\frac{t}{t_{ref}}\right)^k\right)$	$N > 1.8 \times 10^6$

to two shorter than the corresponding life at constant amplitude, depending on the loading spectrum used. A wider set of variable amplitude tests were carried out in the UKOSRP-II programme using different stress spectra. In general, these showed only small differences with the constant amplitude results, but these were within the inherent degree of scatter in the fatigue results. However, the design SN-curves are based on constant amplitude tests which is assumed to also cover the variable amplitude effects.

6.7.5 The Impact of the Development of SN-Curves

Some of the oldest fixed steel offshore structures in European waters will either be designed without any fatigue assessment or with a fatigue assessment using the X or Q curve. As we now assume that the T' curve is a more accurate representation of the safe life of a tubular offshore structure, it is of relevance to compare these S–N curves, as shown in Fig. 6.13 for thicknesses of 16 mm.

For joint thicknesses around 16 mm the Q curve is generally similar or more conservative than the T' curve in air. Further, for joint thicknesses around 16 mm the Q curve is not significantly unconservative for conditions in seawater with cathodic protection. However, for free corrosion conditions there are parts of the Q curve that are providing significantly longer fatigue lives compared to the T' curve for free corrosion. The most important point is, however, that the Q curve doesn't have any reduction for increased thickness and the Q curve will become significantly unconservative for thicker wall thicknesses.

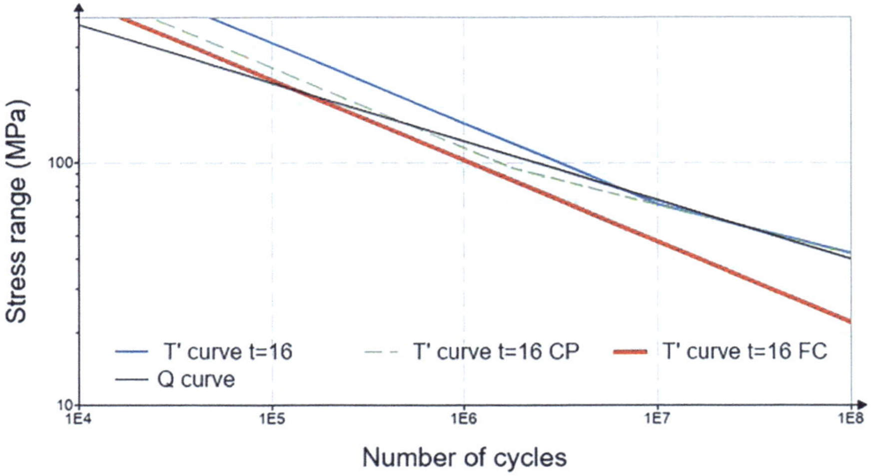

Fig. 6.13 Comparison of Q curve and T'-curves

Since a comparison between S–N curves based on a constant stress range values can be misleading, a representation of a relevant stress range distribution to compare the Q curve and the T' curve is favourable. Here, a closed form fatigue analysis is performed where the stress range distribution is represented by a Weibull distribution, which allows for a closed form calculation of the fatigue damage (DNV 2021). This closed form fatigue analysis can be written as a function of the S–N curve parameters, the shape parameter of the Weibull distribution and the maximum stress range (Smax) in the expected design life. The shape parameter of the Weibull distribution has been shown to be typically in the range between 0.7 and 1.1 for many offshore steel structures in the North Sea and Norwegian Sea (Almar-Næss 1985). For static jacket structures the shape parameter (h) may be expected to be in the lower range whilst for dynamic jackets and particularly floating structures will be expected to be in the upper range. Figure 6.14 shows the ratio of the resulting damage from the Q versus the T' curves for different thicknesses and scale parameters as a function of Smax.

In Fig. 6.14 it is clearly seen that damage according to Q curve is somewhat similar to that of the T' curve for thickness t = 24 mm independent on the Weibull shape parameter, but slightly unconservative for Smax values between approximately 120–400 MPa. Further, it can be seen that the Q curve is providing damage significantly lower than that of the T' curve for larger thicknesses. Hence, structures designed to

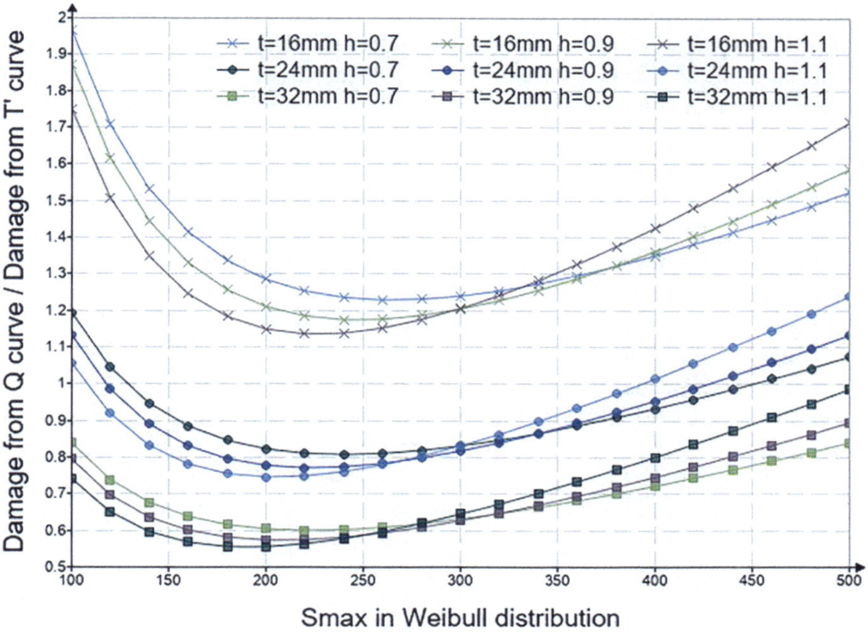

Fig. 6.14 Ratio of resulting damage from the Q and T' curves for different thicknesses and scale parameters as a function of Smax

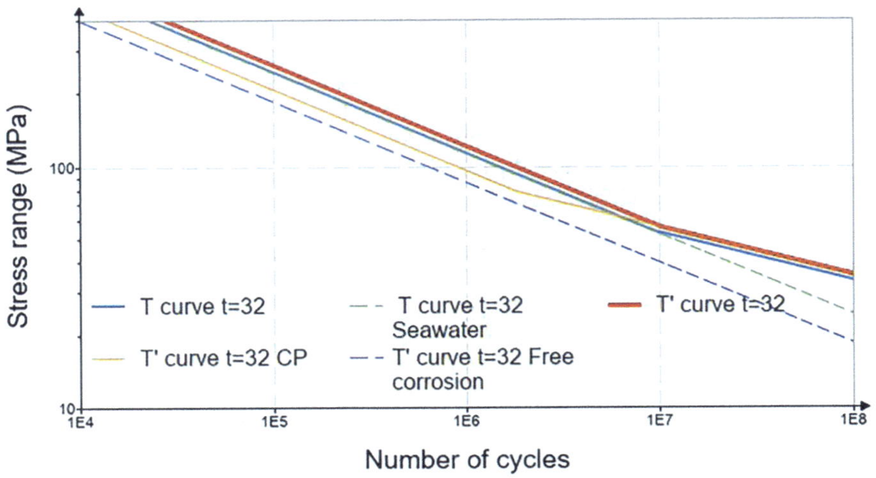

Fig. 6.15 Comparing the T and T' S–N curves for a common thickness of 32 mm

the Q curve can be expected to have less fatigue life than what the modern T' curve would indicate, particularly for wall thicknesses in excess of 24 mm.

As mentioned earlier, with the T curve in 1984 the thickness effect and the effect of the environment were introduced. Although the formulation of the thickness effect was different in the T curve, the effect on the calculated fatigue life is the same as for the T' curve. A common thickness of 32 mm is chosen to compare the T and T' curves in Fig. 6.15.

For the "In Air" T and T' curves there can be seen a slight shift to improved fatigue life with the T' curve. Hence, the T curve will be marginally conservative for all cases.

The T' curve in seawater with "Cathodic Protection" (CP) provides shorter fatigue lives than the T curve for high stress ranges (70 MPa and higher). In contrast, for low stresses the T' curve introduced a second slope which will improve the fatigue life significantly for joints.

Figure 6.16 shows the ratio of the resulting damage from the T versus the T' curves in seawater with CP for different thicknesses and scale parameters as a function of Smax.

As can be seen from Fig. 6.16, a detail designed by the T curve (in seawater) will have an unconservative estimate of the damage compared to the T' curve being used for low values of Smax in the Weibull distribution. For larger Smax > 300–400, the T' curve is giving similar or longer estimates than the T curve.

For "Free Corrosion" there is a more significant, but constant shift towards shorter fatigue lives with the T' curve. This shift implies that all estimates of fatigue lives by the T' curve will be 74% of those determined by the T curve.

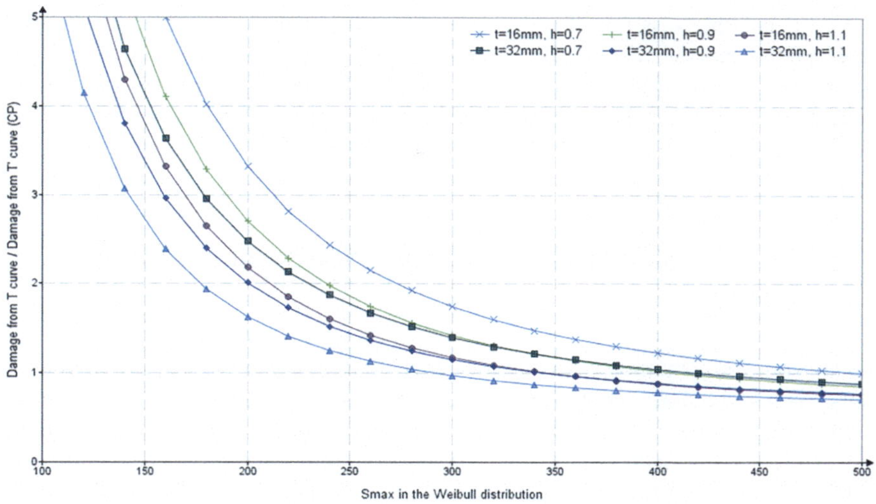

Fig. 6.16 Ratio of damage from T versus T' curves for seawater conditions with cathodic protection for different thicknesses and scale parameters

6.7.6 The Safety Factors for Fatigue

The accuracy of a fatigue life calculation is significantly affected by uncertainties both in the estimation of the cyclic loading, the resulting cyclic stress range distribution and the fatigue endurance as described by S–N curves. The present regime for safely accounting for these uncertainties is to use a design S–N curve (two standard deviations below the mean curve) and the use of a safety factor on the calculated damage. In general, it is required that:

$$D \leq \frac{1}{DFF}$$

where D is the calculated damage by a Palmgren–Miner summation and DFF is the Design Fatigue Factor (safety factor). At present, the DFF will vary from 1 to 10 in standards, depending on the criticality of the member and whether it is possible to inspect the welded detail.

Early standards for design of offshore structures had a slightly different formulation, but related to the present formulation the requirement in for example the 1974 version of DNV's "Rules for the design, construction and inspection of fixed offshore structures" (DNV 1974) was:

- DFF = 1 for structures above a horizontal plane 2 m above high astronomical tide (HAT).
- DFF = 2 for structures below a horizontal plane 2 m above high astronomical tide (HAT.

Ageing existing structures being assessed for life extension may meet challenges if low safety factors were used in the original design, in situations where new standard require much higher safety factors. For example, the upper part of the splash zone (more than 2 m above HAT) may have been designed with a safety factor of 1, whereas present requirements can be a factor of 10.

6.7.7 The Hot Spot Stress Method and Stress Concentration Factors

The hot spot stress method is widely used in the fatigue design of welded tubulation joints. As noted before, a hot spot is a point of high stress and as a result a location where fatigue crack initiation is expected. If this point coincides with a location of possible defects, such as at a weld toe location, joint failure is expected to start at this location. The hot spot stress is normally calculated as a factor (the stress concentration factor—SCF) times the nominal stress (the stress undisturbed by the geometric change as a result of the joint) in the brace.

Since the early 1970s there has been a significant effort devoted to the development of accurate SCFs for welded tubular joints under various loading conditions. Various parametric formulas, expressed with non-dimensional joint parameters for different types of loading and boundary conditions have been developed both experimentally and numerically. In practice, the tubular joint SCF is usually calculated from parametric equations given in standards such as ISO 19902 (ISO 2007), API RP2A (American Petroleum Institute 2000) and DNV-RP-C203 (DNV 2021) or from case-by-case finite element analysis. The standard and code-based SCFs are expected to provide upper-bound values for use in design and life extension assessments. SCFs determined from case-by-case finite element analysis have shown to provide values both on the safe side and to the unsafe side (CoV ~ 12–15%).

These improved SCFs can have a significant impact on the calculation of the fatigue life.

6.7.8 Fatigue of Single Sided Closure Welds

To reduce the likelihood of fatigue cracking full penetration welds made from both sides are generally recommended for platform construction supported by detailed NDE during fabrication. However, single sided welds are commonly used for connecting braces to chords in tubular joints for offshore structures, particularly in large jacket installations. A typical solution is the brace to stub weld and also for circumferential brace joints, as shown in Fig. 6.17. Single sided welds are often used between members with different thicknesses which has added complications.

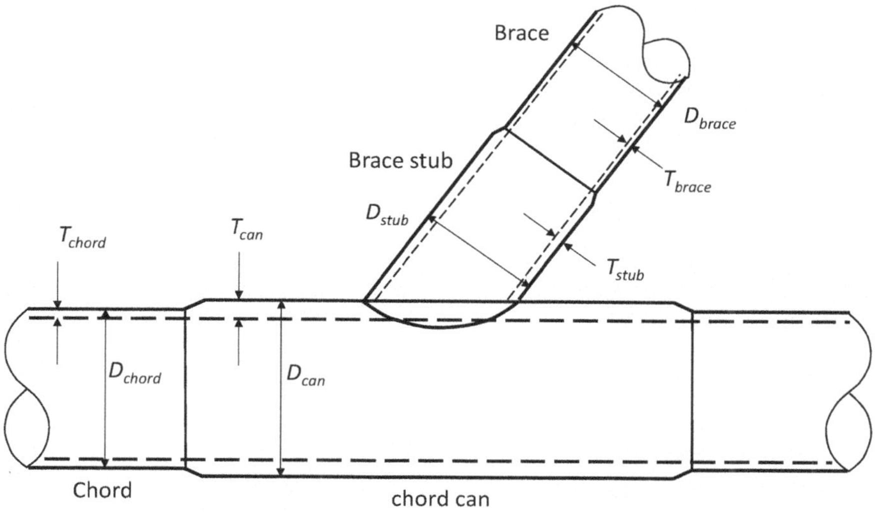

Fig. 6.17 Closure welds

Fatigue cracking in single-sided welds may initiate at the weld toe but may also initiate at the weld root. However, as inspection of the root during operation is difficult or even impossible, it is essential that the design of such single-sided welds ensures the weld toe (which can be inspected) to be the location most prone to fatigue cracking. Hence, the stress concentration at the internal weld root needs to be made sufficiently low compared to that of the external weld toe.

However, historically there have been several cases of fatigue failures of members containing single-sided closure welds. In some cases, through-thickness cracking has been detected in time to avoid member failure but in other cases members have become completely severed (Department of Energy 1990). In some cases, a closure window may be constructed to allow access to internal welding at a joint, for example close to the joint can. Such windows are also normally closed by single sided welds and therefore also vulnerable to fatigue cracking.

In single sided welds small cracks, slag inclusions and other defects cannot be removed by work on the inside of the weld. As such these potential weld root imperfections are important to the fatigue performance of these joints and difficult to locate by normal inspection methods. Ultrasonic flaw detection is seen as unreliable as a test for weld roots during fabrication. However radiographic testing has shown some success (Heshmati et al. 1986) but can only be used in fabrication yards.

Typical imperfections found in root welds, based on Heshmati et al. (1986) and Stacey et al. (1997) are listed below:

- Undercut on butt welds (which will have an adverse effect on fatigue performance),
- Lack of penetration (where the weld metal fails to reach the root,

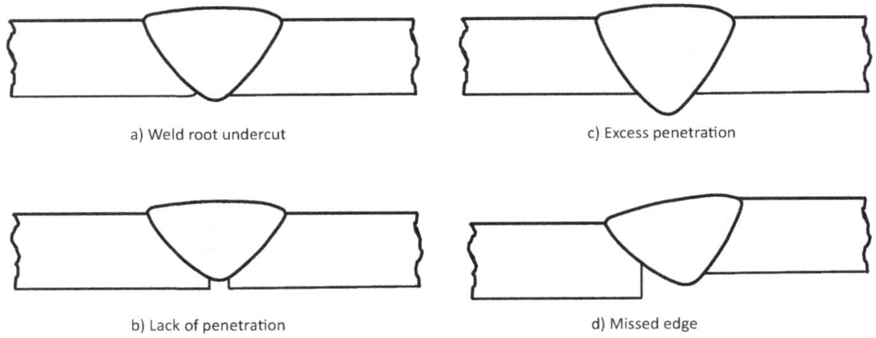

Fig. 6.18 Examples of weld root imperfections (based on Stacey et al. 1997)

- Excess penetration (on the inside of the weld causing a highly localised region of higher stress),
- Missed edge (in misaligned single sided welds with a small root gap part of the weld may be missed from the root run.

Examples of these are shown in Fig. 6.18:

Stacey et al. (1997) did a review of fabrication data in two fixed installations in the North Sea. The two installations were installed in 1991 (structure A) and 1986 (structure B), see Sect. 5.3.8. The study covered a range of joints configurations including single sided welds. The results for single sided welds are shown in Table 6.7 for the two different structures (A and B). Reportable defects relate to the fabrication specification and are required to be both reported and repaired. Repairable defects are those above the specified reference level during UT inspection. These defect s are also required to be reported and repaired. The maximum defects height was 3 mm for structure A and 5 mm for structure B. Initial defects lengths can be very large with a maximum defect for structure B of 7068 mm. The size of the deepest defect at the weld root was approximately 3 mm (Stacey et al. 1997).

The use of a backing strip has been used to improve weld quality. A strip is inserted, bolted up and fitted prior to welding ensuring a more satisfactory weld root. Root side of single sided closure welds are normally assessed using the F2 S–N curve and if a backing strip is used the F curve is recommended (DNV, ABS).

For single sided welds with thickness transitions, there is a significant difference in the stress concentration factors (SCF) at the weld toe if the transition is flush outside or flush inside, see Fig. 6.19. Due to the SN-curve being less severe for the weld toe on the outside, single sided closure welds in brace-stub connection should be designed such that any thickness transition is placed on the outside, i.e. with flush inside (ISO 2007; Lotsberg 2016).

In an assessment, it is vital to determine whether:

- Brace to stub thickness transition are flush inside or flush outside.

Table 6.7 Statistical data for sized defects in single-sided welds

		Reportable defects		Repairable defects	
		A	B	A	B
Total number of defects		29	21	52	87
Defect height	Number of sized defects	4	17	6	52
	Minimum (mm)	1.5	1.5	2.0	2.0
	Maximum (mm)	1.5	2.5	3.0	5.0
	Mean (mm)	1.5	1.9	2.3	5.0
	Standard deviation (mm)	0	0.2	0.5	2.1
Defect length	Number of sized defects	25	21	45	87
	Minimum (mm)	3	15	4	8
	Maximum (mm)	85	7068	3650	1430
	Mean (mm)	28	735	160	201
	Standard deviation (mm)	19	1541	80	226

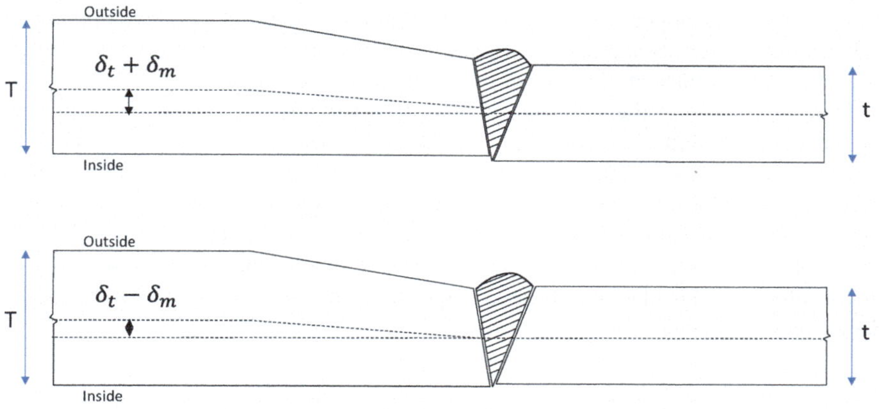

Fig. 6.19 SCF for flush inside weld roots

- If flush outside the weld root is the most fatigue prone part of the weld and careful evaluations are needed in an assessment for life extension.
- If flush inside the problem is reduced and the fatigue crack is normally expected to start at the weld toe and can, hence, be determined by inspection.

- Whether backing strip was used.
- Type of inspection used during fabrication.

6.7.9 Implications of Fatigue Damage

The primary implications of fatigue damage are reduced ultimate structural strength and an increasing rate of fatigue crack growth. For example, an initial through-thickness fatigue crack in a tubular joint can reduce the static strength by up to 40% in bending (Stacey et al. 1996; HSE 1999b) and the remaining fatigue life of the joint is significantly shortened (Mohaupt et al. 1987). Fatigue cracking can lead to an increased chance of brittle or ductile fracture and the possibility of water ingress into tubular members in fixed structures and plated marine hulls. The consequences of fatigue failure therefore need to be properly understood in the context of managing of ageing installations.

Through-thickness cracking can be followed by member severance and loss of stiffness in the local structure. This will lead to load redistribution which will cause other components to be more heavily loaded, possibly resulting in more rapid fatigue cracking in these areas. Hence, multiple cracking may occur and, depending on the level of redundancy, ultimately the structural could fail. Thus, as both fatigue life predictions and component strength are affected by load redistribution (Noordhoek et al. 1987), it is important that due consideration is given in the development of the structural integrity management plan to the possibility of total member failure occurring after a through thickness crack and its consequences.

It is apparent that the influence of load redistribution on fatigue life can lead to unexpected failures, as the fatigue of the intact structures does not account for this load redistribution after fatigue failure. It is also possible that fatigue cracks may initiate and propagate at fabrication defects which are not necessarily located in those joints identified from the structural analysis as being critical. A lack of information on such defects will give an incorrect view of the structural integrity of the installation. This places additional emphasis on the need for an understanding of the system performance.

6.7.10 Fatigue Crack Growth

After a crack has initiated by fatigue in a structural component or joint further application of cycles of stress can cause the crack to grow. The rate and mode of fatigue crack growth are key aspects in determining the life of the component or joint when the crack will have grown to a limiting or critical size. The amount of fatigue crack growth per stress cycle was found by Paris and Erdagon (1963) to depend on the range of the stress intensity factor raised to a power of around three beyond a lower bound threshold value (ΔK_{th}). This can be expressed as:

$$\frac{\mathrm{d}a}{\mathrm{d}n} = A(\Delta K)^m$$

where m is approximately 3, $\Delta K\left(\sim \Delta\sigma\sqrt{\pi\,\Delta a}\right)$ is the stress intensity range for a crack of depth a for a change in stress of range $\Delta\sigma$ and A is a material parameter. The constant A was determined by fatigue testing pre-cracked specimens and measuring the crack growth as a function of the number of stress cycles.

Early testing was done in air (Paris and Erdagon 1963). Later tests were conducted in sea water where free corrosion was found to have a significant effect increasing the fatigue crack growth rate compared with the rate in air (Scott and Silvester 1975). Tests in seawater with cathodic protection also had increased crack growth rates. Evidence for this was from a programme of fatigue crack growth testing of pre-cracked plates in Phase 1 of the UK Offshore Steels Research Programme (Department of Energy 1988).

6.7.11 Effect of a Sea Water Environment and Hydrogen Uptake

When there is a crack in steel in a seawater environment combined with cyclic stress, the amount of fatigue crack growth per cycle is increased compared to that of the same crack in air. The reasons for this increase were investigated in an extensive programme of research by the National Physical Laboratory and the results were published in HSE OTH 325 (HSE 1992a, b). The research concluded that the generation and absorption of hydrogen atoms due to corrosion reactions was a possible electrochemical process in enhancing fatigue crack growth over a wide range of cathodic potentials including free corrosion. However, HSE OTH 325 (HSE 1992a, b) states that localised anodic dissolution of the crack tip could, in principle, explain enhanced crack growth under free corrosion conditions at low stress intensity factor ranges if the crack tip dissolves at its maximum rate. Under free corrosion and at intermediate protection potentials the primary source of hydrogen atoms was thought to be more likely from strain enhanced reduction of water *at the crack tip* into hydrogen and hydroxyl ions. Increasing the mean stress and the stress intensity factor range increases the amount of hydrogen entering at the crack tip and thus raises the crack growth rate.

At cathodic potentials more negative than about – 900 mV the primary source of hydrogen atoms is predicted to be increasingly from cathodic reduction of water at the *external* surface of the steel and the diffusion of atomic hydrogen towards the stress concentration at the crack tip. The increase in the fatigue crack growth rate is then determined by the diffusion mass transport rate of atomic hydrogen, the movement of dislocations, and by the nature and density of voids and other hydrogen trapping sites (HSE 1992a, b). Hydrogen embrittlement of the material around the crack tip allows the crack to grow incrementally in small steps controlled by the rate at which the material becomes embrittled.

There has been a debate about the relative contributions to crack growth from anodic dissolution at the crack tip and hydrogen embrittlement at different potentials.

Dissolution may be the primary mechanism determining crack growth from plain specimens under free corrosion or for deeper cracks at low stress intensity factor ranges. However, hydrogen embrittlement cannot be ruled out as the main factor determining the crack growth rate over all potentials, including free corrosion (HSE 1992a, b).

The presence of hydrogen sulphide from sulphate reducing bacteria (SRB) can promote HE and cause a substantial increase in the crack growth rate over that for steel in clean seawater depending on the concentration of sulphides and level of cathodic protection (see Sect. 6.2.7). At negative potentials, however, the effects of hydrogen sulphide may not be significant because the locally high pH will supress bacterial activity (HSE 1992a, b).

The generation, absorption and diffusion of hydrogen into the material has implications for fatigue crack growth testing of standard fracture mechanics specimens. Test times should be sufficiently long as to establish steady conditions of hydrogen charging along the crack front of the specimen. Tests on tubular joints where hydrogen absorption is from the external surface must allow enough time for a steady hydrogen concentration to be developed through the thickness. Cathodic potentials more negative than – 900 mV can cause a hydrogen gradient in tubular joints but results from small scale specimens under the same conditions where the hydrogen concentration is more uniform should be conservative.

6.7.12 Fatigue Crack Growth Laws

Before 1980, there is no evidence that fatigue crack growth was used in the design of offshore structures or for assessing their expected life. This changed when the early BS PD 6493 was published (British Standards 1980) and provided guidance on methods for assessing the acceptability of flaws in fusion welded structures. The standard recommended a generic lower bound formula for fatigue crack growth in offshore steels in air that designers and engineers could use for engineering critical assessments of the life of joints containing actual or postulated cracks.

The original generic formula for fatigue crack growth in air was still present in BS PD 6493 in 1991 (British Standards 1991). A separate recommendation for fatigue crack growth in a seawater environment was added in the form of changes to the Paris constants for an upper bound fit a complex set of data which was poorly described by the Paris Law. The A parameter for a seawater environment was 7.7 times higher than A for the air environment and larger than the environmental penalty of 2 to 4 measured in S—N endurance tests on welded joints.

In 1998 the UK Health and Safety Executive (HSE) commissioned RN King to undertake a review of the fatigue crack growth data (HSE 1998a, b, c) with a view to recommending fatigue crack growth laws which were more representative of the behaviour of offshore structural steels under different environmental conditions. This included a literature review of fatigue crack growth rates (FCGR) and fatigue thresholds in both structural and high strength ferritic steels.

The environments considered in the King review were air and seawater. The seawater environments considered both free corrosion and an applied cathodic protection potential of – 850, – 1050 and – 1100 mV Ag/AgCl. King found that the effect of cathodic protection at a value of – 850 mV was a marginal increase in the rate of fatigue crack growth compared with freely corroding. However, fatigue crack growth rates at – 1050 mV cathodic protection and – 1100 mV were approximately double those at – 850 mV, see Fig. 6.20. This was put down to the effect of hydrogen ingress because of the CP process.

After examining the data, King found a more accurate description of crack growth using a bi-linear two stage growth law to relate the crack growth per cycle (da/dN) to the cyclic stress intensity factor ΔK rather than the more conventional Paris Law with a single slope. In addition to the external environment, the fatigue crack growth law was also found to depend on the ratio between the minimum and maximum stress intensity factors defining the R-ratio and value of mean stress of the loading cycle (HSE 1998a, b, c). Examination of data for the fatigue threshold stress intensity factor range (ΔK_{th}) indicated that the recommendations of PD6493:1980 (British Standards 1980) remained satisfactory for ferritic steels.

Consequently, King recommended mean and design values for the crack growth constants, C and m for both Stage 1 and Stage 2 of fatigue crack growth for ferritic steels including those with higher strength having a maximum yield stress of up to

Fig. 6.20 Summary of fatigue crack growth rates for R > 0.5 and $\Delta K = 400\ \mathrm{N\ mm}^{-\frac{3}{2}}$ for different environments (based on data in HSE 1998a, b, c)

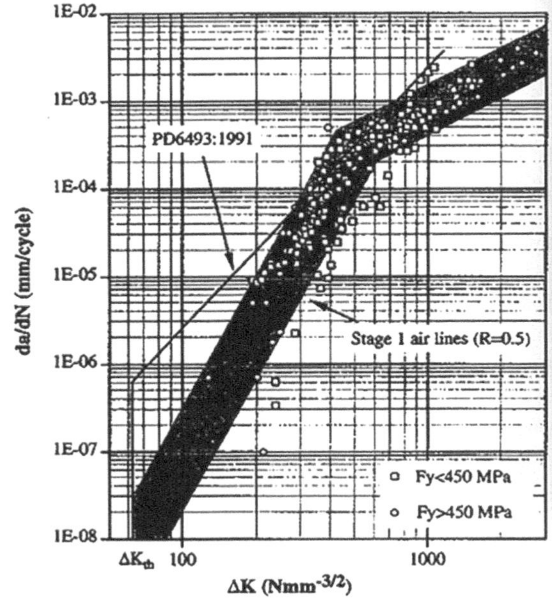

Fig. 6.21 Fatigue crack growth rates for ferritic steels tested in sweater at − 1050 to − 1100 mV with an R-ratio larger than 0.5 (HSE 1998a, b, c)

600 MPa. He made separate recommendations for the high and low R-ratio regimes. An example of these results is shown in Fig. 6.21, which shows that the fatigue crack growth line from PD 6493:1991 under-estimates the measured growth rates at higher values of stress intensity factor range under condition of high R ratio and in seawater with cathodic potentials of − 1050/− 1100 mV.

The new, two-stage fatigue crack growth curves offered improved accuracy compared to BS PD 6493:1991 (British Standards 1991) and were incorporated into the later revisions to BS PD 6493 and subsequently into the British Standard for the assessment of defects in metallic structures, BS 7910 (British Standards 2019).

In addition to the range of the stress cycles, the initial crack depth is also an input needed in analyses of crack growth and fatigue life. The initial crack depth could be a measured crack in the material or an assumed maximum crack size not found by inspection. Typical values for the initial crack depth of fabrication defects such as weld undercuts that may not be reliably detected by magnetic particle inspection range from 1.5 to 4 mm depending on whether the joint is machine ground or as-welded and of complex geometry (British Standards 2019).

King in his study (HSE 1998a, b, c) assumed an initial defect size of 0.15 mm × 50 mm for comparison with S–N data, as this size was considered typical of intrinsic flaws found in weld toes. Larger defect sizes were also investigated. The study indicated that plate joints are more sensitive to defect size than tubulars. It was stated that this was because in plates a larger proportion of the fatigue life is occupied in the early stages of fatigue crack growth where larger defects are more damaging.

The revised recommendations for the fatigue crack growth rate for ferritic steels were validated by performing a number of fracture mechanics based fatigue analyses in conjunction with experimental fatigue test data under constant amplitude and variable amplitude loading. These analyses showed reasonably good correlation between fracture mechanics predictions and the experimental endurance data in both air and seawater environments. The predicted lives in seawater with cathodic protection were shorter than those in air and the results for a CP level of − 1050 mV were always lower than those for − 850 mV.

6.7.13 Effect of Fatigue Cracks on the Integrity of Structures

Fatigue cracks can seriously affect the integrity of the structure if the in-service inspection and maintenance regime fails to detect and repair significant fatigue cracks before the load carrying capacity of the cracked joint or member is seriously impaired. In a fixed tubular structure, where there may be a degree of redundancy from other members, the load carried by a single joint containing a growing fatigue crack may be progressively redistributed elsewhere in the structure. There is then a higher possibility that fatigue cracks may develop elsewhere at locations carrying a higher cyclic stress. Where fatigue cracks are widespread in a fixed structure, the integrity of the whole structure to resist extreme loads can be significantly reduced.

The importance of a through wall crack to an offshore structure depends on the further consequences. In a fixed tubular structure, a through wall crack or complete severance of the joint may be tolerable when load is redistributed and the resulting fatigue life can be accepted. Where the structural element is a fluid retaining boundary, such as a buoyancy or ballast tank of a semi-submersible or other floating installation, then the consequences of a loss of watertight integrity can be severe. Here a single through wall fatigue crack may be sufficient to compromise safety if there is significant water ingress.

For the assessment of fatigue crack growth and associated probability of the fatigue crack reaching a critical size, a structural reliability analysis (sometimes also called a probabilistic fatigue analysis) is a commonly used approach. This analysis would be based on fracture mechanics and fatigue crack growth considering the size distribution and frequency of fabrication defects. Such defect information should in principle be established by thorough investigation into the inspections performed during fabrication, in addition to the sentencing criteria (repair criteria). However, in practice such distributions and frequencies are often taken from published standards, based on research of the type described previously.

Such an analysis should also take the type, quality and efficiency of in-service inspections into account. The analysis should also consider variabilities in material properties and loads. The analysis will determine the probability of a structural limit state exceedance.

6.7.14 Fatigue of High Strength Steels

Fatigue test data and guidance (both for SN-type and fracture mechanics analysis) have mainly been developed from tests on medium grade steels, typically BS 4360 50D (British Standards 1990). Higher strength steels with yield strengths up to 700–800 MPa are also used in offshore structures, particularly in jack-ups and mooring systems. Generally, it has been assumed that in-air SN-curves for medium strength steels are also applicable for higher strength steels. However, there is evidence that the effect of seawater on the fatigue performance of higher strength steels is detrimental due to a greater susceptibility to hydrogen cracking.

The fatigue behaviour of higher strength offshore structural steels (defined generally as those with a yield strength greater than 420 MPa) has been the subject of several studies. A review (Standard Norge 2015) of published work from Europe, USA and Japan compared fatigue crack growth data for a range of higher strength offshore structural steels with data for conventional BS 4360 50D structural steels (British Standards 1990). The data considered parent plate and weld HAZ materials and the influence of environment, i.e. air, free corrosion, cathodic protection potentials ranging from − 850 to − 1100 mV (SCE).

In support of the data on SN-curves, the fatigue crack growth data for parent materials with yield strength in the range 400 to 600 MPa suggest similar or even slightly improved fatigue performance both in air and in a seawater environment under free corrosion compared with that of BS 4360 50D steel (HSE 1997). At the free corrosion potential in sea water, the environmental reduction factor (ERF) of higher strength steels compared with air is in the range of 3–5 at low stress intensity factor range decreasing to around 2.5 at higher values of stress intensity range, see Fig. 6.22. This performance is in line with that for structural steels such as 50D with 350 MPa yield strength.

With cathodic protection of − 850 mV, Fig. 6.22 indicate that the effect of a sea water environment is again more pronounced at low stress intensity factor ranges with an ERF of 4 (HSE 1997). More negative overprotection potentials (i.e. − 1100 mV) can cause fatigue crack growth rates to be increased significantly under high mean stress with an ERF in the range of 4–10 compared with those observed in air. This behaviour is observed both for conventional 50D type and higher strength structural steels. Highly negative protection potentials are associated with reduced fracture toughness due to hydrogen embrittlement.

A review of fatigue crack growth rate curves for high strength steels was performed on behalf of BSI Committee WEE/37 for BS PD 6493:1991 and the results are summarised in (HSE 2002). Data for fatigue crack growth for various higher strength steels was obtained in a seawater environment with cathodic protection potentials around − 1100 mV Ag/AgCl. The steels tested included:

- X65 and X70 pipeline steels
- HT 80 plate material
- HY 130 with a yield strength of 993 MPa.

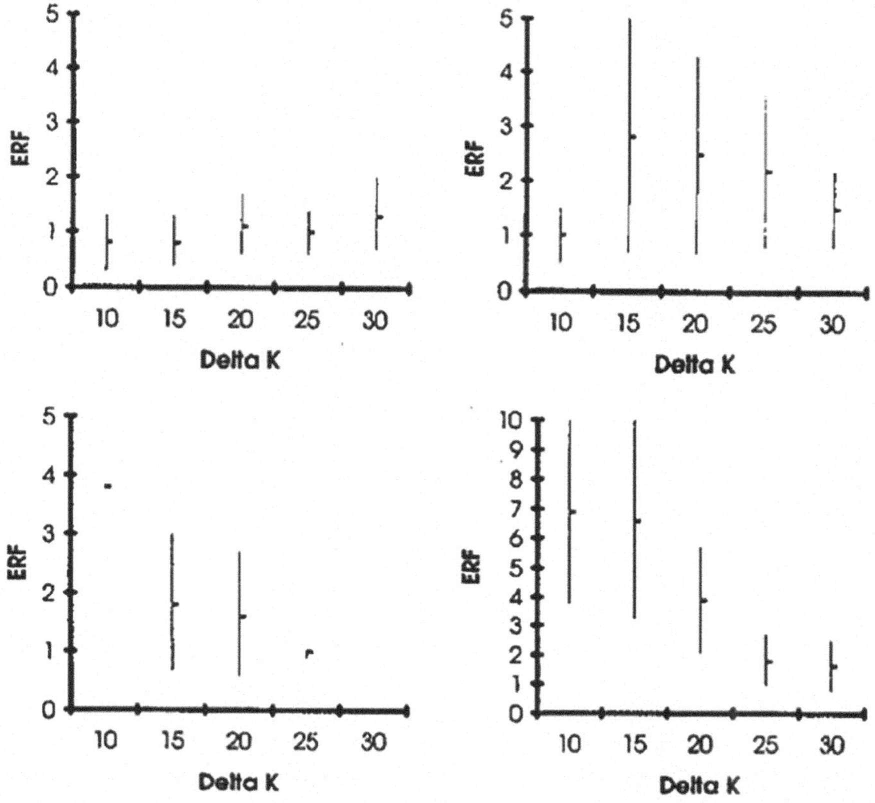

Fig. 6.22 ERF values for HS steels 400–600 MPa as a function of stress intensity range and applied potential (HSE 1997). Top left is showing data for in air, top right for free corrosion, bottom left for − 850 mV cathodic protection and bottom right − 1100 mV

These materials were representative of the types of materials used in jack-up legs at the time. Examination of the da/dN versus DK relationship showed high strength steels to behave in a generally similar manner to medium strength structural steels (Stacey and Sharp 2021).

A limited amount of work has been carried out on fatigue crack growth rates in the heat affected zone of welds on higher strength offshore steels used for jack-up applications. Figure 6.23 shows bounding curves for weld HAZ in 500–600 MPa steels (HSE 2003) where the effect of highly negative potentials can be clearly seen. There is a suggestion in HSE RR 105 (HSE 2003) that fatigue crack growth rates in HAZ are lower than in the parent plate material, however, the experimental conditions were not well defined and it is probable that welding residual stresses may have contributed to this observation.

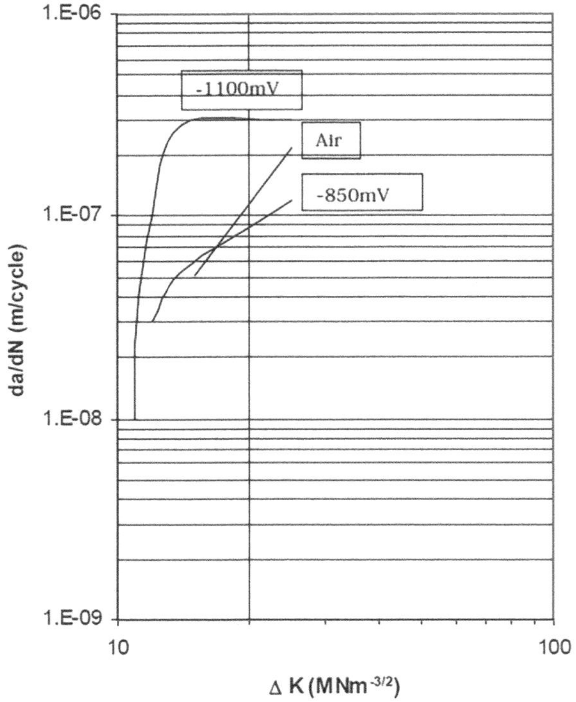

Fig. 6.23 Crack growth rates at different CP potentials compared to in air (HSE 2003)

The presence of active sulphate reducing bacteria (SRB) can have a very marked effect on crack growth rates (Stacey and Sharp 2021), see also Sect. 6.2.7. This has been demonstrated for BS 4360 50D steel by Cowling et al. (1986).

Robinson and Kilgallon at Cranfield University also examined this effect for higher strength steels using sediments collected from the base of a platform in the North Sea (Robinson and Kilgallon 1995). The study showed that absorption of hydrogen was considerably enhanced when SRBs were present and this appeared to be correlated to the total sulphide concentration rather than to microbial numbers. This higher concentration of hydrogen led to higher crack growth rates by at least an order of magnitude and also resulted in much lower threshold stress intensity values.

The work above generally indicates that the fatigue crack growth rate of parent and welded higher strength steels in the range 400–600 MPa is comparable to that of medium strength structural steels. However, the data are limited, and that further data are required for air and seawater with cathodic protection. The effect of active environments on the performance of high strength steels is an important consideration and more work is required to understand the mechanisms that apply.

The results suggest that an essential element of maintaining the integrity of higher strength steel structures is maintaining cathodic protection potentials within a narrow band. Some codes and standards require protection potentials to be within the range of – 770 to – 830 mV (Ag/AgCl)). This is found to be difficult to achieve in practice (HSE 2003), see Sect. 3.5.1.

It should be noted that the data reviewed above are for relatively modern higher strength steels and therefore may not necessarily be representative of all higher strength steels used offshore. It would be expected that, in some cases, higher strength steels used in early structures may have inferior fatigue performance to that of modern steels. This possibility needs to be considered in an assessment of materials in an ageing or older offshore structure.

6.7.15 Fatigue of Castings

The fatigue performance of castings was investigated in a joint industry project (funded by HSE and five oil companies) and reported by Ma and Sharp (1997). In addition, OTH 390 (HSE 1992a, b) addresses the fatigue of castings in relation to the production of the fourth edition Guidance Notes (HSE 1994) and its introduction of an S–N curve for castings. Data on the fatigue performance of cast steel tubular joints were also reported by Herden et al. (1986).

Four areas were investigated in the joint industry programme which were:

- material properties of cast steels,
- stress concentration factors in cast nodes,
- the S–N relationship for cast steel,
- defects in cast nodes.

An extensive literature survey was undertaken and a total of 234 tests on the fatigue performance of castings was produced. In addition, a series of tests on cruciform and real cast nodes was undertaken as part of the programme. Totally, these included approximately 140 cast pieces, 27 V-shaped specimens, 60 cruciform tubular T and DT joints and 15 large scale joints. Most of these tests were in air, although there were also some cruciform and V-shaped specimens tested in seawater under both free corrosion and with cathodic protection. Most of these tests failed at short lives (due to relatively high cyclic stresses) and it was considered that the effect of the environment may not have fully developed (Ma and Sharp 1997). In general, the results showed a factor of two lower fatigue life than that of tests in air. In the absence of more data, it was recommended that this reduction factor of two should be applied to calculated fatigue lives in the higher stress region for cast nodes in seawater environment. The recommendation is presently included in the latest ISO 19902 (ISO 2007). This standard also emphasizes that a finite element analysis should be performed to locate the areas of maximum stress range in the casting which are different from those in welded joints. The brace to casting circumferential butt weld also needs to be checked as it can be the most critical location for fatigue cracking.

A fracture mechanics study was undertaken to establish the likely effect of defects on fatigue life. This showed that the size of defects that might be undetected by either MPI or UT could reduce the crack initiation period by 50%. In addition, it was concluded that the presence of sub-surface defects reduced the crack propagation life by a factor between 3 and 5, compared to a specimen without defects (HSE 1992b).

The available design fatigue life could therefore be only one tenth of the experimental fatigue life (from the S–N curve). On this basis it was recommended that a reduction factor on the mean curve of four standard deviations should be applied for design, which is consistent with a reduction factor of ten on mean life. The recommended design SN curve is given by ISO 19902 (ISO 2007):

$$\log N = 15.169 - 4 \log S$$

6.7.16 Fatigue Performance of Bolts

During the 1990s, TWI in the UK conducted a large programme of tests on the fatigue performance of bolts. The tests were to study the fatigue performance of bolts of strength Grades 8.8, 12.9, B7 and L7 with different diameters, manufacturing methods (rolled or cut threads), and in air and sea water environments under free corrosion and different cathodically protected conditions. The tests also examined the effect of the mean stress in the bolt (pre-tension) on fatigue performance for selected cases (HSE 1998c).

The results showed that the fatigue performance of Grades 8.8, B7 and L7 bolts, tested in sea water under free corrosion and cathodically protected conditions, experienced reductions on fatigue life relative to in-air performance. A reduction factor of 3.0 on the life obtained from the S–N fatigue design curve in air of standard strength bolts curve was able to account for the effect of sea water with adequate cathodic protection in the range − 850 to − 1000 mV. However, the report recommended that the underwater fatigue performance of bolts made of higher strength-grade steels should be specifically addressed via a validated test programme, as should the effect of hydrogen embrittlement.

A later analysis of the TWI test data by MATSU (1997) found that there was a difference in fatigue performance of rolled threads relative to that of bolts with cut threads depending on whether the threads were rolled before or after heat-treatment. Bolts that are rolled before heat treatment have similar fatigue performance to that of bolts with cut threads. In contrast, bolts that are rolled after heat treatment had improved fatigue performance by a factor of between 3 and 5 on fatigue life in both air and sea water. Therefore, when assessing the fatigue life of a bolted structure for life extension and continued service, it is important to know whether the threads were cut or rolled and the heat treatment.

Differences in thickness and mean stress also need to be considered when assessing the fatigue life of older or ageing offshore structures. The TWI tests on Grades 8.8, B7 and L7 bolts with diameter varying between 19 and 44 mm found that the fatigue life reduced with increasing bolt diameter. A correction factor f on stress range for the fatigue life of bolts of diameter d mm ($20 < d < 60$ mm) when using standard fatigue curves that are based on bolts of 20 mm diameter is given by the relationship.

$$f = \left(\frac{d}{20}\right)^{0.3}$$

In addition, the TWI test data clearly showed a reduction in fatigue life as the mean stress from pre-tensioning the bolts is increased. In practice, relaxation of the pre-tension during the early life of a structure will mean that this effect is not generally significant.

Assessing the fatigue performance of bolts used in offshore environments can be based on the fatigue curves given in either the EN 1993-1-9 (EN 2005) or DNV-RP-C203 (DNV 2021a) modified for different thread manufacturing routes, thickness and environment. The fatigue curves in these standards are largely based on the testing of uncoated bolts up to 50 mm diameter in air. While corrections can be made for the differences in thickness and environment there was a lack of relevant data about the actual fatigue performance of rolled hot dipped bolts up to M72 diameter in a marine environment.

Consequently, TWI carried out a programme of fatigue testing of hot dipped galvanised M72 bolts in air and seawater environments. The results are given in TWI Core Research Report (Johnston 2021). That report found that the most suitable S–N fatigue curve for the assessment of life extension and continued service was the DNV GL RP C203 Class G curves (which is equivalent to FAT 50 in Eurocode 3) corrected for thickness using the reference thickness of 25 mm (not the 30 mm reference thickness in Eurocode 3 or the 20 mm suggested by MATSU (1997). While DNV RP C203 Class G or FAT 50 curves are recommended for hot dip galvanised bolts given the detrimental effect of galvanising on fatigue performance, the higher classification of FAT 71 could be considered for bolts in air if galvanising is not performed (DNV 2021b).

6.8 Wear, Erosion and Tearing

Wear can occur whenever solid material surfaces rub against each other and result in material removal and thickness loss from one or both materials in contact. The rate of loss can be increased where the surfaces are in a corrosive environment, where wear exposes fresh surfaces for corrosive attack. Examples of areas on offshore structures where wear can be a problem include the following:

- Mooring lines associated with floating structures, either of chain or of rope,
- Conductors, flow lines and caisson pipes in contact with guides,
- Areas where visiting supply vessels can rub against the structure during loading/unloading.

Mooring system components are particularly vulnerable to wear on contacting surfaces such as between chain links, shackle bolts and at locations where chains are in contact with fairleads. Significant loss of material can occur, sometimes in short periods of time measured in months, as a result of the rubbing of chain link surfaces. There is usually a subsequent loss of load-bearing capacity as the cross-sectional area of mooring chain links becomes too low to carry the applied load.

The amount of wear will depend on surface properties such as roughness and hardness and on the local environment, and the degree of movement. There will be a significant variability in wear due to changes in contact forces, movements and environmental conditions. Seawater has a corrosive effect which may lead to higher synergistic wear rates than the sum of wear and corrosion as independent processes. Protective deposits can be removed by wear thus enhancing the rate of corrosion. Corrosion may also increase surface roughness which may increase the amount of wear.

To minimise wear, friction should be reduced by ensuring that contacting surfaces are relatively smooth without sharp edges or protrusive weld deposits in areas of contact. Reduction in wear can also be achieved by increasing the hardness on one of the contacting surfaces. In practice this combination of material hardness is difficult to achieve. An example of wear is shown in Fig. 6.24.

Erosion is the physical removal of surface material due to numerous individual impacts from a flow of solid particles, liquid droplets or the implosion of gas bubbles (cavitation) onto a material surface. Erosion is a time dependent degradation mechanism but can sometimes lead to very rapid failures. In its mildest form, erosive wear appears as a light polishing of the upstream surfaces, bends or other stream-deflecting structures. In its worst form, considerable material loss can be obtained.

Tearing is a plastic mechanism in which the material is strained beyond its ductility limits and is for example associated with for example impact loading.

Fig. 6.24 Wear and tear in chain links (unknown origin)

6.9 Stress Corrosion Cracking (SCC)

Stress corrosion cracking (SCC) is a type of corrosion-induced failure that occurs in metals under the combined action of tensile stresses when exposed to a corrosive environment. Cracks due to SCC are often identified by their multiple initiation sites, (tree-like) branching and stepped brittle appearance. The combination of tensile stress and the corrosive environment creates a synergistic effect that makes the material more susceptible to cracking than either factor alone. SCC can lead to unexpected and sudden failure of ductile metals from accelerating growth, coalescence and ultimate loss of load carrying capacity. As further described below, stress corrosion cracking (SCC) is mostly an issue for stainless steels and to some extent high-strength structural steels (Fig. 6.25).

For stress corrosion cracking (SCC) to occur three factors need to be present:

- A sufficiently high tensile stress
- A corrosive environment, such as those containing chloride ions, sulphur compounds or alkaline solutions.
- A material whose microstructure is susceptible to SCC, such as stainless and high-strength steels.

Stress corrosion cracking (SCC) generally initiates at zones of stress concentration, such as surface discontinuities, pre-existing cracks and corrosion pits. SCC typically initiates and propagates along grain boundaries or other susceptible microstructural features in the material. These cracks can extend rapidly through the material, often with little plastic deformation, leading to brittle fracture. The corrosive environment can also alter the microstructure of the steel, promoting the formation of

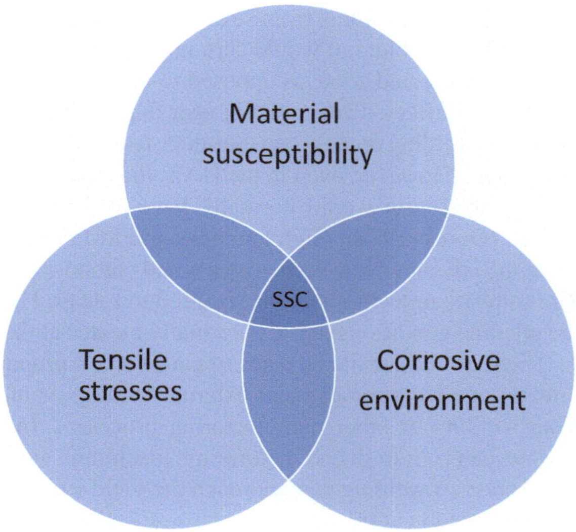

Fig. 6.25 The three factors needed for stress corrosion cracking (SCC)

brittle phases or weakening of grain boundaries, which makes the material more prone to crack initiation and propagation, further exacerbating its brittleness. The stress corrosion cracks can grow either by intergranular or trans-granular mechanisms. In intergranular stress corrosion cracking the crack develops along the grain boundaries of a material, whereas in trans-granular cracking the crack grows through the grains of the material (TWI 2024).

Stress corrosion cracking is a very complex mode of failure and therefore the prediction of it is a difficult task. Several mechanisms for SCC have been suggested that can be characterized as electrochemical and electrochemical-mechanical mechanisms (Revie and Uhlig 2008). The different mechanisms for SCC depend on the nature of the materials under stress and the chemical reactions with the environment they are exposed to.

The susceptibility to SCC of different materials varies, with some materials having a much greater susceptibility than others. Stainless steels, for example, are much more vulnerable than ordinary carbon manganese structural steels in a marine chloride bearing environment. Failures of austenitic and duplex stainless steel from chloride stress corrosion cracking that occurred in process plant offshore prompted a detailed study by the Health and Safety Laboratory (HSE 2011). SCC of austenitic steels continues to be a major threat where these steels are used in offshore.

Certain types of carbon steels, especially those with higher carbon content or those exposed to aggressive environments containing chlorides, sulphides or alkaline solutions, can be susceptible to stress corrosion cracking. In addition, low-alloy steels, which contain small amounts of alloying elements such as manganese, nickel, chromium or molybdenum, can exhibit varying degrees of susceptibility to SCC depending on their composition and microstructure. However, in 1981, after over ten years' experience of offshore structures in the North Sea, Harrison (1981) declared that as far as was known to him there had been no failures of carbon manganese structural steels or welds from stress corrosion cracking in normal sea water. If there were to be a problem, it would certainly have occurred in the high hardness regions of the heat affected zone, as opposed to the weld metal or parent steel.

The heat-affected zone (HAZ) near the welds is generally seen as susceptible to SCC. The welding process can introduce residual stresses and alter the microstructure of the steel. However, even in the HAZ, the controls to minimise hydrogen cracking during fabrication would normally have also been sufficient to prevent SCC in-service (Gooch 1974). High hardness and SCC have not generally been a problem since the advent of low carbon steels and cathodic protection in offshore structures. The sulphate reducing bacteria required to release H_2S from pollutants are killed in the alkaline conditions that CP normally creates at the metal surface (HSE 1998a).

The required tensile stresses to cause crack propagation may be in the form of directly applied stresses from external loading or in the form of residual stresses from welding or other manufacturing processes. In addition to welding, residual stresses can be introduced by forming, machining and grinding. Residual stresses set up because of welding can approach the yield strength of the material if the weld is not post-weld heat treated.

There is more likelihood of SCC occurring in CMn steel structures if the sea water is polluted by sour hydrocarbon products in drill cuttings. An extensive study by Taniguchi and Kataya (1978) showed that in polluted sea water with a pH of 5.4 the critical hardness for cracking was 310 HV. Apart possibly from within corrosion pits, pH values as low as this only occur in sea water if H_2S is present. Such low pH values can also easily occur in carbon steel storage tanks containing sour products associated with stagnant water. Under these conditions it may be difficult to keep hardness levels low enough (for example below the NACE maximum of 243 HV) to avoid cracking in non-heat-treated welds. However, for unpolluted sea water around a platform these conditions are extremely unlikely in the absence of a source of significant contamination.

High-strength structural steels are often susceptible to SCC, especially when exposed to aggressive environments containing chlorides, hydrogen sulphide or other corrosive species. High strength steels are commonly used in tendons, mooring chains and jack-up structures and may be more prone to cracking when chlorides are present in association with high stress, although hydrogen embrittlement is often a more likely cause of degradation (Harrison 1981). However, in high strength steels, SCC should be considered as a possible degradation mechanism in combination with hydrogen embrittlement and hydrogen induced stress cracking. In an assessment of carbon manganese steels in an ageing structure, any detected cracking is unlikely to be the result of SCC alone.

Preventing stress corrosion cracking involves several strategies, including material selection, design considerations, stress mitigation techniques and environmental control measures. For an existing structure the material has already been selected, but an evaluation of its continued robustness is important (see Chap. 10). A robust material with respect to SCC may include corrosion-resistant alloys, the minimizing of tensile stresses through proper design and manufacturing practices, the application of protective coatings or inhibitors in addition to controlling the corrosive environment through maintenance (ensuring correct corrosion protection potentials).

6.10 Detection of Material Changes and Degradation in Steel Materials

6.10.1 Introduction

Testing of materials in existing offshore structures during operation is generally challenging, as the number of feasible techniques for material testing in the field is limited. The methods that are available can broadly be categorised into three types, namely destructive, semi-destructive and non-destructive. In order to provide the relevant information (such as ductility, fracture toughness, proneness to HE and laminar tearing) about a material needed for a life extension assessment, destructive or at least a semi-destructive test is often needed. However, these methods require the removal of parts of the material from the structure and may weaken the structure if the area is not sufficiently repaired.

Non-destructive testing includes several techniques which enable a limited number of material properties to be assessed in special cases. An example of this may be chemical composition analysis can be undertaken using for example optical emission spectroscopy.

Destructive techniques involve destruction of the object being measured and are typically related to research and development. Destructive testing is often far cheaper to implement than non-destructive testing but is limited in practice as destructive testing can only be carried out on structures where these material parts are no longer needed (e.g. samples removed from decommissioned structures). Such testing can include yield stress, fracture toughness properties, chemical composition etc. The information from such tests can be useful for other structures from the same time period and particularly for structures with materials from the same manufacturer. Various studies have been reported on components from offshore structures that have been in service for several years (Sharp and Ersdal 2021).

Semi-destructive techniques are comparable to destructive techniques in that they are less invasive, as only a small amount of material is removed for testing. This allows the structure to better maintain its integrity. Boat sampling testing (BST) is an example of this method. This non-destructive technique is a surface sampling technique, which has been developed for obtaining, in-situ, metal samples from the surface of an operating component without affecting its operating service life. The shape and size of the sample depends upon the shape of the cutter and the surface geometry of the parent material. Miniature test specimens can be developed from the sample and these specimens can then be subjected to various tests, viz. metallurgical evaluation, micro-hardness evaluation, to confirm the integrity and assessment of safe operating life of the component.

6.10.2 Measurements of Residual Stress

Knowledge about the residual stresses may be needed as input to the fracture mechanical assessment of a detail or a defect. An example of such situations could be to determine whether the post-weld heat treatment has been performed and was effective. In addition, the amount of residual stress can be relevant to the assessment of stress corrosion cracking (SSC) and hydrogen induced stress cracking (HISC).

In theory, residual stresses in a metallic material in an existing structure can be measured, but the process will be relatively difficult. The most likely method to be used would be hole drilling, which is the most commonly used technique for measuring residual stresses in structures in operation. Stressed material is removed by drilling a small blind hole in the area of interest and the material surrounding the hole finds a new equilibrium leading to a displacement near the hole which can be measured.

In laboratory situations, also neutron and x-ray diffraction are possible techniques, but these are not relevant for an existing structure. Neutron diffraction uses neutrons to measure the crystal lattice spacing in an object, allowing the residual stress to be determined from the lattice spacing. Neutrons can penetrate deeply into a material allowing measurements beneath the surface. X-ray diffraction allows for surface residual stress measurement by also measuring the lattice spacing locally. X-rays can only penetrate the object surface by a few hundred microns.

To the knowledge of the authors, diffraction methods have only been used to assess residual stresses in single mooring chain that have been recovered from operational mooring lines and analysed in a lab environment.

6.10.3 Elemental Composition of Steels

Optical emission spectroscopy (OES) is a widely used analytical technique used to determine the elemental composition of a broad range of metals. The part of the electromagnetic spectrum which is used by OES includes the visible spectrum and part of the ultraviolet spectrum (wavelengths in the range 130 nm up to around 800 nm.) To obtain material for analysis an excitation device such as an electric spark or arc source is used. A high voltage pulse ionizes the atmosphere between the tip of the counter-electrode and the sample surface, causing it to become conductive forming a plasma which heats the material to a high temperature, causing it to be vaporized and ionized. This causes emission of element-specific light in the ultraviolet (UV) and visible range. The generated light is captured by an optical system which separates it into its spectral wavelengths. These can be measured using a high-resolution grating to separate the light into element-specific wavelengths. A detector then measures the intensity of light for each wavelength, which are unique for each element. The intensity measured is proportional to the element in the sample. The measured intensities are then processed by a computer system, using a predefined calibration enabling the elemental concentrations to be analysed. The sample's elemental composition can then be calculated based on a calibration with certified reference material. Optical Emission Spectroscopy can be available as a traditional laboratory, or cart-mounted system.

OES has many advantages: it's fast and relatively easy to use, it can measure a wide range of elements and concentrations in many different types of metallic materials. These include elements such as carbon, sulphur, phosphorous, boron and nitrogen. It claims to be very accurate when measuring low levels of trace and tramp elements, and it's fairly low-cost compared to other techniques.

X-ray fluorescence (XRF) is a non-destructive technique for measuring elemental composition from magnesium (Mg) to uranium (U), from parts per million to 100% (Beckhoff et al. 2006). Handheld XRF analysers are portable devices that can be used on location for immediate, lab-quality results to determine material composition. Energy dispersive X-ray fluorescence identifies metals and elements in an object by detecting their signature XRF emission energies. Powered by a low wattage x-ray tube, X-rays hit the sample, causing it to fluoresce and send X-rays back to the analyser. The energy spectrum is measured by a detector which identifies which elements are present and the amount of each. The elements that are the lightest in the periodic table below magnesium (including carbon and nitrogen) are not easily detectable by XRF as the X-rays are too weak to be detected and get counted. However, some suppliers claim to be able to detect carbon and nitrogen using this method.

6.10.4 Hardness Tests

Hardness testing indicates the ability of a material to resist indentation and gives an indication of strength, wear resistance and toughness. The process involves the application of a constant load on a rounded or pointed indenter to create an indentation in the material surface. The depth of penetration is then measured to provide an indication of hardness. It can be a stand-alone test but is often carried out in conjunction with other mechanical tests as an overall assessment of material properties.

There are several hardness methods in use which are Brinell, Rockwell and Vickers hardness methods, each with their own standards and benefits (CEN and ASTM material testing standards). Choosing the right hardness test is usually governed by the size and geometry of the sample, the area to be tested and ease of application. Portable testers are available for use in the field.

Hardness is a quasi-mechanical property, as it is a function of several other mechanical properties such as elastic modulus and yield strength, which all influence how far the indenter penetrates under a given load, and how much the material springs back when the indenter is removed. In general hardness correlates loosely to yield strength and ductility, with a high scatter but can provide some information on yield strength.

Hardness tests might be used on an offshore structure to provide an indication of the yield strength and toughness where these properties are unknown. A particular example of where it might be used is to indicate the properties of in-situ heat affected zone material when its toughness is unknown. Such information could be useful if cracks had been detected in such material.

6.11 Remediation of Ageing Effects in Steel

6.11.1 Approaches to Remediation

The remediation of ageing effects in damaged or degraded steel components during operation is generally challenging and expensive. Access to damaged or degraded areas is usually difficult and the remediation techniques are difficult to apply. Operators of fixed offshore structures will normally prefer to leave the damaged areas in situ and seek alternative means for altering the load path by adding strengthening members to the structure. Where a mobile structure can be brought into dry dock, there is more possibility for remediation to be carried out.

6.11.2 Weld Toe Improvement for Increased Fatigue Life

In most types of welded joint, fatigue cracks tend to initiate at the weld toe. This is because of the combined effect of the following features associated with the weld toe (see also Sect. 6.7):

- a stress concentration associated with the change in profile of the weld surface at the fusion boundary where the weld metal fuses to the parent material and
- the presence of undercut and micro-cracks and other flaws at the weld toe.

The combination of high stress with micro-cracks and other inherent flaws can act as a source for fatigue crack initiation and subsequent fatigue crack growth.

In the fabrication of offshore structures, most welds were left in the as-welded condition with little attempt to optimise the weld profile and remove microcracks. In addition, these welds were normally left without any post weld heat treatment to reduce tensile residual stresses. Where such welds have proved to be a source of fatigue cracking, or where analysis has determined that welds are susceptible to fatigue damage, the main approach through which their fatigue life can be extended is from the application of various techniques designed to improve the condition of the weld toe.

Weld toe improvement techniques rely on two main principles:

(1) Removal or reduction in size of weld toe flaws and, as a result, a reduction of the severity of the weld toe stress concentration by dressing the weld toe and producing a smoother interface between the weld and parent material. The main techniques of weld toe dressing are weld toe burr grinding and weld toe remelting using a TIG or plasma welding torch.
(2) Introduction of compressive residual stresses at the weld toe by plastic deformation of the weld toe region. The beneficial effect of introducing compressive residual stress at the weld toe is that an applied tensile stress has to exceed the residual stress before it became damaging. The main techniques are hammer-peening, needle peening, shot peening and high frequency ultrasonic peening. The beneficial effect of the compressive stress decreases as the applied tensile stress approaches the yield strength of the material disappearing altogether at when the maximum applied stress exceeds yield.

The benefit of these techniques when applied to the joints of offshore structures was investigated during the UKOSRP 1 programme. The results of this research are summarised in the background document (HSE 1992b) to the fourth edition of the HSE Guidance Notes (HSE 1994) which subsequently recommended the use of these techniques to extend fatigue life where this was necessary. Detailed guidance on the application and benefit of using these techniques is provided in Annex F of BS 7608 (British Standards 1993) and the DNV Recommended Practice for fatigue design DNV-RP-C203 (DNV 2021).

According to BS 7608 weld toe improvement techniques are applicable to the treatment of weld toes on members at least 6 mm thick in any arc welded steel structure that is subject to fatigue loading. Due to a lack of experimental data for extra high strength steels, the specified benefits for fatigue performance only apply to structural steel and stainless still up to a maximum specified yield strength of 960 MPa. It is, however, reasonable to expect that, in principle, weld toe improvement techniques would also benefit the fatigue performance of welds of higher strength steels and members less than 6 mm thickness, although in the absence of relevant published data the benefit should be quantified by specific testing.

It is important to note that weld toe improvement techniques only address fatigue cracking initiating at the weld toe. Fatigue cracking can also initiate in other regions of the weld, such as from defects at the weld root and heat affected zone. For these situations, weld toe improvement cannot be relied upon to increase fatigue life.

In addition to weld toe dressing and the introduction of compressive stress at the weld toe, three other approaches to extend the fatigue life of welds should be mentioned. AWS D1 (American Welding Society 2000) for the welding of steel suggests that weld profiling is a way to improve fatigue life at the weld toe by optimising the shape profile of the weld as a whole. Another approach is by drilling a hole either through the tip of a growing crack as a means to arrest further crack growth, or by drilling a hole offset from the crack growth as a means to divert a growing crack away from the weld toe. If the fatigue cracking is extreme, the whole weld can the ground out and the entire joint rewelded. These different approaches to extend the fatigue life of welds and their relative benefits are now discussed.

Weld toe dressing

The fatigue life of a weld from cracks initiating at the weld toe may be extended dressing the weld toe by either burr grinding a shallow smooth pre-determined radiused groove at the weld toe or by remelting the weld toe by TIG or plasma dressing. These techniques remove inherent flaws and reduce the stress concentration at the weld toe. The removal of prior fatigue damaged material from the weld toe effectively resets the surface of the material to its condition at the start of life with the added benefit that the future rate of fatigue damage is reduced and the time to initiate a new crack increased.

Procedures for burr grinding are given in standards such as BS 7608 (British Standards 1993) and DNV-RP-C203 (DNV 2021) and other documents. Weld toe grinding requires skill and practice and there are dangers of damaging the weld or removing too much material if it is not carried out properly. It is therefore important that grinding operatives have been suitably trained and are qualified for the task.

The final ground surface is normally inspected for any further defects caused or exposed by grinding using a surface NDT technique, such as MPI. Weld toe burr grinding inevitably results in reduced thickness of weld and the cross-section of the parent material. Care should be taken to ensure that the minimum specified weld throat thickness is maintained. The acceptability of anticipated reduced section thickness may need to be assessed by analysis and if necessary, material can be reinstated by depositing weld with post weld heat treatment and dressing-out when possible.

The benefits of burr grinding were quantified within UKOSRP I (Department of Energy 1988). Tests in air on burr ground butt welds in 25 and 38 mm thick plates showed that at high cyclic stress the mean fatigue life was at least 2.2 times that of as-welded joints tested in the same programme (see data in Fig. 6.26). Later data for 30 and 40 mm thick T-butt welds gave improvement factors of 2.2 and 2 respectively that were approximately constant over all stress ranges investigated (HSE 1992b). These data formed the basis of the improvement factor of 2.2 that was recommended in the HSE Guidance Notes.

The results from tests on butt welds in 25 and 38 mm plates suggested that an even greater benefit might be obtained at lower cyclic stress levels. The data indicated that a change in slope in the appropriate S–N curve for the untreated weld according to the clarification of the weld could be introduced. This benefit was subsequently incorporated in BS 7608 through a recommendation that the design fatigue strength (stress range) for 10^7 cycles could be increased by a factor of 1.5 and that the S–N curve could be increased from 3 by rotating it about this point to a slope of 3.5. The benefit applied to all forms of weld dressing including TIG dressing and hammer peening.

Similarly, by 2012 DNV GL C203 also recognised the beneficial change in the slope of the S–N curve from weld toe grinding as shown for weld detail category D.

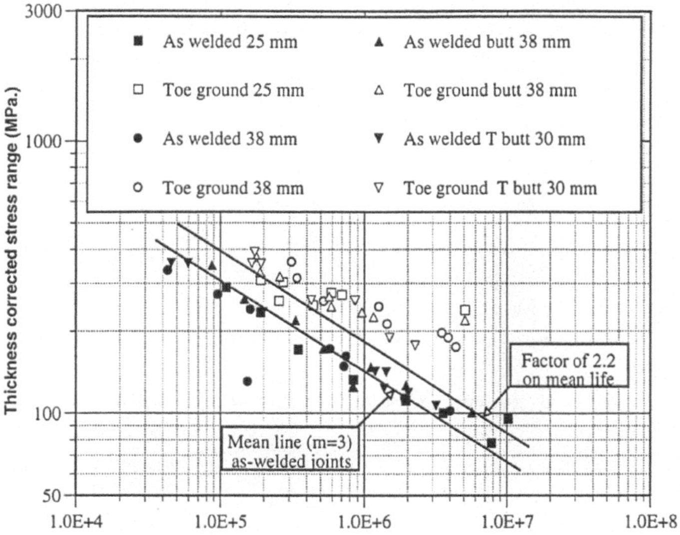

Fig. 6.26 Endurance test for T-butt and cruciform joints with weld toe improvements (HSE 1992b)

Unlike BS 7608, the DNVGL RP differentiates between peening (most beneficial) and burr grinding. The most recent edition of DNVGL RP C203 allows an improvement factor on fatigue life of 3.5 for burr ground welds and 4 for peened welds for welds with a minimum specified yield strength greater than 350 MPa.

Tests on ground plate joints under free corrosion conditions in seawater reduced the beneficial effect of weld toe grinding (Department of Energy 1988). It was presumed that this was due to due to corrosion pitting introducing new stress raisers within the ground region, although in some tests a substantial benefit remained (de Back and Vaessen 1984). As expected, the application of cathodic protection tended to restore the benefit by eliminating the pitting.

Fatigue data for ground tubular joints within UKOSP1 was extremely limited comprising only 6 tests in air (Department of Energy 1987), 9 in sea water under free corrosion conditions (de Back and Vaessen 1984; Hara et al. 1986; Wylde and Yamamoto 1988) and 2 with cathodic protection (de Back and Vaessen 1984). Although the numbers of tests were insufficient to quantify the effect of weld toe grinding, the results were in general agreement with those for butt welded plates.

An alternative method of improving fatigue life by weld toe dressing is by using a TIG or plasma welding torch. Inherent flaws are removed by remelting the material at the weld toe and the local stress concentration is reduced by producing a smooth transition between the weld and parent material. BS 7608 provides further information on this technique (British Standards 1993).

The benefit of weld toe dressing is that it provides grounds for reducing the fatigue classification of the weld and enables to use of a more favourable fatigue design curve resulting in a longer fatigue design life. However, it only affects the fatigue life associated with cracking from the weld toe: the possibility of cracking at other features of the weld (e.g. weld root) is not influenced and should always be considered. In a corrosive environment such as seawater, weld toe dressing cannot be assumed to be effective if pitting corrosion subsequently occurs in the dressed region.

Peening

Weld toe peening aims to introduce compressive residual stresses at the surface of the weld by plastic deformation. This can be achieved by repeated hammering of the weld toe region or directing high velocity shot at the weld toe. Various methods are available including hammer peening, needle peening high frequency ultrasonic peening and shot peening and are described in more detail in BS 7608 (British Standards 1993).

A body of data that showed that hammer and shot peening could enhance the fatigue performance more than other weld improvement techniques (HSE 1994). These benefits remained in seawater and under variable amplitude loading. The testing of hammer peened joints in UKOSRP I showed that, in air, the fatigue strength of the welded connection was similar to that of the parent plate material.

Generally, fatigue failures in the specimens were seen elsewhere than in the weld toe area. In contrast to weld toe dressing the improvement in fatigue performance depends on the maximum and minimum stresses of the cycle and any inherent weld toe flaws and prior fatigue damage cannot be assumed to have been removed.

The magnitude of the improvement was found to depend upon various factors such as the procedure adopted and the yield strength of the steel. Consequently, the Review Panel for the Fatigue Guidance (RPFG) recommended that whilst a benefit should be allowed for peening, the magnitude of the benefit could only be demonstrated on a case-by-case basis by reference to experimental data relevant to the procedures and steel type used (HSE 1992b). Hence, it was not considered appropriate to include a value in the Guidance Notes.

Hole drilling

If a fatigue crack has already initiated and growing, the life of the joint can sometimes be extended by drilling holes through the wall of the component. However, work performed by PJ Tubby at TWI (Department of Energy 1989) indicated that holes at crack tips were not able to significantly arrest cracks in tubular joints before re-initiation occurred elsewhere.

Later research has indicated that drilling a hole through the component wall in a position close to the crack tip can divert the crack away from the weld toe into the hole with relatively good results (Atteya 2023). However, this work also has shown that a new crack can initiate at the weld toe beyond the hole, causing some limitations to this type of repair (Atteya 2023).

Weld profiling

According to AWS D1, weld profiling is believed to offer enhanced fatigue endurance due to a reduction in the local notch stress adjacent to the weld. However, from UKOSRP the experimental justification for this benefit was mixed. In UKOSRP 1 (Department of Energy 1988) plates welded with an improved profile did show increased endurance in both air and sea water with cathodic protection, but no such improvement was found for tubular joints in the UKOSRP 2 tests (Department of Energy 1987). In conclusion, there appeared to be no support for a beneficial effect of weld profiling and as a result no benefit was suggested in the Guidance Notes (HSE 1994).

Rewelding

In extreme circumstances it is possible to grind a cracked weld away and to reweld it. This is a difficult operation as the rewelding may introduce further defects and residual stresses that can be the cause of new cracking.

6.11.3 Mitigation of Wall Thinning from Corrosion and Erosion

Wall thinning is an effect of corrosion and erosion. As noted earlier, corrosion can result in loss of material from exposed surfaces when a structure is not properly protected by an external coating or electro-chemical cathodic protection. Erosion can cause loss of materials from components rubbing together. In addition, scour can in some cases also give rise to wall thinning where flowing particles of sand or other particulate material abrade the surface, see Sect. 6.8.

Corrosion is often found at sea level in the splash zone and topsides where cathodic protection is limited, whereas erosion often occurs where flowlines and caissons are restrained by guides and between the links of heavy-duty anchor chains, while scour occurs at the base of the structure near the seabed. Wall thinning due to corrosion can be a particular problem for ship-like structures and semi-submersibles where water-tight integrity is an important consideration.

In ageing structures, measures taken earlier in life to protect the material from corrosion or hydrogen damage may have become ineffective. For example, surface coatings applied during manufacture may have degraded exposing larger or isolated areas of bare metal. These sites can lead to aggressive corrosion and significant local wall thinning. In addition, sacrificial anodes (see Sect. 6.6.3) can over time be depleted and if not replaced can leave the structure inadequately protected or unprotected.

Before mitigation is considered, it is usual to determine the extent of the wall thickness loss using appropriate non-destructive testing methods, such as ultrasonic testing. With this information, an assessment can be made of the criticality of the damage to the integrity of the component and its role within the structure.

Where pitting and wall thinning have limited structural significance, isolated shallow pits can be ground out while locally thinned areas can be dressed to a smooth surface. In both cases the affected area should be protected from further corrosion, electrochemically or by a coating. Material lost in larger areas of pitting or wall thinning with greater structural significance can be reinstated using epoxy composite wrapping or weld overlays. The latter needs particular care to avoid cracking by optimising the welding procedure and ensuring dry conditions.

Different options for epoxy composite wrapping and their applications are given in Sharp and Ersdal (2021). DNV-RP-C301 (DNV 2015) provides requirements and guidelines for the design, fabrication, operation and qualification of bonded repairs for steel structures. Alternatively, the American Bureau of Shipping has a document specifying requirements for bonded and composite repairs of steel structures (ABS 2022).

Even larger areas of wall thinning can be covered with patch plates. These can either be welded around the edges or bonded adhesively to the base material. Particular attention is required to the design of such plates to ensure that the stresses produced by the shift in the neutral axis remains within code limits. For welded patch plates, a fatigue assessment may be necessary to ensure that cracking from the toe of the repair weld around the plate does not occur.

Where there is gross loss of material and wall thinning such that the member may fail under extreme loading conditions an alternative load path is required. Bolted clamps can be designed that encircle the member and provide additional strength to the affected area. Further details can be found in Sharp and Ersdal (2021). An example of the use of such clamps was the bolted grouted repair clamp that was used to strengthen a nodal joint on the Viking AD platform, see Annex 2.

6.12 Assessment of Steel Materials for Life Extension

Steel production, the requirements for this production and the understanding of materials behaviour have changed substantially over the years of extracting offshore oil and gas from the seabed, largely driven by developments in the North Sea. The standards related to the design of early offshore structures did not fully acknowledge the effects of the factors above on the long-term integrity of the structure. Therefore, any assessment of steel materials after long-term exposure to the North Sea or other similar environment should be considering the history and records of the materials in relation to knowledge that now exists about the performance of materials throughout their life that is summarised in this book.

As we have seen so far in this book, the durability of steel materials in an offshore structure and their suitability for continued service beyond original design life and into a period of life extension is dependent on many factors, including:

- factors determined at the design and fabrication stage:
  - the original standards which were specified for the steel,
  - the chemical composition and levels of impurities and alloying elements,
  - the specific production process and the strength of the steel,
  - welding and any other joining methods and use of heat treatment (pre-heat and post weld),
  - the original construction quality, level of defectiveness and numbers of repairs to welds,
  - the inspection processes applied during fabrication
  - the resulting properties and condition after fabrication,
  - any protective coatings that might be applied,
- factors determined during in-service:
  - the environment to which the steel is exposed,
  - how the steel has been used and loaded,
  - the degradation during use (how the dimensions and properties of the steel might have changed during use),
  - the effectiveness and potential of any electrochemical (cathodic) protection over time,

– the inspection processes applied during in-service,
– the extent the material has been repaired or modified.

This book has shown how the strength and other properties of steel are dependent on their specification, chemical composition and production process. Specifications have changed over the years with older steels tending to have lower properties with greater variability than their modern counterparts. A check on the tensile properties on the original mill certificates for the steel used could indicate whether the structure has sufficient strength for life extension when some members may be carrying higher stresses.

While issues with lamellar tearing and low through thickness properties of thick sections were mostly understood by the mid 1970s, this knowledge was only progressively disseminated. Therefore, the specification of steel for early structures should be checked to determine whether these effects had been properly addressed by appropriate material specification. Otherwise, there is a possibility that lamellar tearing and low through thickness properties could be a problem during life extension when fatigue cracks or higher stresses due to thickness loss could be prevalent.

Welding and joining processes have evolved over the past fifty years and modern welds are likely to contain fewer fabrication defects than was once the case. Because fabrication defects can initiate fatigue cracking and fatigue crack growth during service, the standard of construction quality and the number of fabrication defects detected by manufacturing and in-service inspections are important considerations in a life extension assessment. Poor construction quality places greater importance and reliability on the in-service inspections to detect defects in locations that would be a concern before they become critical.

Some areas of welding are more likely to contain defects than others and these should be given special attention. These areas include:

- stop-start positions,
- welds with unfused surfaces,
- single sided welds such as closure welds where the roots cannot be dressed or inspected,
- weld repairs made during fabrication or in-service,
- fillet welded attachments.

These can all be sites for crack growth during extended service.

When the early structures were fabricated, inspection technology was at a more rudimentary state than it is today. While most critical welds were subject to surface examinations and radiography, the use of ultrasonic testing to detect lack of weld sidewall fusion and root and HAZ cracking was not that common. Therefore, a review of the reliability, coverage and defect sentencing criteria of the manufacturing could be a good indicator of level of potential defectiveness of the structure.

Testing methods and requirements have changed over the years. Steels and welds for more modern structures are likely to have fracture toughness test data at the relevant temperatures, thicknesses and locations including the weld metal and coarse grained and other HAZ regions. The data available for older structures are likely to be more limited and appropriate advice from a fracture mechanics specialist is recommended, particularly where there is the possibility of local brittle zones.

Materials for older structures are more likely to have only Charpy V test data of the parent material and weld as a quality check. Unfortunately, the correlations to fracture toughness can be difficult. Fracture toughness testing on representative material and welds may be beneficial where there are known cracks requiring engineering critical assessment for life extension. This may not be justified for locations in structures where the stresses are known to be low. Therefore, operators assessing life extension of structures are advised to review what fracture toughness data exists in preparation for the possibility of needing to assess a significant crack that may be revealed during this period.

Operators of ship-shaped and other floating structures should be aware that hulls and tanks could be constructed of Grade A class ship steel, particularly where there have been conversions from tankers. Older structures made from these types of steels are unlikely to have guaranteed Charpy energy or other fracture properties and therefore may be more susceptible to fracture. Critical locations where cracks have been found may therefore need special attention.

The offshore sector has long recognised the threat of wall thickness loss due to corrosion within a marine environment. Protective measures such as coatings and cathodic protection have usually been applied to limit the loss of material. However, in-service maintenance of these measures is required because coatings can degrade exposing bare metal and cathodic protection anodes become consumed. A review of the maintenance of the coating and the cathodic protection could indicate areas of the structure where material may have been lost resulting in higher stresses, initiation of fatigue cracks and potential for water ingress. Such a review is prudent before any period of life extension identifying areas where excessive loss of material or severe pitting due to corrosion could have occurred.

An assessment for life extension should normally assess the likelihood of hydrogen embrittlement and hydrogen cracking of the steel and their effect on fatigue crack growth and fracture toughness and tolerance to cracks. While high strength steels, such as may be used in jack-ups and mooring chains, may be more vulnerable, this is not always the case. Untempered martensitic and other vulnerable microstructures are more vulnerable to embrittlement.

Weld metal hydrogen cracking could also have resulted from welding in the open, either during fabrication or while the structure is in operation, where damp electrodes or moisture could have been used. This could have been avoided with appropriate storage of consumables and pre- and inter-pass-heating and records of such weld procedures should be reviewed. In-service conditions, such as the presence of sulphate reducing bacteria, can also lead to increased hydrogen uptake into the material. Further, special attention should be given to safety critical and high strength steels where the cathodic potential has historically been through periods where it has been highly negative (< -1000 mV SCE).

Where hydrogen embrittlement could be considered a possibility on account of the material or welding specification or in-service conditions, operators should search for any relevant test data from constant stress, slow strain rate or fracture toughness tests.

Where such data does not exist and the consequences from fracture of a crack are high, retrospective testing of representative material can be recommended. Where critical parts of a structure are judged to be in a hydrogen embrittled state with low tolerance to cracks, careful consideration is necessary of whether life extension is possible. However, hydrogen embrittlement remains an issue for structures constructed from high strength steels, although evidence of robust material selection can minimize the problem.

How the structure is used and loaded can change the state of a material in different ways. Members and components in contact or rubbing, such as the links of mooring chains or caisson and flowlines passing through guides can become worn and abraded. Material can be removed through scouring from flowing particles of sand or sediment. Areas where there is H_2S present, such as product storage tanks and in the vicinity of drill cuttings, may be susceptible to sulphide stress cracking and other forms of stress corrosion cracking. Where these effects are likely to be present, further investigations are recommended.

In areas of the world where the weather is hostile, the greatest threat to a structure in a period of life extension is likely to be from accumulated and widespread fatigue damage. This book has covered the prediction of fatigue damage and fatigue crack growth in some detail and shown that understanding of the fatigue performance has had a continuous improvement through experience, research and development. Despite modern methods the prediction of fatigue damage can never be wholly reliable because of the intrinsic variability of the microstructural factors underlying fatigue and because the cyclic loading spectra in members and joints are rarely known sufficiently accurately and can be influenced by occasional peak loadings. In regions and structures where the possibility of widespread and accumulated fatigue can be foreseen, an appropriate inspection and monitoring programme during life extension is invariably necessary.

Structures in locations subject to hurricanes and typhoons can become damaged during these extreme events. Climate change is expected to make such events to occur more frequently. Loads may be sufficient to cause plastic strain, hardening and load cycle fatigue and ratchetting. Reassessment of structures subject to extreme events is a code requirement and needs to take these ageing mechanisms into account.

Thus, the assessment of steel materials in ageing structures is a fundamental part of the process for ensuring a safe structure for life extension, including gaining the necessary acceptance. While the performance of materials in structures during the original design life (typically 25 years) has been validated from long experience and can be predicted with reasonable confidence, life extension takes the operator into a regime where knowledge is not so extensive and a more cautious approach becomes necessary. However, the understanding of material's behaviour, the available testing methods, the technological and diagnostic tools now available are sufficiently good for long lifetimes to be possible.

References

ABS (2022) Bonded and composite repairs of steel structures and piping. ABS, November 2022

Almar-Næss (1985) Fatigue handbook. Tapir, Norway

American Petroleum Institute (2000) API RP2A recommended practice for planning, designing and constructing fixed offshore structures, 21st edn. Washington D.C, USA

American Welding Society (1972) Structural welding code-steel, AWS D1.1

American Welding Society (1979) Structural welding code-steel, AWS DI. 1-79

American Welding Society (2000) AWS structural welding code—steel. ANSI/AWS D1.1-2000, Miami, USA

Atteya M (2023) Crack arresting using cutout repair in tubular offshore joints. PhD Thesis, Stavanger University, Norway

Barker NN (1996) A review of hydrogem assisted cracking in high strength steels, Report for HSE by Marine Technology Support Unit, MaTR 236

Beckhoff B, Kanngießer B, Langhoff N, Wedell R, Wolff H (2006) Handbook of practical x-ray fluorescence analysis. Springer, 2006, ISBN 3-540-28603-9

British Standards (1980) PD 6493 Guidance on methods for assessing the acceptability of flaws in fusion welded structures. British Standardisation Institution

British Standards (1982) BS 6235 code of practice for fixed offshore structures. British Standards Institution, London

British Standards (1990) BS 4360 British Standard specification for weldable structural steels. British Standardisation Institution

British Standards (1991) PD 6493 Guidance on methods for assessing the acceptability of flaws in fusion welded structures. British Standardisation Institution

British Standards (1993) BS 7608 Code of practice for fatigue design and assessment of steel structures. British Standardisation Institution

British Standards (2019) BS7910 Guide to assessing the acceptability of flaws in metallic structures

Buitrago TY, Healy J, Chang BE (1993) Local joint flexibility of tubular joints. Offshore Technol Conf 1:405–416

Cole IS, Brook R (1993) Fractography of crack growth in stress corrosion cracking and corrosion fatigue growth in a number of offshore steels. In: Proceedings 3rd international offshore and polar engineering conference, Singapore, pp 296–302

Cole IS, Brook R, Howard I (1988) A comparative study of the performance of three steels for offshore applications. In: Proceedings on conference fatigue of offshore structures, London, September, pp 167–184

Coudreuse L, Renaudin C, Bocquet P, Cadiou L (1997) Evaluation of hydrogen assisted cracking resistance of high strength jack-up steels. Marine Struct 10(2–4):85–106

Cowling MJ, Hancock JW, Appleton RJ, Gall DS (1986) Fatigue crack growth in biologically active environments under realistic load sequences. In: Topp DA, Dover WD (eds) Final report, Cohesive programme on fatigue of offshore structures, July 1983–June 1985, University College London 1986

de Back J, Vaessen GHG (1984) Effect of plate thickness, temperature and weld toe profile on the fatigue and corrosion fatigue behaviour of welded offshore structures: Part I and II. Final report on ECSC Convention 7210-KG/601 (F7.4/81, Stichting Materiaalonderzoed in de Zee, May 1984

Department of Energy (1977) Offshore installations: guidance on design, construction and certification, 2nd edn. HMSO, London

Department of Energy (1984) Offshore installations: guidance on design, construction and certification, 3rd edn. HMSO, London

Department of Energy (1987) United Kingdom offshore steels research project—Phase 2—Final summary report (UKOSRP II), OTH87 265, HMSO London, UK

Department of Energy (1988) United Kingdom offshore steels research project—Phase 1 Final report (UKOSRP I), OTH88 282, HMSO London, UK

Department of Energy (1989) OTH 89 307 Fatigue performance of repaired tubular joints, Written by P.J. Tubby for the Department of Energy, HMSO, London, UK
Department of Energy (1990) Single sided closure welds interim guidance. HMSO, London, UK
Djukic MB, Sijacki Zeravcic V, Bakie G, Sedmak A, Rajicic B (2014) Hydrogen embrittlement of low carbon structural steel. Proc Mater Sci 3(2014):1167–1172
DNV (1974) Rules for the design, construction and inspection of offshore structures, Det Norske Veritas, Høvik, Norway
DNV (2006) Material risk—ageing offshore installations, Report for the Petroleum Safety Authority Norway, REPORT NO. 2006-3496, DNV, Høvik, Norway.
DNV (2015), DNV-RP-C301 Design, fabrication, operation and qualification of bonded repair of steel structures, DNV, Høvik, Norway.
DNV (2021a) DNV-RP-C203 Fatigue design of offshore steel structures, Høvik, Norway
DNV (2021b) DNVGL ST 0126 Support structures for wind turbines
DNV (2022) Review of available material on usability and suitability of high-strength mooring chain (PTIL)—usability and suitability of high strength mooring chains—Final report, DNV report P261.1C-54693, Rev. 0, Høvik, Norway
DNVGL (2015) DNVGL-RP-B101 Corrosion protection of floating production and storage units, DNVGL, Høvik, Norway
DNVGL (2018) DNVGL-RU-OU-0300 Fleet in service, DNVGL, Høvik, Norway
DNVGL (2019) Bolted joints—study on bolt incidents, Work performed by DNVGL for the Petroleum Safety Authority, DNVGL report 5333, Høvik, Norway
EFC (1995) Guidelines on materials requirements for carbon and low alloy steels for H2S containing environments in oil and gas production, European Federation of Corrosion Publications Number 16, Institute of Materials
Efthymiou M (1988) Development of SCF formulae and generalised influence functions for use in fatigue analysis. In: Recent development in tubular joints technology; Anugraha Centre, Egham, UK
EN (2005), EN 1993-1-9 Eurocode 3: Design of steel structures—Part 1–9: Fatigue
Energy Institute (2017) Guidelines for the management of microbiologically influenced corrosion in oil and gas production
Ersdal G (2001) An overview of ocean currents with emphasis on currents on the Norwegian continental shelf. Norwegian Petroleum Directorate, Stavanger, Norway
Ersdal G, Sharp JV, Stacey A (2019) Ageing and life extension of offshore structures: the challenge of managing structural integrity. Wiley
Force (2023) Hydrogen assisted cracking of jack-up installations, work performed for the Petroleum Safety Authority Norway
Gibstein MB (1978) Parametric stress analysis of T joints. In: Proceedings of the European offshore steels research seminar, Cambridge, UK, 27–29 November 1978
Gooch TG (1974), Stress corrosion cracking of welded joints in high strength steels. Welding J 53(7):287–306
Gurney TR (1979) Fatigue of welded structures, 2nd edn. Cambridge University Press, Cambridge, UK
Haagensen P (2013) Tekna Fatigue course 2013, Norsk forening for stålkonstruksjoner
Hara M, Kawai Y, Narumoto A, Matsumoto S (1986) Corrosion fatigue strength of 490 MPa class high-strength steels produced by the thermos-mechanical control process, OTC 5311. In: Proceedings 18th offshore technology conference, Houston, 5–8 May 1986
Harrison JD (1981) Material selection considerations for offshore structures, Welding Institute Report 150/1981, TWI, Cambridge, UK
Herden DC, Tebbett IE, Wood T (1986) New data on the fatigue performance of cast steel tubular joints, OTC-5310-MS
Heshmati E, Guy RG, Lewis G, Levitt G (1986) Analysis methods and inspection procedures for single-sided closure welds in offshore structures, Paper Number: OTC-5351-MS offshore technology conference, Houston, Texas, May 1986

HSE (1991) Hydrogen cracking of legs and spud-cans on jack-up drilling rigs—a summary of results of an investigation, written by Abertheny K, Fowler C M, Jacob R and Davey V S for HSE Offshore Technology Report OTH91 351, HSE 1991
HSE (1992a) OTH 325, Crack-tip electrochemistry in relation to corrosion fatigue of offshore structural steels in seawater—technical summary and implications. HMSO London, UK
HSE (1992b) OTH 92 390, Background to new guidance for steel joints and connections for steel offshore structures. HMSO London, UK
HSE (1994) Fourth edition of the guidance notes for the design and construction of offshore structures. HMSO London, UK
HSE (1995) Offshore installations—guidance on design, construction and certification. HSE Books, 1995, amendment no. 5. HMSO London, UK
HSE (1997) OTH 532 A review of the corrosion fatigue behaviour of structural steels in the strength range 350–900 MPa and associated height strength weldments, prepared by Cranfield University for the HSE, HSE Books, London, UK
HSE (1998a) OTH 555 A review of the effects of sulphate reducing bacteria in the marine environment on the corrosion fatigue and hydrogen embrittlement of high strength steels, prepared by MJ Robinson and PJ Kilgallon for HSE, HMSO, London, UK
HSE (1998b), OTH 511 A review of crack growth rates in air and seawater, written by R N King, 1998
HSE (1998c), OTO 97 067 Fatigue test on steel bolts, Prepared by PL Wells, MaTSU, 1997, for HSE, HSE, HMSO, London, UK
HSE (1999a) OTO 1999 056 A review of the effects of microstructure on the hydrogen embrittlement of high strength offshore steels, prepared by MJ Robinson and PJ Kilgallon for HSE, Research Strategy Unit, Bootle, UK
HSE (1999b) OTO 99/054 Static strength of cracked tubular joints—new data and models, prepared by TWI for the HSE, Research Strategy Unit, Bootle, UK
HSE (2002) Fracture mechanics assessment of fatigue cracks in offshore tubular structures, Offshore Technology Report OTR 2000/077, 2002. Health and Safety Executive, London, UK
HSE (2003) RR 105, Review of the Performance of high strength steels used offshore, prepared by Cranfield University, HMSO London, UK
HSE (2004) RR 207 Review of low cycle fatigue resistance. HSE Books, London, UK
HSE (2011) RR902 Chloride stress corrosion cracking in austenitic stainless steel: assessing susceptibility and structural integrity. HSE Books, London, UK
ISO (2007) ISO 19902:2007 Petroleum and natural gas industries—Fixed steel offshore structures. International Standardisation Organisation
Johnston C (2021) Fatigue performance of large bolts. TWI Core Research Project 1155/2021, TWI, Cambridge, UK
King RN, Stacey A, Sharp JV (1996) A review of fatigue crack growth rates for offshore steels in air and seawater environments, OMAE 1996
Kuang JG, Potvin AB, Leick RD (1975) Stress concentration in tubular joints. In: Proceedings of the offshore technology conference
Loidl M, Kolk O, Veith S, Göbel T (2011) Characterization of hydrogen embrittlement in automotive advanced high strength steels. Mater.wiss. Werkst.tech 42(12):1105–1110. https://doi.org/10.1002/mawe.201100917
Lotsberg I (2004) Recommended methodology for analysis of structural stress for fatigue assessment of plated structures. In: Proceedings of OMAE-FPSO 2004, OMAE specialty symposium on FPSO integrity, ASME. Houston, USA 2004
Lotsberg I (2016) Fatigue design of Marine structures. Cambridge University Press, New York, USA
Ma A, Sharp JV (1997) Fatigue design of cast steel nodes in offshore structures based on research data. Proc Inst Civ Engrs Wat Marit & Energy 124:112–126
Maddox SJ (1991) Fatigue strength of welded structures. Cambridge University Press, Abingdon, UK

Marshall PW (1974) General consideration for tubular joint design. In: Proceedings of the welding in offshore construction international conference, Newcastle, UK, 26–28 February 1974
Martin ML, Connolly MJ, DelRio FW, Slifka AJ (2020) Hydrogen embrittlement in ferritic steels. Appl Phys Rev 7:041301. https://doi.org/10.1063/5.0012851
MATSU (1997), Fatigue performance of threaded connections in air and seawater—derivation of S-N curves, MATSU 1997
Moan T, Bekkvik P, Næsheim T (1981) Alexander L. Kielland-ulykken. Norges Offentlige Utredninger, NoU 1981:11, Universitetsforlaget, Oslo, Norway
Mohaupt UH, Burns DJ, Kalbfleisch JG, Vosikovsky, Bell R (1987) Fatigue crack development, thickness and corrosion effects in welded plate to plate joints. In: Noordhoek C, de Back J (eds) Paper TS3, steel in marine structures, Proceedings of the third international ECSC offshore conference, Delft, June 1987. Elsevier
MSL (2004) Assessment of repair techniques for ageing or damaged structures. Project #502, MSL Services Cooperation, Egham, UK
MTD (1990) Marine Technology Directorate Ltd—Design and operational guidance on cathodic protection of offshore structures, subsea installations and pipelines. MTD Report 90/102
Muzghi IA (2024), Fastener corrosion in Arabian Gulf offshore installations, Fastener Corrosion in Arabian Gulf Offshore Installations (osti.gov), visited September 2024
Noordhoek C, van Delft DRV, Verheul A (1987) The influence of the thickness on the fatigue behaviour of welded plates up to 160 mm with attachment or butt weld. In: Noordhoek C, de Back J (eds) Paper TS4, steel in marine structures, Proceedings of the third international ECSC offshore conference, Delft, June 1987. Elsevier
Paris P, Erdogan F (1963) A critical analysis of crack propagation laws. J Basic Eng Trans ASME 85:528–534
Pircher H, Drodten P, Kaiser H-J, Musgen B (1990), Behaviour of high-strength structural steels exposed to seawater. Thyssen Technische Berichte, Heft 1/90:53–67
PSA (2019) Bolted Joints—Study on bolt incidents, DNV report for PSA, Report No. 2018-5333 Rev. 3, Date 2019- 01-15
Revie RW, Uhlig HH (2008) Corrosion and corrosion control, an introduction to corrosion science and engineering, 4th ed. Wiley and Sons
Reynolds AG, Sharp JV (1991) The fatigue performance of tubular joints—an overview of recent work to revise. Department of Energy Guidance. In: Faulkner D, Cowling MS, Incecik A (eds) Proceedings 4th international symposium on integrity of offshore Structures. Elsevier Science, Essex, U.K., pp 261–277
Robinson MJ, Kilgallon PJ (1994) HE of cathodically protected HSLA steels in the presence of sulphate reducing bacteria. Corrosion 50(8):620–635
Robinson MJ, Kilgallon PJ (1995) Cracking of high strength low alloy steels—controlling the damaging effects of sulphate reducing bacteria, Publication 95/100, Marine Technology Directorate Limited, London, 1995
Scott PM, Silvester DRV (1975) Influence of seawater on fatigue crack propagation rates in structural steel. UKOSRP report 3/03. UK Department of Energy 1975
Sharp JV, Ersdal G (2021) Underwater inspection and repair for offshore structures. Wiley, 2021
Smedley P, Fisher P (1991) Stress concentration factors for simple tubular joints. In: Proceedings of the international offshore and polar engineering conference, Edinburgh, UK, 11–16 August 1991
Stacey A, Sharp JV (2021) Integrity management of ageing offshore structures: technical developments and challenges. Paper AIM-ALE2021-76756 in the proceedings of the ASME 2021 international conference on asset integrity management—ageing and life extension AIM-ALE2021 November 30–December 2, 2021, Virtual, Online
Stacey A, Sharp JV, Nichols NW (1996) The influence of cracks on the static strength of cracked tubular joints. In: Proceedings of the 15th international conference offshore mechanics and Arctic engineering, Florence, Italy, June 1996, Vol III, American Society of Mechanical Engineers, New York

Stacey A, Nichols NW, Sharp JV (1997) Fatigue performance of single sided circumferential and closure welds in offshore jacket structures. In: OMAE, conference, Japan, 1997

Standard Norge (1984), NS3472:1984 Prosjektering av stålkonstruksjoner—Beregning og dimensjonering. Standard Norge, Lysaker, Norway

Standard Norge (1998) NORSOK N-004 Design of steel structures

Standard Norge (2015) NORSOK N-006 Assessment of structural integrity for existing offshore load-bearing structures. Standard Norge

Taniguchi, Kataya (1978) Critical hardness for SCC in sea water including H2S. J Iron Steel Inst Japan 64

Trethewey KR, Chamberlain J (1995) Corrosion for science and engineering, 2nd ed. ISBN 9780582238695, Pearson Publishing

Troutmann A, Mori G, Oberndorfer M, Bauer S, Holzer C, Dittmann C (2020) Hydrogen uptake and embrittlement of carbon steels in various environments. Materials 13:3604. https://doi.org/10.3390/ma13163604

TWI (1991) A compilation of fatigue test results for welded joints subjected to high stress/low cycle condition—Stage 1, Written by TD Rosenburg, RM Andrews and TR Gurney for the Department of Energy Report OTI 91 552

TWI (2024) Stress corrosion cracking—what factors cause and prevent it? https://www.theweldinginstitute.com/Stress-Corrosion-Cracking-What-Factors-Cause-and-Prevent-It, visited August 2024

Visser W (1975) On the structural design of tubular joints. J Eng Ind 97:391–399. Houston, TX, USA, 4–7 May 1975

Wang D, Hagen AB, Fathi PU, Lin M, Johnsen R, Lu X (2022) Investigation of hydrogen embrittlement behaviour in X65 pipeline steel under different hydrogen charging conditions. Mater Sci Eng A 860(2022):144262

Wordsworth AC, Smedley GP (1978) Stress concentration at unstiffened tubular joints. In: Proceedings of the European offshore steels research seminar, Cambridge, UK, 27–29 November 1978

Wylde JG, Yamamoto N (1988) Seawater corrosion fatigue tests on steel tubular joints, paper 7921.01/87/589.2. The Welding Institute, March 1998

Chapter 7
Cementitious Materials in Offshore Structures

7.1 Introduction

Cementitious materials such as concrete (reinforced and pretensioned) and grout are used as materials in many major offshore structures. Concrete is also used widely as a weight coating on offshore pipelines, but these are outside the scope of this book.

Grout is initially mostly used as a joining material between steel elements and as a filling material in foundations, but also as a repair material later in life for both steel and concrete structures. Historically, the piles in fixed offshore steel structures (jackets) were inserted into the legs of the jacket and the annuli were filled with grout to secure load transfer. Later, the piles were driven through pile sleeves, which had also been joined to the main structure by grouting. In addition, a few examples of particularly large jacket structures had to be fabricated in two parts, subsequently joined together at site by grouting. Later in the life of these jacket structures, repairs have in some cases been necessary. Two common repair methods include grout filling of members and joints in addition to grouted clamps and sleeves (See Sharp and Ersdal 2021). Grout is also used to repair cracks and spalling on concrete structures.

The use of concrete in offshore structures has a long history, including war time forts, ships and tanks. In the oil and gas industry, the use of concrete offshore structures began in the 1970s primarily in UK and Norwegian waters. Later, offshore concrete structures have gained a wider use, including in Danish, Canadian, Russian and Australian waters. Examples of offshore concrete structure types for oil and gas production include:

- Gravity-based structures (GBS), for example the Norwegian Statfjord platforms and the UK platform Brent C. There are several types of GBS designs, including the Condeep design, Doris design and Arup's concrete gravity substructure (CGS). At the time of writing there are approximately 40 concrete gravity-based structures (GBS) servicing various oil and gas fields in the world,
- Tension leg platforms (TLP), for example the Norwegian platform Heidrun where the hull is made in concrete material,

© The Author(s), under exclusive license to Springer Nature Switzerland AG 2025
J. V. Sharp and J. Wintle, *Assessment of Materials in Ageing Offshore Structures*,
https://doi.org/10.1007/978-3-031-84489-8_7

- Semisubmersible platforms, for example the Norwegian platform Troll B where the hull is made in concrete material,
- Jarlan walls enclosing concrete tanks, for example the UK platform Ninian and the Norwegian concrete platform Ekofisk T,
- Steel–concrete hybrid,
- Articulated tower (used as a loading column on Maureen) where the cylindrical tower is built in concrete.

Most of the early concrete offshore structures are now in a life extension phase.

More recently, concrete has also been used as a base for offshore wind turbines, mainly in Sweden, Denmark and Norway. Some of these are gravity based in shallow water depths. An example of a wind turbine installed on concrete bases is the 40 MW Middel Grunden farm in Denmark, with 20 conical shaped gravity foundations installed in 2001 in a water depth of < 10 m. The design had the added benefit of being able to resist ice flows. The Hywind Tampen in Norway is a floating spar type concrete structure used as base for the turbines. For the floating wind turbines with concrete sub-structures, the weight of the concrete is very important in providing stability.

As shown in Table 7.1, the first concrete offshore structure to be installed was the Ekofisk tank in the Norwegian sector in 1973, installed on the Ekofisk oil field as a storage unit. The first to be installed on the UKCS was the GBS on Beryl Alpha in 1975. Other significant examples of offshore concrete structures include Hibernia and Hebron in Canada, Sakhalin in Russia and Wandoo and Bream in Australia (Department of Energy 1986a; Fernandes et al. 2016; Kviterud 2024). As indicated in Tables 7.1 and 7.2, water depths for these offshore concrete structures have varied from 30 to 350 m.

As can be seen in Tables 7.1 and 7.2, concrete structures are massive, weighing up to 830,000 tonnes, with concrete volumes varying from 23,000 up to 240,000 m^3 and the quantity of reinforcing steel from 5800 to 80,000 tonnes. In addition, high strength prestressing tendons are installed in both horizontal and vertical ducts and then grouted. The weight of prestressing tendons was as high as 8000 tonnes. Anchorages for the tendons are built into the walls of the concrete components. Grouting is undertaken to protect the tendons against corrosion and to permanently "lock-in" the tendon pre-tension, thereby removing the long-term reliance upon the end-anchorage systems.

Due to the dominance of GBS type structures, this book will use these as basis for the discussion on concrete as a material in ocean structures. However, most aspects will also apply for other types of concrete structures.

The main parts of a Condeep type are:

- Shafts, also called legs, columns and towers,
- Cells, also called caissons and base, used for oil and water storage,
- Steel to concrete transition at the interface to between the shafts and the topside,
- Shaft to base junction,
- Skirts and its interaction with the often-grouted foundations.

Table 7.1 Materials in pre-1990 offshore concrete structures

Concrete structure	Date	Water depth (m)	Overall weight (tonnes)	Volume of concrete (m^3)	Concr. frade (f_{ck})	Weight of reinforcement (tonnes)	Weight of prestressing (tonnes)	Comments
Ekofisk (N)	1973	72.5	235,000	80,000	C45	10,000	3300	Doris tank
Beryl A (UK)	1975	117	200,000	62,000	C45	8500	820	Condeep
Brent B (UK)	1975	139	165,700	65,000	C45	10,700	1270	Condeep
Frigg CDP1 (UK)	1975	98	194,000	56,000		7700	2600	Removed
Brent D (UK)	1976	142	177,800	65,000	C50	14,000	1500	Condeep
Frigg MP2 (UK)	1976	94	203,000	57,000		8300	2600	Removed
Frigg TP1 (UK)	1976	**104**	130,000	70,000		5800	500	Removed
Dunlin A	1977	151	228,600	90,000			2200	Condeep 4 shafts
Frigg TCP2 (N)	1977	104	164,000	60,000	C45	13,500	533	Removed
Statfjord A (N)	1977	145	320,000	87,000	C50	12,000	2600	Condeep
Cormorant A (UK)	1978	150	294,700	131,000		13,930	1100	Condeep 4 shafts
Ninian Central (UK)	1978	**143**	400,000	142,000		27,000	4000	Condeep 1 shaft
Brent C (UK)	1978	141	287,500	107,800		11,400	1400	Condeep 4 shafts
Urbanas (3x) Brazil	1978	**15**		10,000	C35			
Statfjord B (N)	1981	145		140,000	C55			Condeep
Tarsiut Island	1981	26		8400				4 hollow caissons
Schwedeneck A	1983	25		3600	B45			Removed
Schwedeneck B	1983	16		2000	B45			Removed
Statfjord C (N)	1984	145		135,000	C55			Condeep
Gulfaks A (N)	1986	135		125,000	C55			Condeep
Gulfaks B (N)	1987	142	173,000	100,000	C55			Condeep
Oseberg A (N)	1988	105	575,000	120,000	C60			Condeep

(continued)

Table 7.1 (continued)

Concrete structure	Date	Water depth (m)	Overall weight (tonnes)	Volume of concrete (m^3)	Concr. frade (f_{ck})	Weight of reinforcement (tonnes)	Weight of prestressing (tonnes)	Comments
Gulfaks C (N)	1989	217	830,000	240,000	C65	80,000	3500	Condeep
Ravenspurn (UK)	1989	43	120,000	10,000	C50	2750	450	CGS 3 shafts

UK United Kingdom, N Norway, G Germany
GBS Gravity base structure (condeep), *CGS* Caisson concrete gravity substructure

Important parts of any concrete structure are also the steel reinforcement and prestressing cables in ducts and anchorages. The cathodic protection system provided for the steel attachments also provides protection for the reinforcement.

Condeep GBSs were considered ideal to support the very high topside weight of the processing facilities needed for many North Sea structures. Condeep GBSs could also withstand the extreme environmental forces in the North Sea and had the advantage that they could be constructed in the region using local resources. Another major advantage was the ability to store oil in the cells, as in the early days the North Sea lacked an extensive pipeline infrastructure to transport the crude oil to shore.

Most concrete structures rest on the seabed on a large base, often with skirts which penetrate the seabed protecting the structure against horizontal and overturning forces (Fig. 7.1). The size of the caisson and skirts depends on local soil conditions. The topside steel structure is usually supported on one to four concrete shafts extending from the caisson to above the surface of the sea. One or two of the shafts are usually used for drilling supporting, conductor frames and well-conductors. Some shafts may be water filled or dry and normally outfitted mechanically. Alternative designs such as the concrete gravity structure (CGS) has also been used (Fig. 7.2).

Offshore concrete structures are usually constructed using slip-forming. This is a method of construction in which concrete is poured into the top of a continuously moving formwork. As the concrete is poured, the formwork is raised vertically at a speed which allows the concrete to harden before it is free from the formwork at the bottom. Slip forming offers an important advantage of eliminating most horizontal construction joints, thus improving the resistance to seawater, particularly at depth. This method has been used mainly to construct towers and caissons.

Table 7.2 Materials in post-1990 offshore concrete structures

Concrete structure	Date	Water depth (m)	Overall weight (tonnes)	Volume of concrete (m^3)	Concrete grade (f_{ck})	Weight of reinforcement (tonnes)	Weight of prestressing (tonnes)	Comments
Sleipner A (N)	1993	82	463,000	75,000	C65			
Draugen (N)	1993	251	232,000	75,000	C65			1 shaft Condeep, K500TS
NAM F3 (Ne)	1992	43	91,000	23,300				CGS 3 shafts
Troll A (N)	1993	303	683,000	43,000	C70/C75	18,000	2500	Condeep
Heidrun (N)	1994	350		66,000		27,000	4100	TLP
Troll B (N)	1995	340		46,000	C75			Semi
Harding (UK)	1995	109	120,000	36,745	C50			Concrete base
West Tuna (A)	1995	61		29,862	C50/C60			CGS 3 shafts
Bream B (A)	1996	61			C50/C60			CGS 1 shaft
Wandoo (A)	1996	54		28,000	C50			CGS 4 shafts
Hibernia (C)	1997	80		165,000	C85	105,000	8000	Condeep 4 shafts
South Arne (D)	1999	60	153,000	28,000	C65			CGS 1 shaft
Malampaye (P)	2000	43		35,000	C50			CGS 4 shafts
Sakhalin energy (R)	2005	48						
Sakhalin-1 (R)	2005	30						GBS
Adriatic (I)	2008	29					3860	
Sakhalin Arkutun (R)	2012	33		52,000		27,000		
Hebron (C)	2017	109		132,000	C65	40,000	3400	GBS

UK United Kingdom, N Norway, C Canada, A Australia, Ne Netherlands, R Russia, D Denmark, I Italy, P Philippines

GBS Gravity base structure, *Condeep* Concrete offshore structure for deep water, *CGS* Concrete gravity structure, *TLP* Tension leg platform, *Semi* Semisubmersible platform

Fig. 7.1 Example of a Condeep concrete gravity-based structure (GBS)

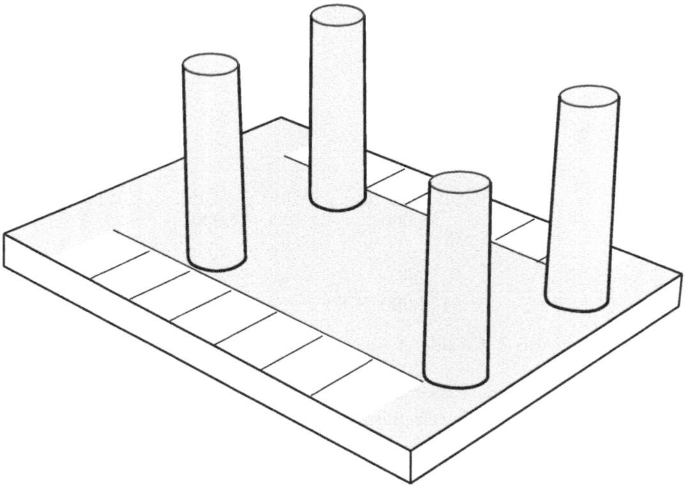

Fig. 7.2 Schematic illustration of a concrete gravity structure (CGS)

7.2 Concrete Materials for Offshore Structures

7.2.1 Introduction

Concrete used in structures is composed of aggregates, cement and water. Natural aggregates are obtained from rolled alluvial material or from crushed rocks of very different natures (silicon, limestone, granite etc.). In mortar only sands with a grain diameter less than 5 mm are used. For concrete, fine gravel is also used. Typical aggregates used in offshore structures were crushed granite and limestone (HSE 1997a). The reduced space between the rebars normally induces the use of reduced size aggregates and the need for enhanced workability. The usual maximum aggregate sizes are 10 to 15 mm.

Normal concrete is suitable for various components of the structure. When the earliest of these platforms were installed, for example the Ekofisk Tank in 1973, the highest strength class of concrete allowed according to Norwegian Standards was used. This strength class was then called B450 and had a cube strength of about 45 MPa and is now called C 45 (Holand et al. 2000). All concrete used various additives for example to improve workability, see Sect. 7.2.4. In addition, several of the earlier structures used small amounts of materials such as blast furnace slag and pulverized fly ash (PFA), see Sect. 7.2.3.

Concrete properties have improved considerably since construction of the first offshore concrete structures in 1968 (Ekofisk). The compressive strength of cubes at that time was about 40 MPa and the concrete consisted of cement, aggregate (coarse and fine), water and plasticizer for example Lignosulfonat which acted as both set retarder and water reducer.

Since then, several large research programmes have been performed to further develop and improve the concrete constituency (cement binder, aggregate, and admixtures). As a result of this the concrete grades have been increasing, as shown in Tables 7.1 and 7.2. For examples, Statfjord B (delivered in 1981) with a required strength of 55/44 MPa (cube/cylinder strength), Gullfaks C (delivered in 1989) with a required strength of 65/52 MPa (cube/cylinder strength) and Hibernia (delivered in 1997) with a concrete grade of 85/68 MPa (cube/cylinder strength). The latest offshore structure, Hebron, has obtained a compressive strength of about 85 MPa at 28 days (cube strength).

Typical mix designs for offshore structures were summarised by Fosså and Widianto (2019) and show the differences with time, as shown in Table 7.3.

However, when investigating specific documentation from the fabrication of individual platforms, it can be seen that silica and fly ash were used at an earlier stage than Fosså and Widianto (2019) indicates. Examples of such materials in early concrete structures are shown in Table 7.4.

As structures increased in size due to deeper water developments, concrete of higher strength grades was needed. This need initiated the development strength

Table 7.3 Typical concrete mix designs used for offshore structures (Fosså and Widianto 2019)

	1975	1990	2017
Concrete (kg/m^3)	460	410	350
Silica fume (%)		2	6
Fly ash (%)			20
Sand (kg/m^3)	780	940	940
Coarse aggregate (kg/m^3)	1000	945	945
Super plasticizer (l/m^3)	4	6	3
Water cement ratio	0.41	0.40	0.36
Slump (mm)	120	240	> 240

Table 7.4 Concrete materials in early offshore concrete structures (Department of Energy 1986a; ELF Aquitaine Norge 1977)

Platform	Concrete mix	Additives	Aggregate
Frigg MCP01 (1976)	Four different mixes used: – two mixes W/C ratio 0.45 max, – two mixes W/C ratio 0.4 max – 450 Water cement ratios:	Sika BV40 (0.5%) Sika Ret. (0–0.3%), Sika AER (0 or 0.005%)	Material deposit at Halden, Norway
Frigg TCP2 (1977)	Cement (PC 300) 480 kg/m^3 Sand (0–10 mm) 790 kg/m^3 W/C ratio 0.6 Concrete grade C45	Betokem LP: 1 L/m^3 Betokem L: 0.1 L/m^3	Coarse aggregate (10–25 mm granite from Ørlandet, Norway) 970 kg/m^3
Dunlin A (1977)	Ground blast furnace slag mixture, three design codes were complied with (DnV CP110, FIP/CEB)		No information
Ninian Central (1978)	Mix was 50 N/mm^2 (CS), Cement content 375 kg/m^3 with 105 kg/m^3 of PFA or 425 kg/m^3 when no pfa used	Cormix P1 super plasticiser: 1.3 l/ m^3, Cramix retarder P2: 0.065 to 2.6 l/ m^3	Glacoid morrain
Cormorant A (1978)	Concrete mixes to DnV, CP110, FIP/ CEB		No information
Brent C (1978)	Concrete grades 50 and 50PF (PF indicated addition of pfa)	Plastocrete between 1 and 0.5%	Limestone aggregate

grades such as C 80–85 which was used in later concrete structures. These high-strength concretes were used to achieve higher compressive strength and better durability associated with reduced permeability to seawater. Important factors for the development of these higher strength grades were (Holand et al. 2000):

- High strength cement,
- Well controlled aggregate grading,
- Admixtures such as superplasticisers and retarders,
- Strict procedures to produce high quality products.

Later, cement replacements such as PFA were introduced on a regular basis to improve performance, see Sect. 7.2.4.

7.2.2 *Chemical Composition of Concrete*

The most common type of cement is Portland cement made from lime, silica, alumina and iron oxide and has an oxide composition as shown in Table 7.5.

These are ground into a powder with a grain size of the order of 10 to 50 μm and burnt, which is a very energy intensive process responsible for large CO_2 emissions. When mixed with water they form a paste that sets and hardens due to hydration reactions. The process of hydration renders these constituents into a structural material. The major compounds that form during the manufacturing process are shown in Table 7.6.

Table 7.5 Typical composition Portland cement (Neville 1973, Table 1.2)

Oxide		% content
CaO	Lime (C)	60–67
SiO_2	Silica (S)	17–25
Al_2O_3	Alumina (A)	3–8
Fe_2O_3	Iron oxide (F)	0.5–0.6
MgO	Magnesium oxide	0.1–4.0
K_2O, Na_2O	Alkalies	0.2–1.3
SO_3	Sulphate	1.3–3.0

Table 7.6 Typical compounds of cement

Name of compound	Formula	Abbreviated formula	% by weight
Tricalcium silicate	$3CaO.\ SiO_2$	C_3S	55
Dicalcium silicate	$2CaO.\ SiO_2$	C_2S	20
Tricalcium aluminate	$3CaO\ Al_2O_3$	C_3A	10
Tetracalcium aluminoferrite	$4CaO.\ Al_2O_3.\ Fe_2O_3$	C_3AF	8
Other			7

Cement starts to set when mixed with water, causing a series of complex hydration reactions. The early stages of hydration produce heat which can be a problem in causing cracking in thick sections. Tricalcium silicate (C_3S) compound hydrates and hardens rapidly and is responsible for the early strength gain. Dicalcium silicate (C_2S) compound hydrates slowly and is responsible for the later strength gain. These tricalcium and dicalcium silicates are the most important compounds responsible for strength, making up 70–80% of cement volume.

The hydration of dicalcium silicate (C_2S) produces calcium silicate hydrate (C–S–H) component and calcium hydroxide (CH). The C–S–H component contributes significantly to the strength and durability of the cementitious matrix.

Tricalcium aluminate (C3A) is a compound in hydrated Portland cement (see Table 7.6) which is susceptible to reactions with sulphates and in the worst cases these can cause cracking of the concrete. The C3A content of the cement was initially seen as an issue offshore, because of the possibility of reactions with sulphates present in seawater which diffuse into the concrete. Hence, cements with low C3A content were preferred; In BS 4027 (British Standards 1980) which was a Standard for sulphate resisting cement the maximum C3A content was limited to 3.5%. However, it became known that the C3A combines with the chloride ion (in seawater) to form an insoluble compound that effectively reduces the concrete's permeability which is an asset. It was found that an increase in the C3A content resulted in an increase in chloride binding capacity which is beneficial to limiting corrosion of the reinforcement. Therefore, cements with moderate levels of C3A (typically about 8%) were subsequently specified for cements used for offshore construction (HSE 1997a).

7.2.3 Cement Replacements Materials

Concrete can be improved by cement replacement materials, typically various industrial waste products. The three typical waste products are pulverised fuel ash (PFA), blast furnace slag (BFS) and condensed silica fume.

Pulverised fuel ash (pfa) is a product from coal fired power stations. The use of pfa in concrete began in the 1950s mainly for reasons of heat and cost reduction. This increased with the introduction of a processed quality controlled pfa (Pozzolan) becoming commercially available. The first standard for the use of PFA, BS3892, was issued in 1965 (British Standards 1965). Ninian Central used PFA in the concrete mix (see Table 7.4). The ash is in the form of small spheres which are prevented from escaping via a chimney by electrostatic precipitation. The constituents of typical fuel ash from UK power stations is given in Table 7.8. The average particle size is 10μm and the size distribution is normally a little coarser than Portland cement. The material was recognised as an admixture but not as a cement replacement in the first edition of BS CP110 in 1972 (British Standards 1972). In 1985, BS 6588 was published for Portland-15–35% pfa cement (British Standards 1985a, b) and in 1996

BS 6610 was published for Pozzolanic cement with 35–50% pfa (British Standards 1996).

Blast furnace slag is a waste product from blast furnaces. The use of ground granulated blast furnace slag (GGBFS) as a cement replacement material began in the mid 1960s. Dunlin A used ground blast furnace slag in the concrete mix (see Table 7.4). Continuous production of a GGBFS known as Chemsave began in 1969 and the first Agrement Certificate for this product was issued in 1972. A British standard, BS5328, was issued in 1976 (British Standards 1976). GGBFS is used extensively for reasons of cost, sulphate resistance and heat reduction. Granules are produced by rapidly cooling the molten slag with air and water to give particles a few millimetres in diameter which are mainly glassy in structure. These need to be ground to a fine powder before use. The CaO and SiO_2 constituents are capable of being activated by Portland cement to form silicate hydrate, enhancing the pozzolanic reaction. The degree of pozzolanic activity of slag is determined by its glass content and the degree of vitrification (Department of Energy 1990a). The first approval certificate (Agrement certificate) for the use of blast furnace slag was issued in 1972 and a British Standard 6699 on this topic was published in 1986 (British Standards 1986).

Condensed silica fume is a by-product from the extraction of silicon. It needs to be condensed and precipitated to enable collection. This leads to sub-micron particles of nearly pure SiO_2 which are significantly smaller than Portland cement particles.

As shown in Table 7.4, the Ninian concrete structure had a cement content of 425 kg/m^3 when no PFA was used and otherwise with cement and PFA with 375 kg/m^3 and a 105 kg/m^3 of PFA. BS 6235 (British Standards 1982b) includes the use of pozzolanic materials with a note to say that the durability of the concrete should be considered carefully. CP110 (British Standards 1972) recommends the possible use of PFA in the range of 25–35%.

Modern standards like ISO 19903 (ISO 2006) state that when fly ash or other pozzolanas are used as additions, their content should normally not exceed 35% of the total weight of cement and additions. The content of silica fume used as additions should normally not exceed 10% of the weight of Portland cement clinker.

Permeability is reduced with the addition of both pfa and granulated blast furnace slag. Table 7.7 shows a significant reduction in permeability for a cement paste with additions of PFA and blast furnace slag with a water-cement ration 0.5.

Table 7.7 Effective diffusivity of chloride ions at 25 °C in three cement pastes with w–c ratio of 0.5 (Holden et al. 1983)

Type of cement	Diffusivity × 10^8 (cm/sec)
OPC	4.47
OPC with 30% pfa	1.47
OPC with 65% blast furnace slag	0.41

Table 7.8 Chemical composition of cementitious materials % (HSE 1996)

	Ordinary Portland cement OPC	Fly ash	Blast furnace slag	Silica fume	Liapor*
Silicon dioxide SiO_2	21	38	35	94	55–65
Calcium oxide CaO_2	63	4	45	1	7% max (with MgO)
Aluminium oxide Al_2O_3	4	13	10	0	15–25
Ferric oxide Fe_2O_3	2	40	1	0	7–15%
Magnesium oxide MgO	3	0	8	1	See CaO
Other products	7	5	1	4	Max 16%

*Liapor is a German aggregate based on expanded clay used by Norwegian contractors on several Norwegian platforms

7.2.4 *Admixtures*

Concrete admixtures are natural or manufactured chemicals or additives added during concrete mixing to enhance specific properties of the fresh or hardened concrete. They are usually added to modify the time it takes for the concrete to set and harden and improving the workability of freshly mixed concrete. In addition, the benefits are enhanced strength, increased durability and better resistance to adverse environmental conditions. Acceleration and retardation of the setting is generally achieved by altering the kinetics of the chemical reactions of different phases in the first few hours of hydration (Domone and Jefferis 1994). These materials are introduced during the mixing process.

Some common types of admixtures and their roles are listed below:

- Water-Reducing Admixtures (Plasticizers), also known as plasticizers or water reducers, reduce the amount of water needed in a concrete mix without compromising its workability. They improve the flow and cohesion of the mix, making it easier to place and finish. They are particularly useful in high-strength and high-performance concrete.
- Retarding Admixtures delay the setting time of concrete, extending the time available for mixing, placing, and finishing. This is especially beneficial in large construction projects or in hot weather conditions where rapid setting may be a challenge. Their use leads to improved workability.
- Accelerating Admixtures speed up the setting and early strength development of concrete. This is useful in cold weather construction or situations where a faster rate of strength gain is desired. Common accelerators include calcium chloride

and non-chloride accelerators. The benefits include early strength development, and an enhanced overall construction schedule.

- Air-Entraining Admixtures introduce microscopic air bubbles into the concrete mix. These bubbles improve freeze–thaw resistance by providing space for water to expand when it freezes. The main benefit is enhanced durability in climates with freeze–thaw cycles, reduced damage from scaling and spalling.
- Superplasticizers are a type of high-range water-reducing admixture that provides significant workability without increasing the water content. They are often used in self-consolidating concrete or in situations where high fluidity is required.
- Corrosion-Inhibiting Admixtures contain corrosion inhibitors that help protect reinforcing steel from corrosion in aggressive environments, leading to increased service life especially in marine or chloride-exposed conditions.
- Shrinkage-Reducing Admixtures help mitigate the drying shrinkage of concrete, reducing the likelihood of cracking. They provide improved durability especially in situations where shrinkage cracking may be a concern.

The use of admixtures requires careful consideration of mix design, compatibility, and dosage. EN 934 (EN 2001) is the harmonised European standard for admixtures. Part 2 covers concrete admixtures whilst part 4 covers admixtures for grout.

It is important that admixtures are used in a controlled manner as part of the overall concrete mix design. It is normal to make trial mixes to ensure required properties are obtained. BS 5075 (British Standards 1982a) was issued in 1982 covering admixtures. This was later replaced by EN 934 (EN 2001), a standard in several parts. Of particular interest for this book is part 4, which covers admixtures for grout for prestressing tendons, in addition to parts 1 and 2 for more general requirements. In addition, BS 8443 (British Standards 2005) provides specifications for establishing the suitability of special purpose concrete admixtures which includes underwater concrete.

ISO 19903 (ISO 2003) indicates that only those admixtures with established suitability shall be used for offshore structures. ISO 19903 further requires the cement with additives should be tested under site conditions to verify that the mix will yield the required effects, without impairing the other properties required. Misuse of admixtures can be detrimental to the properties of the hardened concrete. Most offshore concrete structures made use of admixtures in their construction. These included retarders to prevent material sticking to the shutters and plasticizers to aid slip-forming. As shown in Table 7.4 Frigg MCP 01 used three Sika products as additives (water reducing, retarders and air entraining agents). Ninian Central used Cormix admixtures (retarders and super-plasticisers). In addition, Brent C used a plasticiser.

7.2.5 Microstructure of Concrete Material

The microstructure of concrete is complex and plays a crucial role in determining the material's properties and performance. As described above concrete is a composite material composed of cement, water, aggregates and sometimes replacement materials such as pulverised fuel ash (pfa) and other additives. Its microstructure involves the arrangement and interaction of these components at a microscopic level, influencing the overall strength, durability, and other characteristics of the material.

At the heart of concrete's microstructure is the cementitious matrix. As noted earlier, the cement undergoes a chemical reaction with water resulting in the formation of calcium-silicate-hydrate (C–S–H) components and various crystalline phases, such as calcium hydroxide (CH). The microstructure of this matrix is dense and heterogeneous, with fine crystals dispersed throughout.

The C–S–H component is variable and poorly ordered phases without well-defined stoichiometry (amorphous) and contributes significantly to the strength and durability of concrete (the "glue"). The morphology of C–S–H is often described as having a fibrous or needle-like structure with particles interlocking and intertwining, forming a mesh-like structure. This morphology contributes to the material's cohesion and ability to resist deformation.

Aggregates are individual particles bound together by the cementitious matrix. The arrangement and gradation of aggregates impact the overall strength, workability and dimensional stability of concrete. The interface between the aggregates and the cement paste is critical for load transfer and overall performance.

The microstructure of concrete contains voids in the form of pores and capillaries, see Fig. 7.3. Some of these voids are a natural result of the hydration process, while others may arise due to inadequate compaction or the presence of air entraining agents. The size, distribution and connectivity of these voids influence properties such as permeability, porosity and durability. A denser microstructure with fewer and smaller voids generally leads to better resistance against penetration by water and aggressive chemicals, such as chlorides.

As mentioned earlier, various additives and admixtures are introduced to modify the microstructure and enhance specific properties of concrete. Supplementary cementitious materials like fly ash or silica fume can react with calcium hydroxide, producing additional C–S–H components and contributing to a more refined microstructure with improved strength and durability.

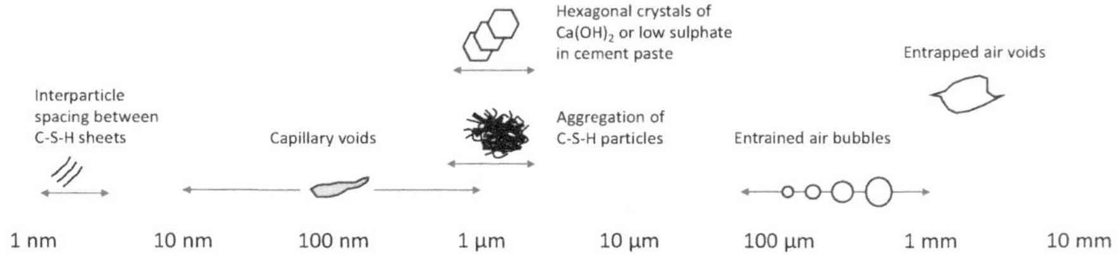

Fig. 7.3 Types of defects in cementitious materials in scale from nanometres to millimetres (based on Mehta and Monteiro 2006)

Apart from the amorphous C–S–H component, the concrete microstructure often includes crystalline phases like ettringite and calcium aluminate hydrates (C–A–S–H). These crystals form during the hydration process and contribute to the overall strength and stability of the material. The arrangement and interlocking of these crystals impact the mechanical properties of concrete.

The region between the aggregates and the cementitious matrix is called the Interfacial Transition Zone (ITZ). This transition zone is distinctively different from both the aggregates and the surrounding cement paste, with different properties and features. This difference is often due to incomplete hydration of the cement particles near the surface of the aggregates. As a result, there may be a higher concentration of un-hydrated cement particles, calcium hydroxide and other by-products in the ITZ. The ITZ typically exhibits higher porosity compared to the bulk cement paste. This increased porosity can be attributed to factors such as incomplete hydration, the difficulty of the cement paste in fully surrounding and coating the irregularly shaped aggregates, and the potential presence of microcracks or voids in this transitional region. The Interfacial Transition Zone is in addition more susceptible to the formation of microcracks due to for example differential shrinkage between the aggregates and the surrounding cement paste. The weaker and more porous nature of the ITZ can lead to a reduced load transfer between the aggregates and the cement paste, resulting in decreased mechanical properties and increased permeability near the interface.

The thickness of the Interfacial Transition Zone (ITZ) in concrete can vary and is influenced by factors such as mix design, curing conditions, type of aggregates and the specific properties of the cementitious materials used. Generally, the ITZ is a few micrometers to a few millimeters thick (Mehta and Monteiro 2006).

The microstructure of the concrete at different scales are illustrated in Fig. 7.4.

Microcracks can develop within the concrete microstructure due to factors such as shrinkage, thermal gradients, and loading conditions. The presence of these microcracks can influence the material's durability, permeability and resistance to external

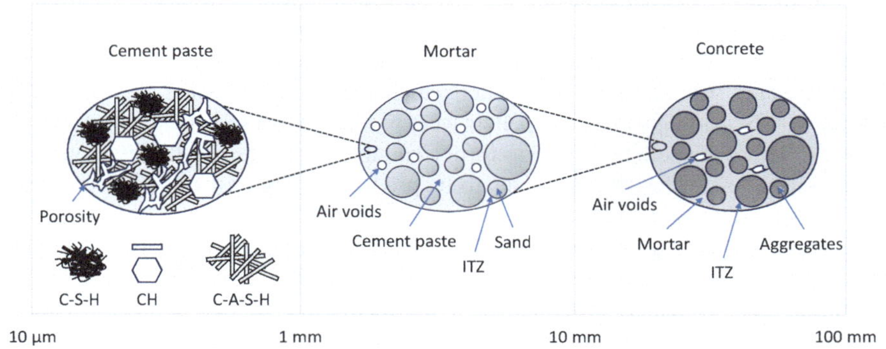

Fig. 7.4 Microstructure of cement and cementitious material at different scales, based on Mehta and Monteiro (2006)

forces. Proper mix design and curing practices aim to minimize the formation of undesirable cracks and defects.

Under deformation, the initial voids and microcracks (that were present before any load was applied) grow through the mechanism of brittle fracture and generate permanent deformations. However, the first stage of deformation is the result of almost reversible movements of atoms and can be considered elastic. Beyond the loads which correspond to the elastic limit, microcracks between the paste and the bigger aggregate grains begin to spread. As a result, permanent deformations occur in conjunction with the elastic ones. For even higher loads, loss of cohesion means that microcracks reach the paste.

Cracks are created in the structure of the grains and thus contribute to the permanent deformation. The microcrack damage become strongly anisotropic. In the final stage of fracture, macroscopic cracks and the final fracture occurs when microcracks join to form a surface discontinuity through the whole component.

7.2.6 Lightweight Aggregate Concrete

High strength lightweight aggregate concrete (LWA) has many advantages for the construction of offshore structures. These include a high strength/weight ratio, good overall durability and resistance to thermal cracking. Light weight aggregates (LWA) are not widely available and hence limit the widespread use of LWA for offshore structures.

Lightweight concrete has been used offshore in several locations. A particular site is the tops of the storage cells to absorb energy from dropped objects and hence minimise damage. It has also been used to reduce the floating draft of offshore structures to allow them to pass over a dry dock cill. Platforms where lightweight concrete has been used include Hibernia, Troll A and B.

Lightweight aggregates are often formed artificially by sintering fly ash (residue from the burning of coal in power stations) mixed with powdered coal or slag from blast furnaces or by sintering clay or shale. The nodules which are formed are graded by size (HSE 1996). Norwegian Contractors (1987) used Liapor aggregate based on clay. The clay was ground to a powder which is then transformed into pellets of similar size by adding water. The pellets were processed in a kiln to a temperature of 1200 °C where the pellets expand to form a suitable aggregate. Different treatments in the kiln allow different grain sizes to be produced, ranging from 4 to 16 mm in size, thus modifying the bulk densities.

Lightweight aggregates have a low-particle relative density because of the cellular pore system. The cellular structure within the particles is normally developed by heating certain raw materials to incipient fusion; at this temperature, gases are evolved within the pyroplastic mass, causing expansion, which is retained upon cooling. Strong, durable, lightweight aggregates contain a uniformly distributed system of pores, developed in a continuous, relatively crack-free, high strength vitreous phase. Pores close to the surface are readily permeable and fill with water within the first few hours of exposure to moisture. Interior pores, however, fill extremely slowly, with many months of submersion required to approach saturation. A small fraction of interior pores are essentially non-interconnected and remain unfilled after years of immersion.

The chemical components of blast furnace slag, fly ash, silica fume and Liapor are shown in Table 7.8 and compared with those in ordinary Portland cement (OPC). It can be seen that the SiO_2 contents in fly ash, blast furnace slag and Liapor are significantly higher than those in OPC. In comparison the CaO_2 content of OPC is much higher than that in both fly ash and blast furnace slag. Silica fume is not surprisingly predominantly SiO_2 (HSE 1996).

The compressive strength of typical high strength LWA is usually determined by the strength of the aggregate rather than by the strength of the cement paste but is unlikely to exceed 75 MPa. The surface of the aggregate is generally rough and provides a good bond with the cement matrix, enhanced by the penetration of the cement into the aggregate. Voids within the aggregate also allow water to penetrate and to be retained within the aggregate particles. This absorbed water benefits the curing process but can lead to damage to the concrete under fire loading or under freeze–thaw conditions.

Electron micrographs of hardened LWA show LWA particles well bonded to the cement paste matrix (Department of Energy 1990). The boundary between the two materials was difficult to identify, compared to the abrupt transition that usually occurs between normal density aggregates and the cement matrix. There is also evidence that there is a chemical bond due to new substances being formed in LWA from reactions between the cement and the aggregate (Department of Energy 1990).

The modulus of elasticity of LWA is slightly lower (Table 7.9) so that minimal strain incompatibilities arise when load is applied to the interface between the aggregate and the cement matrix. This compares with normal density concrete where the aggregates normally have a higher modulus. This means that normal density concretes often fail under load around the aggregates because of micro-cracking, whilst LWA should have fewer problems in this respect. Table 7.9 compares the properties of both LWA concrete based on Liapor and normal density concrete. As can be seen the density of the LWA concrete is 23% less than that of comparative normal density concrete.

Table 7.9 Properties of LWA and normal density concrete (Norwegian Contractors 1987)

Concrete properties	LWA concrete (Liapor)	Normal density concretes
Cement	410 kg/m^3	400 kg/m^3
Density of fresh concrete	1.88 kg/m^3	2.43 kg/m^3
Water-cement ratio	0.4–0.42	0.4–0.42
Compressive strength at 28 days	6–70 MPa	60–70 Mpa
Modulus of elasticity	23,500 MPa	29,000 Mpa
Permeability coefficient	10^{-13} to 10^{-14} m/sec	10^{-13} to 10^{-14} m/sec

Tests have been carried out using an impact layer (fender layer) of low strength LWA on top of normal density concrete demonstrating significantly reduced damage to the normal concrete from simulated dropped objects (e.g. drill piles) by a factor of almost four (Department of Energy 1989). This impact layer is often used to absorb energy from a dropped object and minimising damage on the tops of storage cells. It was found that the lightweight concrete fender layer prevented the scabbing and severe damage which occurred in the un-fendered layer. The fender was shown to be effective in reducing peak impact loads and spreading the load over a larger area of the structural concrete (Department of Energy 1989). It is important that the fender should have low strength and be thick enough to stop the dropped object before it penetrates the surface of the structural concrete.

The permeability of LWC is expected to be low compared to normal density concretes, because of the difference in microstructure due to the bonding between the cement mortar and the aggregates. Specimens immersed in seawater has been tested by several authors (HSE 1996), confirming this low permeability of LWA. However, Table 7.9 indicates that LWA concrete based on Liapor has a similar permeability to OPC. In addition, water penetration tests were carried out on the LWA used for the Heidrun platform. Results indicated an average penetration of 10 mm over a 72 h immersion period (HSE 1996).

Chloride penetration is a major issue for reinforced concrete, as further described in Sect. 8.1.4. Some tests have shown diffusion coefficients which were low compared to normal density concrete (HSE 1996). However, there is some divergence between laboratory test results, compared to field exposure trials (HSE 1996). Test on chloride penetration in LWA using Lytag as the aggregate showed little difference between LWA and normal density concretes after a period of 90 days. However, it was pointed out that longer exposure periods may have shown a larger difference (Department of Energy 1990).

7.2.7 UK Concrete in the Ocean Programme

The UK Concrete in the Oceans programme (Department of Energy 1987a) involved tests on 3 types of concrete, as shown in Table 7.10. All grades were cured in either air, fog or seawater, all for a period of either 28 or 42 days. It was found that the curing regime had considerable effects on the subsequent permeability and chloride penetration.

Within the UK regulation in 1977, the Guidance Notes developed requirements for cement and other components. The cement used should be one of the types listed in Table 7.11, according to the relevant British Standards listed. In addition, latent hydraulic binders such as ground granulated blast furnace slag, pfa or silica fume may be separately combined with ordinary Portland cement in the site mixer. For tricalcium aluminate the C3A content should be not less than 5% and not more than 12% for ordinary and rapid hardening Portland cements.

Table 7.10 Concrete mixes used in UK concrete in the oceans programme (Department of Energy 1987a)

Offshore grade	Cement content Kg m^{-3}	Pfa content Kg m^{-3}	w/c ratio	28-day cube strength N mm^{-2}
Standard	435	0	0.45	70
Lower	305	0	0.7	35
Pfa mix	348	87		70

Table 7.11 Types of cement recommended in the UK guidance notes (Department of Energy 1984)

Type	Standard
Rapid hardening Portland cement	BS 12
Ordinary Portland cement	BS 12
Portland blast furnace cement	BS 146
Low heat Portland blast furnace cement	BS 4246
Portland Pfa cement	BS 6588
Pozzolanic cement with pfa as pozzolan	BS 6610

7.2.8 Concrete Durability

The concrete is generally of a high quality, low permeability to minimise ingress of seawater during the structure's life, as the chlorides can lead to corrosion of the reinforcement. Lightweight concrete (less dense and more absorbent to impact) has also been used to minimise the effect of impact damage on the tops of cells, see Sect. 7.2.6.

To meet durability requirements offshore concrete is generally very high quality, high strength and with features including:

- low water/cement ratio;
- thick covers to the outer layer of reinforcement (typically 70 mm in the splash zone, 45 mm underwater);
- limited permeability to seawater; and
- post tensioned to limit cracking

Early concrete GBS structures had high quality concrete 28 day strengths in the region of 50 N/mm^2. It was found that the addition of small amounts of pozzolanic material such as fly ash (Pfa) or blast furnace slag increased the 28 day strength and also reduced permeability to seawater. Pozzolanic materials in a finely divided form and in the presence of moisture, can chemically react with the calcium hydroxide in cement to form compounds which have cementitious properties, see Sect. 7.2.3. Later offshore GBS structures were built with the addition of pozzolanic materials giving improved properties such as a 28 day strength up to 70–80 N/mm^2. The properties of pfa additions to concrete were investigated in both the Concrete in the

Oceans Programme (Department of Energy 1987a) and in the Concrete Offshore in the Nineties programme (Department of Energy 1990), discussed later.

As noted earlier, the constituent materials for structural concrete are reinforcement, cement, aggregate and water. Admixtures and additions may also be included. Approval of the constituents and reinforcements should be based on material testing where chemical composition, mechanical properties and other specified requirements are tested and checked against applicable international standards and approved specifications.

7.3 Reinforcement and Prestressing Tendons

7.3.1 Reinforcement

As concrete material is weak in tension and shear, reinforcing steel is added to absorb these stresses. In addition, reinforcing steel is also sometimes added to absorb excessive compressive stresses. The amount of reinforcing steel in offshore concrete structures ranges from 5300 tonnes in Frigg MP1 to 80,000 tonnes for Gullfaks C (see Table 7.1).

Steel reinforcement is available in the form of plain steel bars, deformed steel bars, cold-drawn wire, welded wire fabric and deformed welded wire fabric. Typically, offshore concrete structures include deformed bars which are round steel bars with lugs. These lugs or deformations are rolled into the surface of the bar during manufacturing. These deformations create a mechanical bond between the concrete and steel. Steel reinforcement needs to be placed with a sufficient cover to avoid corrosion, which is one of the main degradation mechanisms for offshore structures, see Sect. 8.1.5.

Reinforcing steel need to conform to applicable standards, such as BS 4449 (British Standards 1997) and ASTM A615 (ASTM 2024).

The actual properties of the reinforcing steel may be difficult to obtain when an offshore concrete structure is being assessed. As an example, it is clearly indicated in the re-analysis report of Frigg MCP from 1994, the actual reinforcement used during the fabrication was difficult (or impossible) to obtain. In the Frigg MCP re-analysis they assumed ASTM A615 grade 60, which offers a yield strength of at least 420 MPa (Table 7.12).

Table 7.12 Change in reinforcement quality with time

Period	Typical reinforcement quality	Comments and reference
1970s	KS40, $f_y = 400$ MPa ASTM 615 grade 60, $f_y = 420$ MPa	Frigg TCP DFI resume (ELF Aquitaine Norge 1977)
1990s	B500A-C, $f_y = 500$ MPa	BS4449 (British Standards 1997)
2010s	Grade 500W, $f_y = 500$ MPa	Widianto et al. (2016)

According to Widianto et al. (2016), the Hebron platform used Grade 500W weldable deformed bars per the requirements of relevant Canadian standards (CSA A23.1-09) for reinforcement. T-headed bars were used to improve anchorage, to eliminate hooks and reduce congestion. The minimum concrete cover for steel reinforcement (including stirrups) was 50 mm in the splash zone and 40 mm in the submerged zone. Prestressing strands in the Hebron platform conformed to ASTM A416/A416M (ASTM 2014).

7.3.2 Prestressing Tendons

The use of pre-stressing tendons is fundamental for the design of concrete gravity-based structures. Buoyancy conditions at float-out require the structure to be as light as possible, limiting the use of heavier reinforced concrete. Another important factor is that prestressing is required to prevent cracking in the concrete, which helps to protect the steel reinforcement from corrosion (particularly in the splash zone) and also to keep the structure watertight. The pre-stressing tendons are very high strength steel, usually in the form of strands with yields strengths above 1000 N/mm^2. As such, the material is susceptible to hydrogen induced stress corrosion cracking (SCC) and hydrogen embrittlement (HE), and considerable care is needed in the protection of the strands prior to installation in the structure. The strands are normally embedded in steel ducts and subsequently grouted to provide additional protection.

Prestressing of concrete structures by steel tendons under tension is used to place the larger part of the cross-section of concrete into permanent compression. This minimizes tensile stresses which can cause concrete to crack under externally applied loads, creep and temperature actions. Pre-tensioning (where the tension is applied prior to the setting of the concrete) and post-tensioning (applied after the surrounding concrete has set and hardened) are used. In offshore structures, post-tensioning is the main technique.

As already noted, prestressing tendons are usually installed in pre-formed ducts in the concrete framework and tensioned after the concrete is set and hardened. The ducts are subsequently grouted using Portland cement grouts and the tendons are left in a stressed state while grout is setting. The grout acts to ensure load transfer and improve corrosion protection. Tendon lengths in offshore concrete structures can reach up to 150 m in horizontal or circumferential ducts and between 30 and 130 m in vertical ducts.

Very high strength steels are normally used for the tendons, as significant forces are necessary to put massive cross-sections of concrete into the required degree of compression. A survey of prestressing systems and tendons used in offshore structures was reported by Burdekin and Rothwell (Department of Energy 1980).

Most tendons selected for use in offshore structures were based on British Standards, specified in either CP110 (British Standards 1972) or those included in the first edition of Department Energy Guidance Notes (Department of Energy 1974) as shown in Table 7.13, together with the characteristic yield strength requirements. For the size of tendons used in actual offshore platforms it can be seen from Table 7.13 that the characteristic strength was in the range from 1500 to 2000 N/mm^2.

The actual offshore platforms included in the survey of prestressing systems by Burdekin and Rothwell (Department of Energy 1980) are shown in Table 7.14. Most of the pre-stressing systems used proprietary tensioning methods.

Concrete gravity base structures (GBS) tend to have a large base area and one or more slender towers to support the deck. In the base a number of the structures have horizontal prestressing tendons either in a grid pattern or a radial pattern applied as post-tensioning systems.

In the towers the prestressing tendons are generally vertical. The tendons run for varying lengths with overlaps and termination of the tendons as required by the design criteria. As the towers are subject to significant bending loads, the prestressing systems are part of the structure with the aim to improve the effectiveness of the concrete material. In some of the structures there are also concrete ring beams at different levels on the tower involving horizontal circumferential prestressing tendons.

As noted earlier, typical tendon lengths are up to 150 m in horizontal or circumferential ducts and from 30 to 130 m in vertical ducts. The lengths of vertical tendons used in these offshore structures are generally greater than those for land-based structures. These lengths present difficulties in the grouting process, which are discussed in Sect. 7.5.

Normal practice has been for the tendons to be designed for stressing to 70% UTS (ultimate tensile stress). Allowance for relaxation and creep effects leads to final loadings of about 65% UTS. Since prestressing is designed for maximum storm wave conditions from any direction there is generally an appreciable level of redundancy in the design of the prestressing systems. The same general range of proprietary

Table 7.13 British Standards and their characteristic strength requirements

British Standard	Type of tendon	Type of steel	Characteristic strength requirements
BS 2691:1969 as specified in CP110 and Department of Energy Guidance Notes	Wire 2-7 mm diameter	Cold drawn carbon steel 0.6–0.19% C	2 mm 2020 N/mm^2 3–4 mm 1720 N/mm^2 5–7 mm 1570 N/mm^2
BS 3617:1971 as specified in CP110 and Department of Energy Guidance Notes	7 wire strand 6.4–15.2 mm diameter	Cold drawn carbon steel 0.6–0.19% C	6.4 mm 1816 N/mm^2 10.9 mm 1760 N/mm^2 15.2 mm 1636 N/mm^2
BS 4757:1971 as specified in CP110 and Department of Energy Guidance Notes	19-wire strand 18–31.8 mm diameter	Cold drawn carbon steel 0.6–0.19% C	18 mm 1720 N/mm^2 25.4 mm 1550 N/mm^2 31.8 mm 1483 N/mm^2

Table 7.14 Types of prestressing tendons used in offshore concrete structures (Department of Energy 1980)

Platform	Year installed	Type of tensioning	Weight tonnes	Types of prestressing system used
Ekofisk	1973	Freyssinet	3300	12 × 13 mm tendons 24 × 13 mm tendons
Beryl	1975	Freyssinet	820	12 × 13 mm tendons
Brent A	1975	Freyssinet	1270	12 × 13 mm tendons
Frigg–CDP-1	1975	Freyssinet	2600	12 × 15 mm tendons
Dunlin	1976	Freyssinet	2200	12 × 13 mm tendons
Frigg-MP2	1976	Freyssinet	2600	12 × 15 mm tendons 24 × 15 mm tendons
Frigg -TCP-2	1977	Freyssinet	533	12 × 15 mm tendons, f_y = 1566 MPa
Ninian	1977	Freyssinet	4000	12 × 15 mm tendons
Brent C	1978	CCL	1400	7 × 18 mm Dyform tendons 15 × 18 mm, Dyform tendons
Cormorant	1978	BBRV, CCL	1100	BBRV 85 × 7 mm wires CCL 7 × 18 mm tendons

prestressing systems and materials successfully used in land-based structures has also been used for offshore platforms.

The tendons are made of wire rope in various configurations. Wire rope is composed of metal wires twisted into a helix that forms a composite rope. Larger diameter wire rope consists of multiple strands, as shown in Fig. 7.5. In general, multiple or strand wire cables have been used as tendons.

As previously mentioned, the study by Burdekin and Rothwell (Department of Energy 1980) indicated that the proprietary Freyssinet tensioning system was used in

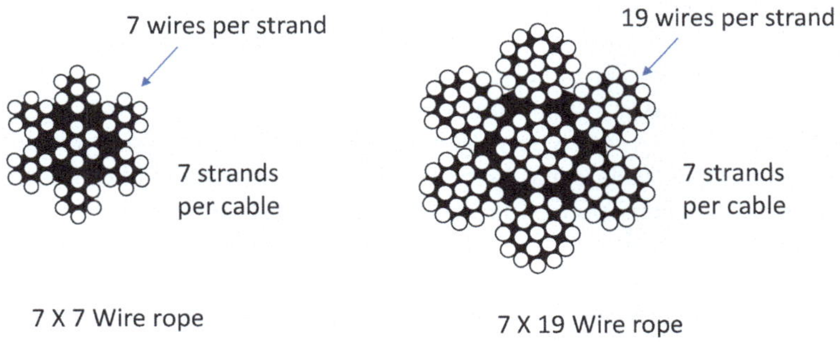

Fig. 7.5 Wire ropes

many of the concrete offshore platforms in the North Sea. In addition, several other offshore concrete platforms have used the Freyssinet system, such as the "Hibernia" in Canada and the Petrobras platform "Brazil" in El Salvador. In the Freyssinet system, high strength steel wires (typical yield strength of 1500 N/mm^2) of 5 mm or 7 mm diameter are grouped (7, 8, 12, 16, 19 or 24) into a strand with a helical spring inside. The spring keeps a proper spacing for the strands. Anchorage devices consist of concrete cylinders with concentric conical holes and corrugations on their surfaces, and a conical plug carrying grooves on its surface. The individual wires are separated and are carried along these grooves at the ends.

As noted earlier, prestressing cables are located in ducts of which there are several different types. Corrugated spiral sheet ducting has been commonly used in diameter 75 to 150 mm but for vertical tendons where the main tower has been slip-formed the ducts have usually been in the form of mild steel tubes, again of diameter 75 ml to 150 mm but of wall thickness 3 to 4.5 mm. The concrete cover to ducts is typically a minimum of 75 mm in the submerged zone and 100 mm in the splash and atmospheric zones.

Anchorages are used to allow the prestressing strands to be terminated and tensioned. The main function of an anchorage is to transfer the prestressing force to the concrete once the stressing process is completed. These usually consist of an anchorage plate which bears onto the concrete over a relatively small area. The tendon is connected to the plate either through wedges or other methods. Anchorages are usually located in recesses which are filled after the tendons have been stressed. The discontinuity between the surrounding concrete and the fill can provide an area of weakness allowing chlorides to penetrate. A successful approach has been to prime the recess surface with a suitable epoxy bonding compound and to use a small aggregate concrete for the fill, followed by an epoxy coating to the concrete surface (HSE 1997b).

As noted above the typical strength of prestressing stands has been in the range from 1500 to 1600 N/mm^2. Such high strength steels are generally susceptible to corrosion in the presence of water and oxygen, and these steels are also sometimes susceptible to stress corrosion if stressed in tension in an unfavourable environment. Stress-corrosion failure occurs by the slow growth of cracks until a critical crack size is reached, overload fracture of the remaining steel cross-section then takes place. Susceptibility to stress corrosion cracking in steels depends upon the type of steel, the stress level, tie presence of defects or stress concentrations and the environment. Stress corrosion cracking of high strength steels is covered in Sect. 6.9.

The difference between this degradation mechanism in prestressing tendons with normal high strength steels is that the former are enclosed in grout filled ducts which should provide a degree of protection from the highly alkaline environment. However, it is accepted that the filling of the ducts is not complete due to the difficulty of grouting long ducts, particularly horizontal ones. Ingress of seawater into the ducts may occur potentially causing local corrosion and local failure of the strand. However, due the bond between the grout and the tendons this loss of local prestress is unlikely to cause wider loss and place the structure at risk. A vulnerable location is the anchorage where the tendons are terminated at the surface. Some protection may be available through the external cathodic protection system. However, some instances of corrosion at the anchorages have been observed requiring repair.

Due to the effect of damaging corrosion, careful storage of prestressing strands on site before installation is necessary including protection from weather and seawater. Protection by coating with a soluble oil is also used.

7.4 Design Requirement for Concrete in Offshore Structures

7.4.1 Historic

Design requirements for concrete offshore structures have developed as the number of installed structures has increased. The early installations were designed to land based codes such as ACI 318-74 (ACI 1974), NS 3473 (Standard Norway 1973) and BS CP110 (British Standards 1972). In addition, FIP Recommendations (FIP 1972, 1974) and DNV standards (DNV 1977) were used. The codes and standards used for particular structures are shown in Table 7.15.

The choice of specified grades of concrete depends on the environmental conditions to which the concrete is to be subjected, see Table 7.16. The submerged and splash zone sections are the most vulnerable, demanding concrete grades of either 40 or 50. The atmospheric zone is less critical and grade 30 is deemed sufficient. In addition, for concrete in the submerged, splash zones a minimum cement content of 400 kg/m^3 should be used and the free water/cement ratio should be 0.4 or less.

Table 7.15 Codes and standards used for particular structures

Structure	Codes and standards used
Frigg TCP2 (1977)	NS 3473, NS 3474 (Standard Norway 1976), DNV (1977)
Ninian Central (1978)	ACI 318, BS CP110, FIP (1974)
Dunlin A (1977), Cormorant A (1978)	BS CP110, FIP (1974), DNV (1977)
Brent C (1978)	BS CP110, FIP (1974)

Table 7.16 Minimum grades of concrete (typical values)

Zone	Exposure condition	Minimum grade
Submerged	Directly exposed to seawater	40
	Directly exposed to crude oil or subject to severe abrasion	50
Splash	Directly exposed to salt water or salt water spray	40
Atmospheric	Directly exposed to marine atmosphere	40
	Protected from direct exposure to marine atmosphere	30

Table 7.17 Specified cover to the reinforcement

Zone	Exposure condition	Minimum cover mm		
		~ 1975	~ 1990	~ 2020
Submerged	Directly exposed to seawater	60[a,b], 50[d]	70[c,e]	50[f]
Splash	Directly exposed to salt water or salt water spray	75[a,b], 65[d]	60[c]	60[f]
Atmospheric	Directly exposed to marine atmosphere	75[a,b], 50[d]	60[c], Ø—5[e]	50[f]
Prestressing tendons				100[f]

[a]Department of Energy 1st ed Guidance Notes (Department of Energy 1974)
[b]FiP 3rd edition (FIP 1975)
[c]BS8110 (British Standards 1985a, b)
[d]ACI 318 (ACI 1974)
[e]NS3473 (Standards Norway 1973)
[f]ISO 19903 (ISO 2006)

Several requirements were made in even early codes and standards regarding materials. These include that seawater should not be used as a mixing or curing water because of the potential hazard of reinforcement corrosion due to presence of chlorides. With respect to aggregates and a possible alkali silica reaction, it has been noted that some aggregates may be susceptible to deleterious reactions with alkalis present in the cement. Appropriate measures should be taken to avoid or minimise this problem. Limits on the level of C_3A were also introduced to minimise the risk from reactions with sulphates, see Sect. 7.2.2.

The specified concrete cover to the reinforcement is another important parameter, again depending on the environmental conditions. Typical values are given in Table 7.17. It can be seen that there is some consistency between the different codes and standards, with the splash zone requiring the largest cover.

7.4.2 Department of Energy Guidance Notes

In 1974, the UK Department of Energy published the first edition Guidance Notes (Department of Energy 1974) to support the Certification legislation. This was very basic but did cover both steel and concrete materials. The second issue (Department of Energy 1977) extended the guidance and the third edition (Department of Energy 1984) further developed the guidance. By 1984, approximately 10 concrete platforms had been installed in the UK sector of the North Sea.

The concrete section in the third edition (Department of Energy 1984) was based heavily on land-based practice supplemented by information from FIP (1975). The fourth edition of the Guidance Notes (HSE 1994) was the last and took account of research from the Concrete in the Oceans programme (Department of Energy 1987a, b) and other sources. The issues addressed in developing the fourth edition were:

- the inclusion of hydraulic binders (e.g. pulverised flue ash—pfa) with the cement,
- the tri-calcium aluminate (C_3A) level,
- the possibility of alkali-silicate with the use of silicate materials in the aggregate.

The addition of pulverised fuel ash (pfa) to the cement had been introduced into some offshore structures due to its ability to reduce permeability and hence limit chloride ingress. A pfa mix was tested in the Concrete in the Oceans programme and its characteristics are discussed in Sect. 7.2.3. There was clear evidence that the resistivity of hardened concrete was significantly higher when a 20% replacement of pfa for cement was used. However, there was no clear evidence that this led to improved resistance to corrosion.

C_3A is a compound in hydrated Portland cement which is susceptible to reaction with sulphates. This led to the development of sulphate resisting cements. This reaction in severe cases can lead to disintegration of the concrete. The sulphate content of seawater is about 250 parts per 100,000 which is recognised to be at the lower level of severity regarding the potential problem with sulphates.

The level of C_3A in the cement was considered in the fourth edition and the 12% maximum level in previous guidance notes retained but with a lower limit of 5% introduced. The argument for a lower limit is that the resistance to chloride penetration in hardened concrete is influenced by the C_3A level. As noted in Sect. 7.2.2, the benefit of a certain level of C_3A in the cement arises from the removal of free chlorides in the pore solution by a binding process for the chlorides penetrating through the cover involving C_3A, which has a positive effect on corrosion resistance.

The potential ASR reaction was also addressed in the fourth edition guidance notes pointing out that the choice of aggregates is important. A reaction between alkalis in Portland cement and forms of silica (ASR reaction), which may be present in certain types of aggregate, can lead to the formation of a gel which absorbs water. The resulting progressive expansive reaction can crack the concrete. There is no proven effective remedy once ASR has begun in the concrete and hence it should be avoided preferably by choice of aggregate. It is understood that most offshore structures were built with granite-based aggregates, rather than those containing silicates. Hence the alkali reaction is unlikely to be a problem with offshore structures.

Regarding concrete mix the fourth edition recommends a C40 grade (a C40 grade has a 28-day strength of 40 N/mm^2) for all areas exposed to seawater with a C50 mix for concrete exposed to crude oil or subject to severe abrasion.

7.4.3 British Standards on Concrete Structures

BS CP110, the structural use of concrete, was published in 1972 (British Standards 1972). It was described as "The Unified Code" bringing together the separate codes relating to reinforced, prestressed and precast concrete. This code dealt with the design, materials and construction of buildings and structures in reinforced concrete, prestressed concrete and precast concrete. It was written in limit state terms. Guidance was given on the specification and workmanship of concrete, the inspection and testing of structures. Design in CP 110 was based on specified concrete grades such as 30 N/mm^2 for post-tensioned prestressed concrete. The code was withdrawn in 1985.

This was followed by BS 8110 (BS 1985b) which became the main Code of Practice for the design of reinforced or prestressed concrete structures. This standard was divided into two parts, namely:

- Part 1, Code of practice for design and construction, was intended for general use.
- Part 2, Code of practice for special circumstances gave more detailed design approaches.

In 2004, BS 8110 was replaced by Eurocode EN 1992 (EN 2004) which is now the recommended code for design of concrete structures. However, for offshore concrete structures, ISO 19903 (ISO 2006) is the specific design code since its introduction in 2006.

It would be reasonable to assume that the UK platforms designed prior to 1985 were designed making use of CP110. Particularly, it is stated that CP110 was one of design codes used for the Cormorant A platform in 1978. Later UK platforms would have used BS8110.

7.4.4 International Federation for Structural Concrete

The Comité Européen du béton (CEB), or in English; European Committee for Concrete was established in 1953 and was a not-for-profit association committed to advancing the use of concrete structures. Similarly, Fédération Internationale de la Précontrainte (FIP), or in English; International Federation for Prestressing, was established in 1952. These organizations played an important role in stimulating research and developing guidance documents for the design, construction and development of concrete. These two organisations merged in 1998 to create Fédération Internationale de béton (fib).

FIP's first edition of "Recommendations for the design and construction of concrete sea structures' was issued in 1972. The second and third editions followed quickly after in 1974 and 1975, reflecting the growth in marine structures in this period. In 1978 it was decided not to concentrate on continuously updating the FIP Recommendations but to concentrate on producing in-depth background information concerning important aspects of offshore concrete structures. These included:

- Foundations of concrete gravity structures in the North Sea (1978)
- Cover to steel reinforcement for floating concrete structures (1982)
- Sea Operations (1982)
- Permanent anchorages for offshore structures (1982)
- Inspection, maintenance and repair of concrete sea structures (1982).

In 1981 FIP decided to review the third edition and a fourth edition was produced and published in 1985 (FIP 1985). It is noted that the purpose of the recommendations is to provide general guidelines on the design of concrete sea structures. They are not intended to take the place of existing statutory regulations. The above background documents are still available, but now from Fédération Internationale de béton (fib).

The FIP Recommendations were used in the design of Frigg MCP 01 and Cormorant platforms.

7.4.5 ISO 19903

The current international standard for offshore concrete structures is ISO 19903 (ISO 2006). This standard was to a large extent based on the Norwegian Standard NS 3473 (Standard Norway 1998) and lists the materials requirements and properties. With regard to cement, it states that only cement shall be used with established suitability with a demonstrated track record for good performance and durability in marine environments, and for good performance and durability in exposure to stored oil or other fluids. The standard also indicates that the tricalcium aluminates (C3A) content should preferably not exceed 10%. However, as the corrosion protection of embedded steel is adversely affected by a low C3A content, it is not advisable to aim for values lower than approximately 5%, see Sect. 7.2.2.

The two main types of cement listed in the standard are:

- Portland cements
- Portland composite cements (with silica, fly ash or slag and minimum 80% clinker.

For the mixing water ISO 19903 states that it should not contain constituents in quantities that can be detrimental to the setting, hardening and durability of the concrete or can cause corrosion to the reinforcement. In this respect salt water (e.g. raw sea water) should not be used as mixing or curing water for structural concrete.

In respect of aggregates, they should have particle sizes, grading and shapes such that they are suitable for the production of concrete. Aggregate should be free from harmful substances in quantities that can affect the properties and the durability of the concrete adversely. There is a possibility of alkali aggregate reaction with certain aggregates and they should be evaluated according to recognized test methods.

Latent hydraulic or pozzolanic materials such as silica fume, fly ash and granulated blast furnace slag may be used as additions (see Sect. 7.2.3). They can help to reduce the permeability of the hardened concrete. The amount is dependent on requirements to workability of fresh concrete and required properties of hardened concrete. ISO 19903 states that when fly ash or other pozzolanas are used as additions, their content should normally not exceed 35% of the total weight of cement and additions.

Admixtures are used to improve several properties (see Sect. 7.2.4). The standard indicates that only those with established suitability shall be used and should be tested under site conditions with the cement and additions to be used to verify that these products will yield the required effects without impairing the other properties required. To improve frost resistance air-entraining admixtures may be used to improve the properties of hardened concrete or to reduce the tendency of bleeding, segregation or plastic cracking.

In terms of the concrete mix ISO 19903 lists requirements for the water-cement ratio which is very important for durability. The concrete should have an effective water/cement ratio not greater than 0.45. In the splash zone this ratio should not be higher than 0.40 and this ratio should also apply to concretes intended for areas that can be exposed to severe frost action (e.g. artic regions). Where pozzolanic or latent hydraulic additions are intended to be used in the production of concrete, in combination with Portland cement these materials may be included in the calculation of an effective water/cement ratio using a formula given in ISO 19903. This includes an efficiency factor for active additions such as silica, fly ash and slag, where the efficiency of the material is considered as a replacement of cement with respect to durability in the given environment. For example, for fly ash this factor varies from 0.2 to 0.4.

The cement content of the concrete is important for durability. ISO 19903 states that in the splash zone, the cement content should not be less than 400 kg/m^3. For reinforced or prestressed concrete outside the splash zone, the minimum cement content should be dependent on the maximum size of the aggregate, as follows:

- a minimum cement content of 360 kg/m^3 for aggregates up to 20 mm
- a minimum cement content of 320 kg/m^3 for aggregates between 20 and 40 mm in size.

The chloride content of the concrete is important with respect to corrosion of the reinforcing steel. ISO 19903 states that the total chloride ion content of the fresh concrete should not exceed 0,10% of the weight of cement in either ordinary reinforced concrete and in any concrete planned to contain prestressing steel.

The final strength of the concrete is also important for durability and ISO 19903 states that for concrete exposed to sea water and stored oil, the characteristic cylinder strength at 28 days should not be less than 40 MPa. In practice higher values have been used.

For the reinforcing steel it is recommended that it complies with ISO 6935 (ISO 2007) or relevant national standards. In general, hot-rolled, ribbed bars of weldable quality and with high ductility should be used.

For prestressing steel, it is recommended that it should comply with the relevant parts of ISO 6934 (ISO 2020) or any relevant national standards. The ducts for the prestressing tendons should be of steel unless other types are specified by design and should be watertight and with adequate stiffness to prevent damages and deformations.

7.5 Grout Materials for Offshore Structures

7.5.1 Introduction

A fundamental property of a structural grout is that it can be poured or pumped to voids in a structure to provide a structural component. Typical applications are grouted connections, pile sleeve connections, grouted repairs, grout strengthening of tubulars, underbase grouting for concrete installations, protection of prestressing tendons in a concrete structure and downhole drilling. Grouts used offshore fall into three main categories:

- Portland cement grouts
- Expansive Portland cement grouts
- High alumina cement grouts (HAC).

Both of the first two may be used with various additives to improve rate of gain of strength and workability, fillers can be used to reduce density and some grouts can include a significant percentage of sand. The American Petroleum Institute has nine classes of oil well cement. Classes A–C have a low C3A content and no retarder. A typical class used offshore in structural applications is Class B. The key features of these grouts are the following:

- Constitution
- Workability
- Time for setting
- Heat of hydration
- Susceptibility to movement during setting (early age cycling)
- Strength and durability in hardened state.

7.5.2 Constitution and Properties

Most grouts consist of ordinary Portland cement mixed with water and with possible additives. The chemical constitution of ordinary Portland cement is described in Sect. 7.2.2. Comparison with Oilwell B grout is shown in Table 7.18. As can be seen there are few differences. Typical grouts used offshore are oilwell types originally designed for downhole drilling purposes.

Curves showing increasing strength with time at a temperature of 8 °C (typical North Sea temperature at depth) for grouts from each of the three categories are shown in Fig. 7.6, which also demonstrates the effect of adding a small percentage of calcium chloride to the Portland cement mix. The early gain in strength performance of the high alumina concrete (HAC) grout is remarkable and enables it to be used where early strength is required (e.g. pin piles).

Shrinkage or expansion of the grout can have a significant effect on bond strength. Tests on grout prisms indicate that both HAC and Portland cement grouts can be

Table 7.18 Oxide compositions of Oilwell B and ordinary Portland cement (HSE 1992)

Oxide	Percentage composition	
	Oilwell B	Ordinary Portland cement
CaO	64.9	65
SiO_2	21.5	19.9
Al_2O_3	3.9	5.2
Fe_2O_3	5.1	3.3
MgO	0.8	1.2
SO_3	2.3	2.6
K_2O	0.4	0.6
Na_2O	0.1	0.2
Other	4.4	6.4

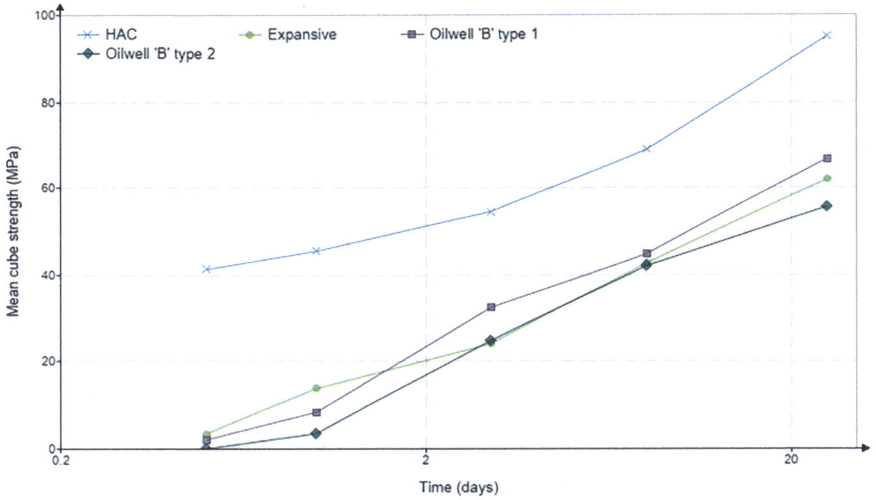

Fig. 7.6 Variation of grout cube strength with time for four grouts, where Oilwell 'B' type 1 includes 1.0% $CaCl_2$ on cement and Oilwell 'B' type 1 includes no $CaCl_2$ (based on Billington and Lewis 1978)

expected to shrink by approximately 0.2%, whereas a typical proprietary post-setting expansive grout might expand by the same amount.

The other property of interest to users is the elastic modulus which is dependent upon age and strength. Elastic moduli plotted against compressive strength for three different grouts is shown in Fig. 7.7. The initial lag of development of stiffness with strength for HAC is noticeable, this effect being common in rapid hardening grouts and concretes.

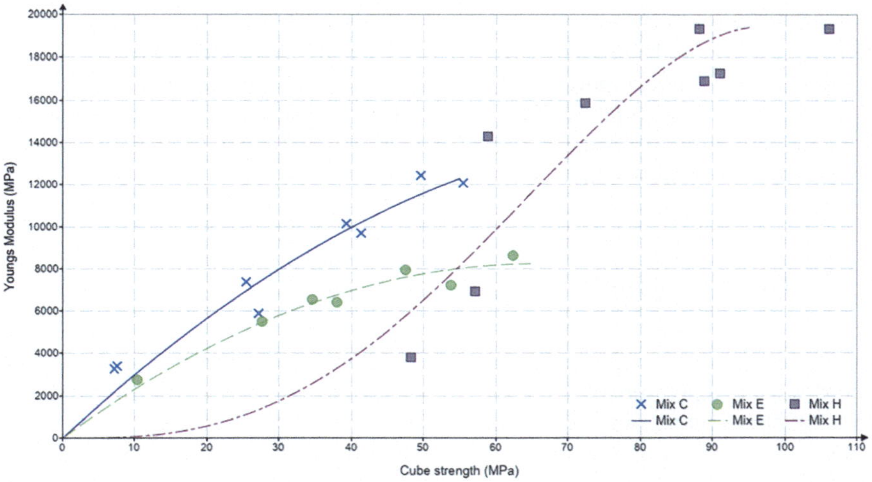

Fig. 7.7 (Y) Variation of elastic modulus with compressive strength for three different grouts (based on Billington and Lewis 1978)

7.5.3 *Standards*

In ISO 19902 (ISO 2008a) it is specified that high sulphate-resisting Portland cement or API standard oilwell cement grouts, mixed with fresh water should be used for grouting of steel structures. It is strongly recommended that sea water shall not be used in cement and grout mixes due to chemical attacks, potential corrosion and other potential adverse durability effects. Water should be free of hydrocarbons and other deleterious matter.

In the composition of the grout mix, the water/cement ratio should not exceed 0.40. It is stated that high alumina cement (HAC) grout shall only be used with extreme care and restricted to below splash zone applications in order to reduce conversion and associated strength loss.

Admixtures may be used to improve the grout slurry properties provided it can be demonstrated that they do not adversely affect the required grout durability and performance.

7.5.4 *Onshore and Offshore Grout Trials*

Onshore trials are required to confirm the compressive strength, unless previous production data was available (ISO 2008a). These trials should consist of at least three separate batches of grout, with a minimum of six cubes being cast from each batch for each age at which strength is to be determined. For strength compliance a compressive strength of 75 mm cubes at 28 days should be used. It is also recommended that strength gain should be measured typically at 30 h, at 3 days and at 7 days.

The grout mixing equipment and mixing times used in the onshore trials should be the same as those which were to be used during offshore grouting to ensure the grout slurry is similarly sheared. Grouting shall not proceed until compliance with the test results has been achieved.

7.5.5 Offshore Quality Control

Specific gravity measurements of the grout slurry prior to pumping also need to be recorded using a pressurised fluid density balance in accordance with ISO 10414-1 (ISO 2008b) or monitoring devices calibrated to this type of balance.

During the grouting operation samples shall be taken from randomly selected batches. The cubes should be cast and tested to a recognized standard and match cured to the in-place grout environmental conditions. At least one cube should be tested from each sample to confirm compliance after 28 days.

7.5.6 Materials Testing of Grout

The relevant properties of grouts are:

- Compressive strength
- Density
- Viscosity
- Fluidity (flow)
- Bleed.

Different tests are available to test these properties which are described below. Some tests are available in the field whilst others require laboratory facilities.

The compressive strength of grout is usually measured by tests on either cubes, prisms or cylinders prepared in accordance with for example EN 196-1 (EN 2016). Moulds are filled with grout which are then cured underwater at a temperature similar to that of the offshore environment. The cubes are removed from the curing tank, dried and grit removed. The cubes are tested using a calibrated compression machine. Typical specifications may require the strength of the grout to be at least 20 MPa at 7 days and 30 MPa after 28 days (Domone and Jefferis 1994). The mean strength of the cubes tested at any age should exceed the specified characteristic strength by a margin depending on the number of trial results.

To provide a continuous record of the density of the slurry whilst being mixed and pumped offshore the mixing plant is fitted with densitometers (Domone and Jefferis 1994). The most commonly used densitometer works on the principle of the attenuation of a beam of gamma radiation passing though the grout in the delivery pipe. The degree of attenuation is related to density by calibration. Offshore there is an additional full-scale trial mix carried out to calibrate the densitometers placed on the mixing plant. This calibration is made with a range of densities either side of the specified design value. The density is also measured using a pressurised slurry density balance which measures weight consisting of a fixed-volume cup on one end of a graduated beam with a counterweight on the other end.

A rotational viscometer consists of concentric cylinders. It is designed to enable viscosities to be measured together with yield and gel strength (the yield stress after a period of resting). However, the method has limitation for stiff grouts. An alternative viscometer which is portable consists of a slowly rotating bob which is lowered into a test cup filled with grout. The torque on the bob is then recorded as an apparent viscosity. A range of viscosities can be measured using different bobs.

There is also a flow through (named Colcrete flow trough) which consists of a funnel with a wide mouth which is then filled with grout. The funnel is mounted over one end of a trough. To carry out the test a stopper is fitted to the funnel. The funnel is filled with grout and the stopper is then removed and the distance the grout travels along the trough is recorded which is a measure of the viscosity.

A grout slump test measures the consistency of a grout batch to see how easily the grout will flow. It is important that the quality of grout used offshore in structural applications (e.g. pile-sleeve connections) is uniform and meets the design requirements. The parameter that is related to the characteristic bond strength and which can be measured offshore is slurry density (Domone and Jefferis 1994). Initially an offshore trial mix is usually required both to demonstrate the required density and to provide test cubes and to confirm that the mix meets the required strength.

These procedures should provide assurance that the quality of the grout delivered offshore meets the design specification. This testing at the delivery stage is important as it is very difficult to measure the quality of the grout after setting in the component.

For grout to be used for prestressing ducts the flow properties are very important. There are several test procedures available for testing flow properties ranging from the simple flow funnel to the more elaborate viscometers (Domone and Jefferis 1994). For a funnel test the funnel is filled with a specified volume of grout or to a specified level and the time is recorded for a specified discharge volume.

An alternative to the funnel test is the plunger test. This involves a tube filled with grout and a plunger is then inserted into the tube and the time taken for the plunger to fall through a specified distance is measured. This is normally repeated three times and average of the latter two tests recorded. As the rheology of grouts is time dependant this needs to be taken into account when making the tests.

The bleeding of the grout is also important, as sedimentation and bleeding can lead to the risk of voids in ducts, particularly for vertical prestressing tendons. With cement hydration much of the bleed water will be re-absorbed back into the grout. In general, after 3 h of hydration the grout will almost have achieved its final volume change due to bleed. Most of the relevant specifications limit the maximum permitted bleed after this time to 2% (HSE 1997b). The degree of bleeding can be measured using a procedure to measure the amount of bleed water expelled from a short column of grout after a period of usually three hours.

In an alternative test the loss in height of the grout is measured at 24h and is therefore a measure of loss of solid volume which is normally limited to 1%.

7.6 Cementitious Materials After Construction

7.6.1 Introduction

A number of different types of defect or blemishes can arise during concrete construction which can affect the strength and the future durability of the concrete. Many of these were described in the Concrete in the Oceans report "Typical Blemishes visible on the surface of concrete underwater" (Department of Energy 1987b), including detailed photographs taken from offshore structures.

Slip forming was a standard method for construction widely used for cells and shafts in offshore concrete structures. In this method, a shutter is continuously fed with concrete and raised in frequent small displacements (lifts) at a rate such that the concrete, when exposed beneath the shutter has achieved sufficient strength to withstand the imposed loads. The slip forming process for such enormous structures typically took several months.

The making of concrete mix for these concrete structures over a period of several months required a consistent quality over this period, which was a challenge. Quality control of the concrete was continuous during the construction, which included a full-scale concrete laboratory testing samples on a regular basis. These tests included for example bleed tests to check the concrete w/c ratio before pouring the concrete mix, in addition to for example cube tests to verify strength retrospectively. However, in practice there were variations in the quality of the concrete (strength and porosity).

In comparison to the uniform appearance of steel surfaces, the concrete surface is variable with different blemishes, defects in addition to foreign elements such as steel attachment plates. All these variations need to be considered in the evaluation of strength, durability and possibly also in life extension. The most important defects are described in Sect. 8.1.

Grout is placed within pre-stressing ducts and also for some types of clamps. In these cases, it is contained and its appearance and its quality are not inspectable until after decommissioning (see for example Sect. 12.2). Grout is also used to repair damage to concrete structures. The surface of such grout can vary and some examples of its appearance are given in the Concrete in the Oceans report "Typical blemishes visible on the surface of concrete underwater (Department of Energy 1987b). In addition, this report also shows examples of leakage of cement grout through gaps along a concrete to shutter interface.

7.6.2 Defects from the Concrete Construction Process

Cracking of concrete can occur in different forms. Settlement cracks can develop due to concrete settlement if there is loss of support during construction due for example to premature removal of formwork. Shrinkage cracks can arise due to the evaporation of water from the concrete mix, increasing in severity as the amount of water in the concrete increases. Cracking can also occur due to the thermal contraction resulting from excess heat generated during setting. This is more likely in large concrete sections using higher cement contents and high placing temperatures and can appear within two to ten days following construction. This cracking can follow the line of the reinforcement and lead to enhanced risk of chloride induced corrosion and carbonisation.

Construction joints are a defect that may occur as it is difficult in practice to obtain a full bond across a cold joint in fixed shutter systems. Cold joints occur when there is an interruption to the supply of concrete for various reasons. The delay allows the concrete to stiffen before fresh material is placed on top. Such defects can be significant and normally need mitigation, such as surface treatments using epoxy or pitch paints.

Low cover to the reinforcement can arise if the reinforcement or formwork is misplaced relative to each other. Adequate cover is an essential requirement to deliver durability, as the cover is needed to delay chloride ingress to the reinforcement (which could cause reinforcement corrosion and subsequent spalling or delamination of the concrete). Particularly in the splash and atmospheric zones the requirement for adequate cover is very important, while underwater where the structure is cathodically protected the requirement for adequate cover may be less important.

Honeycombing is a result of inadequate compaction of concrete during construction or insufficient fine material in the mix causing failure of cement mortar to fill spaces around coarse aggregates. Honeycombing can reduce durability because the reinforcement is less protected. If defects are minor, they can be repaired using cement mortar after removal of the formwork. For larger defects, epoxy bonded concrete may be required.

Voids can arise from inadequate compaction of the concrete by failure of the material to completely fill the framework. Voids can arise from the entrapment of air bubbles in the concrete or on the surface during placement. These commonly occur in vertical cast-in-place walls and columns. Larger surface voids would normally be repaired during construction.

Blistering is the development of hollow bumps of varying sizes on the surface as a result of trapped air under the finished concrete. It can occur due to excessive vibration of the concrete mix, excessive entrapped air in the mix or from excessive water evaporation on the top surface of concrete. The proper use of additives can minimise this problem.

Defects from slip forming can be horizontal ridges from the repetitive movement of the shutter or vertical drag marks from scarring of the concrete due to stones caught against the shutter and pulled vertically.

Extraneous material can be shutter debris due to failure to clean the shutters before placing the concrete. Accidental inclusions such as metallic objects (e.g. nails) can lead to rust staining (see rust stains below).

Rust stains can arise from corrosion of either buried steel object or metallic debris lying close to the surface or from corrosion of reinforcement. Buried steel elements may be exposed to the possibility of corrosion from a defective construction joint, cracks in the concrete or defective concrete due to grout loss.

Exudation is usually a deposit of salts (normally white) formed on a concrete surface as a result of a chemical reaction within the material. This may be because the water used for the concrete mix contained soluble salts. This is unlikely in offshore construction as sea water was not allowed to be used in the mix. As the concrete hardens these soluble salts will migrate to the surface forming salt deposits. Typical sites are construction joints, cracks and points of weakness in the concrete (e.g. grout loss, honeycombing). If seawater flows through a crack or joint, a chemical reaction can occur between the constituents of the concrete and salts in the seawater such as calcium or sodium.

In addition, the following defects may be relevant:

- Scaling, which is the loss of a thin layer of hardened concrete surface, can result from shutter movement after the initial set of the concrete or premature removal of the shutter.
- Delamination occurs when the top layer of concrete hardens before the underlying concrete hardens. Water evaporating from the underlying concrete can form a weak layer, resulting in delamination.
- Removal of shutters can cause damage, where the concrete surface is rough and irregular with coarse aggregate visible.

Repairs and surface treatments during construction (Department of Energy 1987b):

1. Cementitious repairs can lead to a patch similar in appearance to the main concrete although the surface may be finer. The edges of the repair may be marked with a slight lip.
2. Epoxy repairs are used when a high strength material is required with rapid cure times. They can be applied in bands, patches or larger areas. They have a plastic like appearance. They are used typically for surface seals to construction joints to eliminate possibility of water penetration through the joint. They are also used to seal areas of suspect concrete and in some cases as seals around built-in items to prevent corrosion of the item.
3. Sprayed concrete. This is mortar or concrete projected onto the concrete surface. This method can be used for repairs to defective concrete e.g. honeycombing, or to reinstate cover where the reinforcement is too close to the surface. Sprayed concrete usually has good adhesion to the base concrete. Gunite and Shotcrete are proprietary examples of sprayed concrete.

In general, the above defects can have varying impacts on the durability of concrete, as discussed in Sect. 7.2.8.

7.6.3 Voids in Grouted Connections

As noted above, voids can occur due to several causes such as poor grouting techniques, unstable grout exhibiting separation and bleeding and leakage during grouting operations. In ducts for prestressing tendons, voids most commonly occur at changes of curvature and at anchorage in horizontal ducts and upper anchorages in vertical ducts (Gifford and Partners 1997).

Prestressing cables are typically installed in steel ducts which are grouted to ensure load transfer and improve corrosion protection. As noted earlier, tendon lengths in offshore concrete structures can extend to lengths such as 150 m in horizontal or circumferential ducts and between 30 and 130 m in vertical ducts. In the early days of concrete offshore construction, it was recognized that the lengths of both horizontal and vertical ducts were greater than those in many land-based structures. These long lengths can complicate the grouting process, potentially leading to incomplete duct filling. This incomplete filling can result in voids that may cause local corrosion of the prestressing cables (see Sect. 8.1.8 on corrosion protection) and will reduce the bond between the grout and tendon.

Trials with grouting materials and procedures were conducted to avoid bleeding problems in vertical ducts (Department of Energy 1988). Bleeding can occur at the top of such ducts as bleed water rises via multi-strand tendons by capillary action. To address this, a header system was devised where the strands protruded above the top anchorage, allowing any bleeding to occur outside the stressed length. The use of admixtures also helped reduce bleeding effects. Despite good practice, small voids can still form at the top of vertical ducts, necessitating refilling from the top (HSE 1997b).

Void issues are even more challenging for horizontal and circumferential ducts. To ensure complete filling in horizontal ducts, it is crucial to keep the grout fluid for several hours. This is to accommodate sufficient delivery times to making the grout able to fill the complete volume of the duct. Admixtures are beneficial in making the grout more fluid and extending the setting time.

The quality of duct filling cannot be determined by on-site inspection; hence, controlling the grout pumping process is the main criterion for ensuring quality and minimising voidage.

Grout filling and the issue of voids also occurs in the repair of damaged or under-strength steel structural members, where grout filling is used to increase their static strength and fatigue endurance. Complete grout filling has been shown to lead to full recovery of the static strength of damaged members. However, voids can occur at the ends of angled members or in horizontal members (Sharp and Ersdal 2021) reducing the strength of the repaired member. Voids in horizontal members may result in trapped water potentially causing corrosion (Sharp and Ersdal 2021).

Voids can also occur in grouted sleeves and clamps used as repair techniques. Such voids would limit the bond between the grout and the steel and hence reduce the effect of load transfer to the sleeve and clamp. Some small voidage was reported in the bolted clamp which was recovered from the Viking AD platform but was considered unimportant (see Appendix 2 and Sharp and Ersdal 2021).

7.6.4 Early Age Cycling of Grout

As mentioned previously, grouted connections are widely used in support structures, both for offshore oil and gas and also for offshore wind installations. Grouted connections rely on high-strength grout filling the annulus between components, such as between jacket legs and piles and between monopiles and transition pieces.

Early Age Cycling (EAC) is a possible problem for grouted connections due to relative displacement of the components due to cyclic wave, wind and current loading during curing of the grout material. In the early stages after placement of the grout (generally the 24 h period following placement) the grout used in these connections may not have fully cured or hardened. This can make the grout in the connection vulnerable to movements, especially during cyclic loading, and these movements can cause the grout properties to be degraded during its setting.

The degradation of the grout properties can include:

- reduced strength as the grout may not have developed its full strength, leading to reduced load-bearing capacity. This can result in premature failure of the connection under applied loads, particularly on the long-term.
- cracking as the cycling loads on the connection can induce cracking in the grout or at the interface between the grout and the connected elements. These cracks can compromise the integrity of the connection and lead to further degradation over time.
- durability issues as incomplete curing of the grout can also affect its long-term durability, making it more susceptible to environmental factors such as moisture ingress and chemical attack.

Early tests were undertaken to simulate the effect of these movements in offshore oil and gas structures, including the work by DNV and Wimpey for the UK Department of Energy (Department of Energy 1986b). In the work for the Department of Energy, grouted connections with 324 mm pile diameter were subjected to a stiffness controlled cyclic movement for the 24 h during which the grout was setting. The connections were subsequently fully cured and then subjected to static testing. The majority of the laboratory tests were performed with an initial relative movement of $\pm$ 0.35% of the inner tubular (pile) diameter. This level of movement was chosen based on calculations of the anticipated movements between foundation piles and sleeves for a number of relevant offshore oil & gas structures (Department of Energy 1986b). The tests were performed for different sizes of shear keys. Tests with shear keys of $h/s = 0.012$ indicated that these movements did not have any serious adverse effect on their static strength. Further tests with larger values of h/s up to 0.07 were undertaken and reported in Harwood et al. (1996). These investigations found that the movements during setting began to have an effect as h/s increased beyond 0.05.

Two tests were also undertaken by DNV (Harwood et al. 1996) to investigate the condition of the grout after early age disturbance. These specimens were split open following grout setting and detailed examination showed that the grout contained a weak zone adjacent to the pile and shear-keys. This indicated that the weakening of the grout mix was due to a preferential shear zone of weak material. At the time of Harwood et al. (1996) being published there was insufficient data to give a clear indication of the early age movement effect over the full range of h/s. Further tests were commissioned by the Department of Energy to investigate the early age effect over the range of h/s between 0.02 and 0.04.

The test data in general showed that the reduction in static strength of the connections increases as the relative pile-sleeve movement increases. Further, the reduction also increases as the pile shear key density (h/s) increases. Finally, the reduction in static strength increases with the grout strength. The data also showed that there was a large reduction in the fatigue performance of connections that had been subjected to early age cycling (Harwood et al. 1996). As a result, Harwood et al. (1996) proposed an early age movement de-rating factor both for static and fatigue strength, which is now incorporated in ISO 19902 (ISO 2007).

Examination of the grout in a grouted clamp recovered from the Viking AD platform (see Appendix 2) showed that the grout's compressive strength met design requirements. However, shear properties were found to be anisotropic due to the layered nature of the grout, with shear strength along these layers being about one-third of that of un-layered grout. This phenomenon of layering was presumed to be caused by early age cycling (HSE 2000c).

Early age cycling has proven to be a significant problem for offshore wind turbines, in the connection between the sub-structure and the tower. This important technical challenge came to light in April 2010 when a grouting failure occurred allowing the tower to slip down on their monopile foundations (Renewable Energy Magazine 2012). Later investigations and research indicated that this grouting failure could have been caused by early age cycling in addition to the absence of shear keys in the connections. Similar problems have been seen for jacket structures for oil and gas production installations after decommissioning, see Sect. 7.6.5. However, this has so far caused less problems during operation.

As a result, a significant amount of research into this area was undertaken, including a proposed joint industry project by DNV. In addition, investigations were undertaken by Lohaus et al. (2014) and Cotardo et al. (2019) which show various influences of the early age movement on the strength development of the grout. An increased sedimentation of the aggregates and deaeration was shown to occur in areas where large movements had occurred. However, in general; segregation and sedimentation lead to an insignificant decrease of the final strength of the grout material.

Compressive strength reductions of up to 45% were observed, when applying a relative movement of approximately 1 mm. It has to be mentioned, that the movement was applied continuously over a period of 24 h. The strengthening effect of the grout was not considered by a force related reduction of the movements, like it would be expected from real offshore conditions (Lohaus et al. 2014).

In contrast (as noted above), investigations by DNV (1994) and Wimpey (Department of Energy 1986b) in the 1980s and 90s showed severe damage in the shear planes of grouted connections filled with ordinary Portland cement (OPC), depending on the shear key height h to shear key spacing s ratio (h/s ratio).

As already noted, excessive early age cycling may compromise the performance of the grout material and consequently the integrity of for example a wind turbine foundation. At present, DNV-ST-0126 (DNV 2021) limits the effect of EAC by specifying that relative movement during the initial phases of grout curing should not exceed a maximum 1 mm. However, preventing excessive early age cycling displacements during installation of offshore wind turbines can lead to significant additional costs and longer installation times. Hence, there is increasing interest from the offshore wind industry that the 1 mm criterion is relaxed. The database used for evaluating the effect of early age cycling is still limited and often based on small scale specimens which are not necessarily applicable to the increasing size of the offshore wind turbine structures.

As can be seen, there is evidence that early age cycling does have a significant effect on grouted connection in certain conditions. However, the understanding of the effects of early age cycling is a relatively immature topic in the industry and more work has to be conducted to fully understand the phenomena.

7.6.5 *Testing of Grouted Connection from Decommissioned Structures*

The Viking AD platform was decommissioned in 1998 after twelve years in service and one of the grouted strengthening clamps was obtained by HSE for structural integrity assessment (see also Appendix 2). This included a non-destructive assessment of the grout layer between the clamp shell and the jacket tubulars (HSE 2000a). In a second phase of the assessment one half of the clamp was dismantled so that amongst other tests compression tests could be undertaken on samples of the grout (HSE 2000b). The clamp assembly was approximately 8 m long, 10.7 m wide.

On examination it was found that the grout layer between the clamp and the tubular member was uniform in thickness (between 45 and 48 mm) and hence the clamp and the tube were concentric. Following dismantling of the clamp it was seen that the annulus was complete with no large voids present. The bond between the body of the clamp and the grout material was good, although there was evidence of an unidentified crystalline material at the interface. There was no indication of adhesion between the grout and the tubular and a black powdery substance was present at the interface. The grout itself had a layered or slated appearance which was believed to be due to early age cycling. Tests showed that the compressive strength of the grout met the design requirement and that the shear properties were anisotropic being influenced by the layer structure and that the shear strength along a layer was approximately one third of that across a layer.

References

ACI (1974) ACI 318-74 building code requirements for structural concrete. American Concrete Institute

ASTM (2014) ASTM A416/A416M standard specification for low-relaxation, seven-wire steel strand for prestressed concrete

ASTM (2024) ASTM A615 standard specification for deformed and plain carbon-steel bars for concrete reinforcement

Billington CJ, Lewis GHG (1978) The strength of large diameter grouted connections. In: Offshore Technology Conference, Houston, Texas, May 1978. Paper Number: OTC-3083-MS. https://doi.org/10.4043/3083-MS

British Standards (1965) BS3892:1965 Pulverized-fuel ash Part 1: Pulverized-fuel ash for use as a cementitious component in structural concrete. British Standards Institution, London

British Standards (1972) BS CP 110–1:1972 Code of practice for the structural use of concrete—Part 1: Design, materials and workmanship. British Standards Institution, London

British Standards (1976) BS5328 Concrete. British Standards Institution, London

British Standards (1980) BS 4027 Specification for sulphate resisting Portland cement. British Standards Institution, London

British Standards (1982a) BS 5075 1982 Concrete admixtures: specification for accelerating admixtures, retarding admixtures and water reducing admixtures. British Standards Institution, London

British Standards (1982b) BS6235:1982 Code of Practice for Fixed Offshore Structures. British Standards Institution, London

British Standards (1985a) BS 6588:1985 Specification for Portland pulverized fuel ash cements. British Standards Institution, London

British Standards (1985b) BS8110 Structural use of concrete. British Standards Institution, London

British Standards (1986) BS 6699:1986 Specification for ground granulated blast furnace slag for use with Portland cement. British Standards Institution, London

British Standards (1996) BS6610:1996 Specification for pozzolanic pulverized-fuel ash cement. British Standards Institution, London

British Standards (1997) BS 4449:1997 Steel for the reinforcement of concrete weldable reinforcing steel, bar, coil and decoiled product. British Standards Institution, London

British Standards (2005) BS 8443 Specification for establishing the suitability of special purpose concrete admixtures. British Standards Institution, London

British Standards (1985) BS 8110 Structural use of concrete—code of practice for design and construction. British Standards Institution, London

Cotardo D, Haist M, Lohaus L (2019) Early-age movement in grouted joints for offshore applications—determination of the development of grout-stiffness. In: Proceedings 29th International ISOPE, 2019

Department of Energy (1974) Offshore installations: guidance on design, construction and certification, first edition. HMSO, London

Department of Energy (1977) Offshore installations: guidance on design, construction and certification, second edition. HMSO, London

Department of Energy (1980) Burdekin and Rothwell, Survey of corrosion and stress corrosion in prestressing components used in concrete structures with particular reference to offshore applications. HMSO London, UK

Department of Energy (1984) Offshore installations: guidance on design, construction and certification, third edition. HMSO, London

Department of Energy (1986a) Data on existing structures, final report. Department of Energy—Offshore Certification Bureau

Department of Energy (1986b) The strength of grouted pile-sleeve connections, report OTH 86 210, HMSO London, UK

Department of Energy (1987a) Concrete in the oceans programme—coordinating report on the whole programme, OTH 87 248, HMSO, London, UK
Department of Energy (1987b) Typical blemishes visible on the surface of concrete underwater. OTH 87 261, HMSO, London, UK
Department of Energy (1988) OTH 88 289 Grouts and grouting for construction and repair of offshore structures. HMSO London
Department of Energy (1989), The assessment of impact damage caused by dropped objects on concrete offshore structures—OTH 87 240, HMSO, London, UK
Department of Energy (1990) Concrete offshore in the nineties—COIN. Offshore Technology report OTH 90 320, HMSO
DNV (1977) Rules for the design, construction and inspection of offshore structures. Det Norske Veritas, Hævik, Norway
DNV (1994) Fatigue testing of grouted connections, written by Skjolde M and Sele A, Report 94-3243. Det Norske Veritas Industry AS, Norway
DNV (2021) DNV-ST-0126 Support structures for wind turbines. DNV, Høvik, Norway
Domone PLJ, Jefferis SA (1994) Structural grouts. Blackie Academic and Professional, Glasgow UK
ELF Aquitaine Norge (1977) Design, fabrication and installation resume for Condeep TCP2—Concrete structure Frigg field. Veritas report 503080-54/1, Stavanger, Norway
EN (2001) EN 934 Admixtures for concrete, mortar and grout. Part 2. Concrete admixtures. Published 2001, revised 2009. CEN
EN (2004) EN 1992 Eurocode 2: Design of concrete structures. CEN
EN (2016) EN 196-1:2016 Methods of testing cement—Part 1: Determination of strength
Fernandes JF, Bittencourt TN, Helene P (2016) A review of the application of concrete to offshore structures. American Concrete Institute, ACI Special Publication, Farmington Hills, MI, USA (2008), pp 377–392
FIP (1972) Recommendations for the design and construction of concrete sea structures, first edition
FIP (1974) Recommendations for the design and construction of concrete sea structures, second edition
FIP (1975) Recommendations for the design and construction of concrete sea structures, third edition
FIP (1985) Recommendations for the design and construction of concrete sea structures, fourth edition
Fosså KT, Widianto (2019) Concrete mix design development for offshore structures. ACI Spring Convention, Quebec City
Harwood R, Billington C, Buitrago J, Sele A, Sharp J (1996) Grouted pile to sleeve connections: design provisions for the new ISO standard for offshore structures. In: Proceedings of the 15th OMAE conference on offshore mechanics and Artic engineering in Florence, vol 3. ASME, New York
Holand I, Gudmestad OT, Jersin E (2000) Design of offshore concrete structures. Spoon Press, London, UK
Holden WR, Page CL, Short NR (1983) The influence of chlorides and sulphates on durability. In: Crane AP (ed) Corrosion of reinforcement in concrete construction. Ellis, Horwood, Chichester, pp 143–150
HSE (1992) OTH 92 364 Log term durability of grouts in a marine environment, HMSO London, UK
HSE (1994) Fourth edition of the Guidance Notes for the design and construction of offshore structures. HMSO London, UK
HSE (1996) Engineering properties of high strength lightweight concrete, M. Hordyk, HMSO, London, UK
HSE (1997a) OTO 96 050 Guidance for concrete structures—background report, Written by the Offshore Certification Bureau for the health and safety executive, HMSO, London, UK

HSE (1997b) OTO 97 053 Durability of prestressing components in offshore structures, Written by Gifford and partners for the health and safety executive, HMSO, London, UK
HSE (2000a) HSE OTO 2000/057 Assessment of strengthening clamp from Viking offshore platform part 1, HMSO, London, UK
HSE (2000b) HSE OTO 2000/058 Assessment of strengthening clamp from Viking offshore platform part 2, HMSO, London, UK
HSE (2000c) Assessment of strengthening clamp from the Viking offshore platform: Phase I, Offshore Technology report 2000/057, HMSO London, UK
ISO (2006), ISO 19903 Concrete structures
ISO (2007) ISO 6935-1 Steel for the reinforcement of concrete
ISO (2008a) ISO 19902 steel structures
ISO (2008b) ISO 10414-1:2008 Petroleum and natural gas industries—field testing of drilling fluids
ISO (2020) ISO 6934-4:2020 Steel for the prestressing of concrete
Kviterud A (2024) Betongplattform. Wikipedia article, visited August 2024
Lohaus L, Cotardo D, Werner M (2014) A test system to simulate the influence of early age cycling on the properties of grout materials. In: Proceedings 24th ISOPE, Busan, Korea
Mehta PK, Monteiro PJ (2006) Concrete: microstructure, properties and materials, 3rd ed. McGraw-Hill, London. https://doi.org/10.1036/0071462899
Neville AM (1973) Properties of concrete, Pitman Publishing, Sir Isaac Pitman and Sons LTD, Pitman House, Parker Street, London WC2B 5PB, UK
Norwegian Contractors (1987) Structural lightweight—aggregate concrete (LWA—concrete), Norwegian contractors. Technical report 1987
Renewable Energy Magazine (2012) Grouting still a major issue for offshore wind, Grouting still a major issue for offshore wind (renewableenergymagazine.com), visited September 2024
Sharp JV, Ersdal G (2021) Underwater inspection and repair of offshore structures. Wiley
Standard Norway (1973) NS 3473 concrete structures, design rules, Oslo, Norway
Standard Norway (1976) NS 3474:1976 Prosjektering av betongkonstruksjoner—Materialer, utførelse og kontroll, Oslo, Norway
Standard Norway (1998) NS 3473 concrete structures, design rules, 4th ed., Oslo, Norway, 1992 (in English), 5th edition 1998
Widianto JK, Taborda G, Bidne K (2016) Concrete gravity-based structure—construction of the Hebron offshore oil platform. Concrete International, June 2016

Chapter 8
Cementitious Materials—Degradation, Detection

8.1 Expected Evolvement and Degradation of Cementitious Materials

8.1.1 Failure Modes of Concrete Structures

Generally, the properties of concrete improve with age. As an example, compressive strength increases with age due to hydration. However, experience has shown that cementitious materials can also degrade through different mechanism such as chemical, bacterial and physical. In addition, as concrete is a composite material, any deterioration of the reinforcement and prestressing has an important effect on the cementitious material. Observed material degradation and failure modes, with their typical locations in concrete structures are indicated in Table 8.1.

8.1.2 Deterioration of Cementitious Materials

Investigations have revealed that very fine cracks at the interface between coarse aggregate and cement paste exist even before any load is applied to the concrete. These cracks remain stable until the load reaches about 30% of the ultimate load, after which they begin to lengthen and increase in number (Neville 1973). The overall stress under which these cracks propagate is influenced by the water-to-cement ratio of the paste. This marks the stage of slow crack propagation. When the load reaches 70 to 90% of the ultimate strength, cracks begin to open through the mortar (cement paste and fine aggregates) and connect the bond cracks, forming a continuous crack pattern. This is the first stage of crack propagation, and if the load is maintained, failure may occur over time. Failure behaviour is influenced by the heterogeneity of the concrete. Unlike a homogeneous body under a simple stress state, where stress

© The Author(s), under exclusive license to Springer Nature Switzerland AG 2025
J. V. Sharp and J. Wintle, *Assessment of Materials in Ageing Offshore Structures*,
https://doi.org/10.1007/978-3-031-84489-8_8

Table 8.1 Observed material degradation and failure modes of concrete structures (based on ocean structures 2009)

Materials degradation and failure modes	Concrete elements						
	Legs/towers/ shafts—general	Storage cells	Steel concrete transition	Shaft/ base junction	Prestressing anchorages	Splash zone	Foundations (base and skirts)
Chemical deterioration	✓	✓		✓		✓	
Bacterial degradation	✓	✓		✓			
Physical deterioration	✓	✓	✓	✓	✓	✓	
Cracking and delamination of concrete (from corrosion of steel reinforcement)	✓	✓		✓		✓	
Corrosion of prestressing tendons	✓	✓		✓	✓	✓	
Cracking due to fatigue (normally of the reinforcement)			✓	✓			
Damage to concrete and reinforcement from ship impact	✓				✓	✓	
Damage to concrete and reinforcement from dropped objects		✓			✓		
Cracking of concrete due to thermal effects		✓		✓			
Damage to foundations from scour & settlement damage							✓

Description of these material degradation mechanisms and failure modes are further described in this Chapter trajectories are straight or simple curves, concrete exhibits different behaviour due to the presence of aggregates. The material properties vary from point to point, and the bond surfaces between aggregates and paste can form various angles with the direction of the external force. Consequently, the local stress significantly deviates from the nominal applied stress.

Seawater contains a number of chemical ions which can participate in chemical reactions and which could lead to long term degradation of the concrete. These include sodium, potassium and magnesium, as well as sulphate ions. In addition, seawater contains chlorides which can be detrimental to the reinforcement steel, see Sects. 8.1.3 and 8.1.5.

As noted earlier, sulphate attack is a reaction between the sulphates in seawater and calcium hydroxide in the hardened cement paste. The reaction products can be expansive, leading to cracking and crazing. This is a well-recognised problem for marine structures and is usually designed out by, for example, including some pulverised fuel ash in the mix. However, this did not occur with the early installations. Visual examination of the legs should detect any sulphate attack problems.

Carbonation is a slow reaction between the CO_2 in the air and hydroxyl ions in the concrete to produce carbonic acid. This acid reduces the alkalinity of the cover concrete, allowing further carbonation to take place at greater depth. Eventually this leads to the embedded steel losing the protection of the alkaline environment and the possibility of corrosion. However, evidence from tests on bridge decks and other structures shows that with large depths of cover of high quality concrete this is unlikely to be a problem in the air zone of the concrete towers.

The C3A (tricalcium aluminate) content of the cement was initially seen as an issue as it was thought that the C3A would react with the sulphates present in seawater, causing problems and hence cements with low C3A content were preferred. However, it became known that the C3A combines with the chloride ion to from an insoluble compound that effectively reduces the concrete's permeability which is an asset. Therefore, cements with moderate levels of C3A (typically about 8%) were specified, see Sect. 7.2.2.

In some non-offshore concrete structures, alkali-aggregate reactions have been observed. Here, the alkaline nature of the cement reacts with certain aggregates, causing localized damage and loss of integrity. However, as noted earlier, the crushed granite commonly used in many offshore concrete structures has generally not posed this problem. It remains uncertain whether all aggregates used in offshore concrete structures are susceptible to this type of damage.

8.1.3 Cracking of Concrete Due to Exposure at Depth

Many reinforced concrete installations have survived in the sea for many years. However, there is little experience of such installations operating in deep water where there is significant pressure. Offshore concrete structures present unique challenges, particularly concerning the long-term effects of deepwater immersion on strength.

Initially, immersing concrete in seawater at depth induces a beneficial pre-stressing effect, which diminishes as internal pore pressure builds up. This pore pressure develops because concrete is porous, allowing water to infiltrate its pore structure. The time required for this pressure to develop depends on the concrete's permeability, which is typically linked to its strength. Cracking in the concrete further accelerates the establishment of internal pore pressure. In typical offshore installations, this pressure will be fully developed in ageing structures.

As a result of this lack of experience the Concrete in the Oceans programme initiated a series of tests on reinforced concrete beams and cubes exposed for a number of years in Loch Linnhe in Scotland at a depth of 140 m. These were recovered after a period of up to 5 years and their state compared with similar beams stored in a tank on land. The salinity, dissolved oxygen levels and temperatures at Loch Linnhe at this depth are similar to those in the North Sea.

The reinforced beams were constructed in pairs, placed back to back and stressed to hold the cracks open. The overall size of each beam was 1300 × 550 × 250 mm. Two different covers were used, 25 and 75 mm. Two different concrete grades were used with 28 day strengths of 35 and 60 N/mm^2. They were made from cement and aggregate similar to those used offshore.

The strength of these specimens was compared with control cubes maintained for a similar time under shallow fresh water in a laboratory After a short period of one year immersion the two sets of results were similar but after 2.5 years deep immersion the strength of the cubes was significantly different from the control cubes. The control cubes had either maintained or slightly enhanced their strength but the cubes exposed at depth showed a small loss in strength (compared to the control cubes). There was some concern about the testing regime as there were differences in time between testing the control cubes and deep-water cubes after removal from their location. The concern was that internal pore pressure in the deep-water samples had not been released before testing. However thin slices from cylinders exposed at depth showed internal micro-cracking, particularly around the edge of coarse aggregate particles, which was not present in the shallow water exposed samples. This may explain the small loss in strength.

This loss of strength after deep water immersion has been confirmed by other tests (Haynes and Highberg 1976) which involved samples being tested after 6 years at depths up to 1500 m. It was noted that after a short period of exposure the specimens suffered a ~ 10% loss in strength after deep immersion. After 5.6 years in the ocean the concrete showed a ~ 15% loss in strength compared to the control specimens. The authors concluded that low water temperature and pressure slowed the strength gain in concrete compared to the control unpressurised specimens.

Clayton (Clayton 1986) carried out a series of tests on concrete specimens pressurised to 600 bar (equivalent depth of ~ 6000 m) for a short period of only 6 days. Both flexural and compressive strength was measured under pressure on the pressurised specimens and compared with control unpressurised specimens. The mean compressive strength of the pressured cubes was found to be ~ 10% less than the control specimens. These tests showed that the rate of pressurisation was not important, as most of the strength loss occurred after a critical immersion period is exceeded.

The tests also included flexure tests on concrete beams whilst under pressure and also after depressurisation. There was no significant difference between the two sets of results. The mean flexural strength of the beams both under pressure and after depressurisation was 2.2 N/mm^2 which compared to the strength of control tests of 4.2 N/mm^2, showing a ~ 50% loss in strength. It was also found that there was a detectable strength loss after a lower level of pressurisation of 30 bar equivalent to a water depth of 300 m. Specimens were examined microscopically after pressurisation and micro-cracks, mainly between the coarse aggregate and cement paste were observed. This is similar to the observations in the Concrete in the Oceans programme.

In the MATS (1984) programme the durability of concrete was assessed after long periods of exposure in deep water in the North Sea. After two years of exposure some of the specimens were recovered and tested. After a longer period of exposure specimens were also recovered for testing. The tensile strength results showed significant reductions in strength (decreasing from 6 MPa after 2 years to only 3.8 MPa after 8 years. Other work (Hove and Jacobsen 1998) reviewed several test series on the effects of pore pressure and concluded that the effect was small and noted that many tests were undertaken using previously pressurized specimens but tested at atmospheric pressure. This was not the case for the work undertaken by Clayton (1986).

Zhou et al. (2020) conducted tests applying water pressure up to 4 MPa (equivalent to 40 bars, and water depths of 400 m) on concrete specimens, followed by testing. These showed a reduction in compressive strength of ~ 13%. Lower grade specimens showed a larger reduction in strength. The tests showed a correlation between the increment in water content under pressure and the reduction in strength.

As a result, some early offshore concrete structures were designed using a 10% reduction in expected final compressive strength. However, a 10–15% loss in compressive strength was considered unlikely to be important unless there are other issues such as the original concrete not meeting its design specification. The significant loss of flexural strength is of more concerning.

Later work claimed that neither compressive nor tensile strength was significantly influenced by pore pressure. However, the microscopic tests showing some internal cracking around aggregate particles is evidence that for those tests some deterioration of properties was evident, particularly with the flexural tests.

8.1.4 Chloride Penetration into Concrete

Concrete is naturally alkaline, because of the presence of several hydroxides resulting from the reactions between the mix water and the Portland cement particles. This alkaline environment is important in providing protection against corrosion of the reinforcing steel. Loss of this alkalinity can occur through several ageing processes, which include ingress of chlorides, sulphate attack and carbonisation. However, for offshore structures chloride penetration is the most important.

Hardened concrete is a porous material and when immersed in seawater chlorides can penetrate the concrete. When measuring chloride ingress there is a difference whether the amount is quoted as a percentage by weight of cement or concrete. The critical level of chloride ingress in the Concrete in the Oceans project (Department of Energy 1987) is given as 0.4% by weight of cement, while in other cases it is reported as 0.1% of weight of concrete.

Corrosion can occur, provided there is also a sufficient supply of oxygen. The rate of corrosion will then primarily be controlled by the level of chlorides per volume cement, the availability of oxygen and the electrical resistivity of the concrete (see Sect. 8.1.5). It is normally assumed that corrosion of the reinforcement occurs at a certain threshold value of chloride, as noted above.

The presence of chlorides in the concrete affects the concrete resistivity which is also dependant on moisture levels. Measurement of resistivity is useful in determining chloride and moisture levels (see Sect. 8.2.3).

Many studies have been made on the penetration of chlorides into hardened concrete immersed in seawater. An early study in the Concrete in the Oceans programme was undertaken by UK Department of Energy and the industry (Department of Energy 1987). As part of this programme, concrete beams and prisms were exposed in several different environments, including the splash zone and deep water. A specific project entitled "Efficiency of cover" (Department of Energy 1987) was undertaken to see which parameters had the most effect in protecting the steel reinforcements. After several years of exposure measurements were made of chloride ingress. In addition, surveys were made of structures that had been in the sea for long periods and again samples were taken for chloride measurement. In general, the results obtained showed significant variability (Department of Energy 1987). It was found that surface chloride levels were strongly influenced by the condition of the concrete surface and immediate environmental conditions. Chloride surface levels in the splash zone were almost twice those measured in the tidal and underwater zones. Low surface levels were found in specimens exposed in the atmospheric zone. The survey of the Royal Sovereign lighthouse, with a high concrete strength and low w/c ratio of 0.37, showed chloride levels of only 0.09–0.16% by weight of concrete after 10 years in the atmospheric zone at depths of 20–25 mm.

An attempt was made to fit a diffusion type model (e.g. based on Fick's model) to the results from the "Efficiency of cover" project which showed that most of the variables in the Fick's model affected chloride ingress into the concrete to some degree. The lowest grade of concrete (approx. 50 N mm^{-2}) allowed faster ingress of chloride compared to the other two concrete grades tested (Table 8.2). The C_3A content of the cement within the range 7–14% had little effect on chloride penetration but in the lower grades of concrete partial replacement of OPC by pfa reduced the apparent diffusion coefficient by a factor of two to three.

Other tests were performed to simulate immersion in deep water. The deep water was simulated by exposing the specimens to a pressure of 500 psi (34.5 bars), which would represent concrete at a depth of 345 m. The tests at a depth of 20–25 mm, in

Table 8.2 Chloride levels at 20 mm depth—results from exposure tests (shallow immersion in seawater)

Exposure time year	% Chloride by weight of concrete standard grade (0.475 w/c)	% Chloride by weight of concrete low grade (0.69 w/c)
0.5	0.13	0.14
1	0.16	0.27
2.5	0.2	0.31

comparison, showed chloride levels of 0.19% (by weight of concrete) after a period of immersion of only 5 months, indicating the effect of depth.

The Concrete in the Oceans programme (Department of Energy 1987) also investigated chloride penetration of chlorides down cracks. It was found that chlorides penetrated rapidly down cracks of all sizes within relatively short periods of time, reaching levels normally accepted as being necessary to de-passivate the reinforcement. Results showed that the chloride level at all points down static cracks up to 75 mm in depth was approximately 1.25% by weight of cement, irrespective of crack width and length of exposure. However, in many cases corrosion was not found, indicating the importance of other factors such as oxygen levels (see next section).

The Norwegian Research Council undertook a review of the durability of concrete structures in the late 1990s. Part 2.4 of this review concerned offshore concrete structures (Norges Byggforskningsinstitutt 1999). The review took account of the results of visual inspections, measurement of concrete cover, electro-chemical potential and core tests to map the chloride ingress on 6 offshore concrete structures in and above the splash zone (four of these were GBS, one semisubmersible and one concrete tank). All these structures were provided with a cathodic protection system to protect mechanical equipment but also provided some protection of the reinforcement. An overview of the degree of chloride ingress from this report is shown in Table 8.3.

Table 8.3 Overview of identified chloride ingress on concrete structures

Structure	Year of installation	Year in operation at inspection	
Statjord A	1977	16	Was provided with an epoxy membrane on the outer surface of the shafts which seemed to have a significant effect in reducing the chloride ingress
Gullfaks A	1986	7	Moderate chloride ingress (0.1% at 10 mm depth)
Gullfaks C	1989	5	Moderate chloride ingress (0.1% at 25 mm depth)
Oseberg A	1988	8 and 9	Moderate chloride ingress (0.1% at 25 mm depth)
Troll B	1995	2	Low chloride ingress (0.1% at 6–10 mm depth)
Ekofisk Tank	1973	17 and 20	High chloride ingress (0.1% at 40–45 mm depth)

Table 8.3 shows moderate chloride ingress after relatively short periods of between 5 and 9 years at 25 mm depth. Significant chloride ingress after 17–20 years at depths of 40–45 mm was found in the Ekofisk tank. In addition, visible rust stains were identified in areas with low concrete cover for all structures, particularly for Oseberg A. A very low degree of carbonation was also seen on all structures.

DNV undertook a study for PSA entitled "Material Risk—Ageing Offshore Installations" (PSA 2006). This included a section on chloride penetration into concrete. Results from several offshore structures were given for periods ranging from 2 to 20 years. The sites from which cores were taken included both splash zone and at water depths down to 140 m. In the DNV report, a threshold value of 0.15% Cl as weight of concrete was assumed for corrosion of the reinforcement. The results used in this DNV report were taken from several offshore structures for given periods ranging from 2 to 20 years. The sites from which cores were taken included both splash zone and at water depths down to 140 m, as shown in Table 8.4.

Although there are some inconsistencies in the results between Tables 8.3 and 8.4, it can that for the older structures (17–20 years) in some cases the chlorides had penetrated to the reinforcement at a level which was greater than the threshold level (0.15%Cl as weight of concrete). Surprisingly the chloride levels taken from a core at a depth of 140 mm, after 8 years of exposure on the Stafjord A platform were not higher, given the pressure effect of depth. The use of an epoxy coating was shown to work well on the Statfjord A platform in limiting chloride ingress but not on the Heidrun TLP (stated to be a much poorer coating).

According to DNV (PSA 2006) there was sufficient data from the Brent C platform to carry out a probability-based durability analysis. However, it is not stated how the risk of depassivation was calculated and for what location. It was shown that a 10% probability for steel corrosion is exceeded after a service period of approximately 20 years, and a 50% probability is exceeded after ~ 60 years. For the splash zone this level would seem too low, based on other evidence, despite a high level of cover in this zone (typically 75 mm). Where cracks exist from construction or from operation

Table 8.4 Chloride penetration in particular structures

Site	Age	Elevation of sample core (m)	Depth at which threshold value reached (mm)
Statfjord platform A	8	− 140	> 50
		+ 7	25
		+ 7 to 16 m, with epoxy coating	20
Ekofisk tank	17	Not specified	20–35
Brent B platform	20	+ 14.4	40–45
		− 11.5	10–20
Brent C platform	17	Several different locations	16–40
Heidrun	2	With epoxy coating	15–20
		Without epoxy coating	15–20

the penetration of chlorides is much faster and high level of chloride can reach the reinforcement in a few years.

The DNV report (PSA 2006) notes that in the submerged part of offshore concrete structures, a variety of freely exposed metallic components, such as pipes, brackets and supports are in electrical connection with the embedded reinforcing steel. In this case, these components can effectively connect with the large cathodic areas of the embedded reinforcement which are in contact with the oxygen in seawater. This potentials problem can be managed by an effective cathodic protection system for all the freely exposed metallic components (see Sect. 8.1.7).

8.1.5 Corrosion of Reinforcement in Concrete Structures

As previously mentioned, concrete is naturally alkaline which is important in providing protection against corrosion of the reinforcing steel. This protection can remain for long periods, provided it has a substantial depth of high-quality cover. However, the permeability of concrete allows seawater chlorides to eventually reach the steel reinforcement (see Sect. 8.1.4). Activation of the reinforcement with loss of passivity can occur when sufficient chlorides reach the steel surface and if sufficient oxygen is available this will usually lead to corrosion.

Corrosion can progress over time, leading to expansive rust products that often cause the concrete cover to spall (see Sect. 8.1.1), where localized concrete sections can be lost. Once the cover spalls, the reinforcement steel is more exposed and corrosion accelerates unless repaired. The splash zone is especially vulnerable due to abundant oxygen and seawater, making corrosion likely in this area. This type of corrosion is typical in many concrete marine structures and it is recognised that maintaining the integrity of the steel reinforcement is important.

The level of oxygen is important in maintaining the cathodic reaction in the corrosion process. Depletion of oxygen at cathodic sites may lead to polarisation of the cathodic reactions to an extent that the anodic processes at the reinforcement are inhibited. Oxygen diffusion is a key factor in controlling corrosion, particularly underwater and in saturated concrete where corrosion is under cathodic control. Oxygen diffusion was measured in the Concrete in the Oceans programme (Department of Energy 1987). The results showed a wide variation over three orders of magnitude with values either side of the diffusion coefficient for oxygen in seawater (6×10^{-5} cm^2/sec). These values ranged from 1.35×10^{-4} cm^2/sec in concrete at a distance from the face of up to 48 mm in concrete with pfa additions in the splash zone to 2.96×10^{-6} cm^2/sec in high strength concrete underwater.

Corrosion products which form on the reinforcing steel depend on the availability of oxygen. In the splash and air zones where oxygen is plentiful, the corrosion product is typically $Fe_2O_3.H_2O$ or FeO (OH), normally referred to as rust which is expansive and the cause of spalling in concrete structures. Underwater where oxygen is limited the corrosion product is usually Fe_3O_4 (magnetite) which is black and less expansive and is hence less likely to lead to spalling.

Fig. 8.1 Example of intense local corrosion in reinforcement in an underwater previously cracked location, also indicating visible corrosion products which have oxidised on exposure to air (Department of Energy 1987)

Cracking in the concrete cover can exacerbate the ingress of seawater to the reinforcement. Hence, design protocols typically aim to control post-construction and serviceability cracking. Post-construction cracking is typically limited to 0.1 mm in splash and atmospheric zones and 0.3 mm in submerged zones. Most concrete standards include a serviceability limit of 0.1–0.4 mm crack width, due to tensile stresses, depending on the environment. Additionally, cracking of the cover can occur during operation from high static stresses from storms and accidental impact from dropped objects or ship collisions. This cracking can be followed by spalling allowing further water ingress.

Limited oxygen availability underwater restricts corrosion and as indicated by laboratory studies (Department of Energy 1987), the resulting corrosion products underwater are less expansive and usually do not cause spalling of the cover. Intense localized corrosion of the reinforcement in underwater concrete has been observed in concrete test sections, as shown in Fig. 8.1 (Department of Energy 1987; MATSU 1991).

The requirements for intense local corrosion include:

- There is a local breakdown of passivity, such as from cracking and the low concrete resistivity (due to long-term seawater immersion),
- An efficient cathode is also required, which may be in the splash zone where there is sufficient oxygen to support a high local corrosion rate,
- Electrical conductivity through the steel reinforcing network is present which can act as a link between an anode and a remote cathode.

Similar localized intense corrosion of reinforcement has also been observed in bridge decks saturated with chlorides from de-icing salts.

In structures subjected to cyclic stresses, fatigue of the steel reinforcement can occur. This can eventually lead to localized loss of reinforcement, resulting in local tensile stresses in the concrete and potential cracking of the concrete cover. The presence of seawater exacerbates this issue.

Accidental events can cause localized damage to towers (from collisions) or to the tops of caissons (from dropped objects), leading to seawater ingress and local corrosion of the steel reinforcement. It is important to detect and repair such damage promptly to minimize further corrosion.

8.1.6 Development of Concrete Surface Layers

Thin protective layers on the surface of concrete and grout exposed to seawater have been shown to develop under certain conditions (Department of Energy 1990a). This layer mainly consists of aragonite (calcium carbonate, $CaCO_3$) and brucite (magnesium hydroxide, $Mg(OH_2)$), but other compounds have also been observed. This layer is formed as a result of seawater containing a number of chemical ions which can participate in chemical reactions. These could lead to long term changes to the surface of the concrete. These chemical ions include sodium, potassium and magnesium.

Several tests have been undertaken to examine these layers as a result of immersion in seawater. Work by Buenfield at Imperial College (Department of Energy 1990a) found that the brucite layer was around 30 μm in thickness, overlain by a thicker but more slowly developing layer of aragonite. The thickness of the deposited layers appeared to be dependent on the pH of the cement paste pore solution and the permeability of the cement paste. The degree of saturation of seawater with aragonite or brucite was also found to be a function of its pH value. Uncontaminated seawater has a pH within the range 7.8 to 8.3 and is supersaturated with aragonite at pH values greater than around 8.1.

As part of the Concrete in the Oceans programme, concrete prisms and beam, some of which were pre-cracked, were immersed in seawater for periods of up to 6 years (Department of Energy 1990a). These included both shallow water and deep water, the former at Portland harbour (Bincleaves seawater site) and the latter in Loch Linnhe in Scotland at 200 m depth. On recovery of the specimens the surfaces were examined. The white deposit on the crack surface of cores from the Loch Linnhe specimens was found to be generally brucite with some portlandite ($Ca(OH)_2$) and calcite ($CaCO_3$). White and brown deposits were found on specimens exposed in the splash zone at Portland harbour. The white deposits were mainly portlandite which had partly carbonated and also calcite. Although chloride penetration was observed in all specimens no significant strength deterioration was found due to chemical attack after five years of exposure.

Jacobsen (2011) undertook a series of exposure tests on different concretes exposed for up to two years in relatively cold seawater at Rødby Harbour in the southern part of Denmark. The focus of the long-term durability project was to study durability and the evolution of the concrete microstructure with time. None of the concretes showed signs of brucite precipitation at the surface after two years exposure except a concrete containing 15% fuel ash. All the other concretes showed a maximum 1 mm cracked zone at the surface where the paste was more or less decalcified.

Another programme of investigating the effects of marine exposure on grouts was undertaken by Imperial College (HSE 1992). Here, samples were immersed in seawater for a period of up to 5 years, both underwater at the Bincleaves seawater site and in the laboratory. The specimens were then examined in detail using several different methods. The grouts included Oilwell B cement, ordinary Portland cement and also high alumina cement.

Several techniques were used to examine the exposed samples which included backscattered electron imaging and scanning electron microscopy. After exposure brucite and aragonite layers were observed in the laboratory specimens. A back scattered image of oilwell B cement after 7 days in seawater shows a well-defined layer of predominantly brucite, but with traces of aragonite (see Fig. 8.2). It was found that the thickness of this layer at any age was variable but it increased with duration of exposure to seawater from between 40 and 70 μm at 3 days to about 160 to 200 μm after 90 days. Temperature variations were found to have a significant effect on layer precipitation, but for the range of pressures encountered by offshore concrete structures the effects were negligible. Additionally, it was also noted that organic compounds present in sea-water were found to possibly inhibit precipitation of layers.

However, this was not the case for the 5-year-old specimens from Bincleaves. One explanation of this was that natural seawater at this depth contained traces of dissolved polar organic substances which could have been absorbed onto the surface of the grout, blocking nucleation sites and thereby inhibiting precipitation.

The protective effect of these surface layers has been demonstrated to reduce chloride penetration (Department of Energy UK 1987). These layers modify the permeability of the concrete and hence reducing the permeation of chlorides and

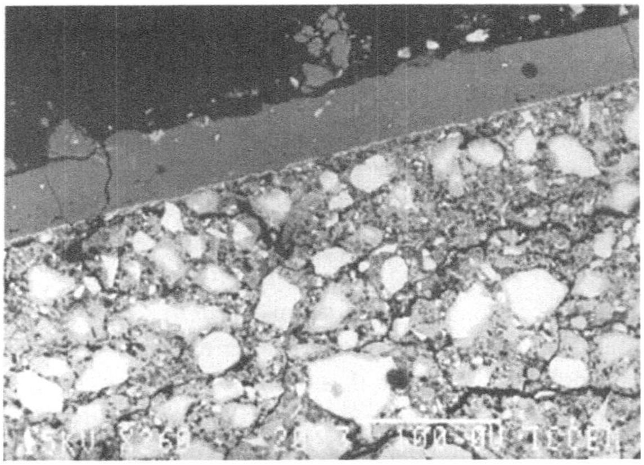

Fig. 8.2 Back scattered electron image of ordinary Portland cement after 7 days showing layer at magnification ×260 (HSE 1992)

oxygen to the steel reinforcement. However, the layers appear to be variable, vulnerable to damage or loss from abrasion and other wear processes, thus limiting their effect in practice.

8.1.7 CP of Concrete Structures

Most offshore concrete structures have a cathodic protection (CP) system. In early structures these were meant to protect the attached steelwork to the concrete columns. However, the steel reinforcement in these early structures was typically electrically connected to the steelwork, despite attempts to isolate the reinforcement. This connection happened particularly with unintended electrical connectivity to flowlines and pipelines as well as other external attachments. This resulted in higher-than-expected drain on the sacrificial anodes, necessitating replacements. In later structures, the reinforcement was deliberately connected to the CP system to benefit from cathodic protection (MTD 1990).

Cathodic protection offers significant advantages in that it protects the reinforcement to some extent, minimising the level of corrosion where seawater has permeated to the steel or where cracking is present.

Design criteria for CP systems in offshore concrete structures recommend a minimum allowance of 1 mA/m^2 for the reinforcing steel (MTD 1990). Maintaining the CP system is essential to minimize corrosion reactions in concrete structures. However, the CP system provides minimal protection for the splash zone, which is most vulnerable to corrosion. This is why higher-than-normal levels of concrete cover are used in this zone.

Seawater reacts with the concrete to form aragonite (calcium carbonate) and brucite (magnesium hydroxide), which can build up at the mouth of cracks and limit seawater ingress, offering some beneficial effects (see Sect. 8.1.6).

8.1.8 Corrosion of Prestressing Tendons

High-strength prestressing tendons are heavily used in offshore structures to ensure that the concrete material remains in compression, see Sect. 7.3.2. Hence, these are required for maintaining the structural integrity.

Tendons are manufactured from high strength steel and are very suceptible to corrosion damage, with potentially serious local loss of prestress. As a result, it is vital that the tendons are protected from seawater. This protection is typically achieved by placing the tendons in ducts with subsequent grouting. The degree to which grouting of the steel ducts has been effective, has led to concerns that seawater can penetrate into the ducts. The concern is particularly relevent for long ducts and ducts with horizontal orientation,

In bridges, corrosion of prestressing tendons in grouted ducts has been observed. Due to the lack of oxygen in these ducts, black rust (Fe_3O_4 or *magnetite*) is formed. The corrosion processes can be written as:

$$3Fe(OH)_2 \rightarrow Fe_3O_4 + H_2 + 2H_2O$$
$$3Fe + 4H_2O \rightarrow Fe_3O_4 + 8H^+ + 8e^-$$

The important element of this process is the production of hydrogen atoms, which are very damaging to the high strength tendons (due to hydrogen embrittlement). This damage is likely to be localised and given the bond between grout and the tendons is intact, the prestressing may be retained overall. To the knowledge of the authors, such corrosion has not been observed in concrete offshore structures. However, the limited inspection capability for tendons will limit any observations of such corrosion.

A review of the durability of prestressing components (HSE 1997) concluded that the earliest concrete offshore structures (built before 1978) were more vulnerable to tendon corrosion. In contrast, later platforms benefited from improved grouting materials and procedures. It was also noted that a significant loss of prestress (~ 40%) in a leg would be required before failure under typical design wave loading could occur. Moreover, such failures would need to be localized in the same section area to pose a danger. In land-based structures, failures have typically occurred near anchorages or construction joints.

8.1.9 Bacterial Induced Deterioration

Bacterial activity in unlined concrete storage tanks containing water and oil at the base of offshore structures can lead to the production of acids, which can attack the concrete. Oil can be introduced at higher than ambient temperatures from the well and there is an oil–water mixture from the operation of displacing the stored oil with seawater. Bacterial activity in the sea-water pillow beneath the oil can produce acids harmful to the concrete.

Aerobic oil oxidising bacteria (OOB) can produce organic acids (e.g. acetic) which according to Khoury and Sullivan in the report OTH 90 320 (Department of Energy 1990c) can lead to reductions in pH values down to about 5. Anaerobic sulphate reducing bacteria (SRB) can lead to significant level of hydrogen sulphide (H_2S) being produced. The presence of sulphur oxidising bacteria (SOB) can oxidise the sulphide to sulphuric acid.

As already noted, SRBs can cause loss of concrete material reducing wall thickness. The rate of loss is dependent on the pH value. Unfortunately, the storage tanks are almost impossible to inspect due to very limited access and hence the level of damage from SRBs is difficult to assess. However, bacterial growth and the presence of H_2S was detected in the legs of the Cormorant A platform after up to 100 days after flooding the leg. Levels of 10^{-4} mg/litre of H_2S were reported (Department of Energy 1990a), together with SRB concentrations in the range of 10^{-2}/ml.

Drill cuttings can accumulate around the lower sections of a concrete platform or on cell roofs. These piles consist of cuttings and oil or water based muds. Up to the early 1980s muds were based on diesel oil, replaced in 1984 by low toxicity oil. The presence of oil and water can encourage the development of bacteria, which are likely to be anaerobic, favouring the growth of SRBs and possible deterioration of the concrete material.

High levels of bacterial activity have led to serious damage to concrete sewers, pipes and storage tanks in the past, in some cases leading to major loss of wall thickness and strength. The major damage would appear to be above the water line, with minimal effects observed below the low water level. A literature review of this type of damage was undertaken by Khoury and Sullivan as part of the London University Concrete in the Nineties (COIN) programme (Department of Energy 1990a).

ISO 19903 (ISO 2006) notes that favourable conditions for SRB growth can be present in un-aerated water in, for example, the water-filled portion of concrete shafts and cells. This acidic environment can cause concrete softening and corrosion of the steel reinforcement. ISO 19903 also notes that inspection of a concrete surface likely to be affected by SRB activity is difficult to undertake in practice but suggests that some information can be obtained by adequate monitoring of SRB activity and pH levels. Again, this is difficult in storage cells.

A programme of tests was undertaken by Khoury and Sullivan (Department of Energy 1990c). This included the casting and curing of about 750 small cylindrical concrete specimens. Half of these were pre-conditioned for eleven months in North Sea seawater to simulate the internal surface of oil storage tanks. The other half were stored for the same time at 100% relative humidity to simulate concrete at the central zone of the storage tank wall, which was assumed to be unaffected by exposure to seawater.

Acidified fluid was then passed over the samples for between 21 and 90 days. Various concentrations of acetic and sulphuric acid were used to give pH values of 8 (baseline), 5 and 3. It was found that the pH had a very significant effect—the compressive strength loss at pH 3, which is highly acidic, was four times that at the less acidic pH 5. Actual bacterial activity in North Sea storage tanks has occasionally been reported to produce a pH of 5 (Department of Energy 1990a).

Compressive strength measurements and changes in the samples' weight showed that acetic acid was far more aggressive than sulphuric acid. Losses in strength after 23 days in pH 3 solutions at 20 °C, were 43% and 11% for acetic acid and sulphuric acid solutions respectively. The strength losses after 90 days immersion in pH 5 solutions were 19% in acetic acid and 5% in a sulphuric acid solution. Weight losses in acetic acid/seawater solutions were also measured at different temperatures in the range 20–60 °C; these losses ranged between 30 and 51%, with the higher loss at the higher temperature.

X-ray diffraction and electron microscopy techniques were used to determine the layers formed by attack from both acids and these showed that the layers contained a high percentage of gypsum (calcium sulphate). Sulphuric acid produced an extensive growth of crystalline gypsum needles. The depth of attack (DOA), which was estimated from optical microscopic photographs of surface sections of the specimens, showed that the reaction products developed on the concrete in layers.

Acetic acid produced five layers, compared to sulphuric acid's three, and penetrated six times as far into the concrete. The layer system was found to be soft (gypsum in the case of the acetic acid) and could be removed by abrasion. The innermost layer appeared to be the reaction zone where the hydration products were depleted, and gypsum deposited.

The layers retarded the rate of acid attack by reducing the inward diffusion of aggressive ions from the outer surface. Periodic removal of the layer during exposure led to nearly double the depth of attack compared with non-removal of the layer. The authors (Department of Energy 1990a) developed an equation for the depth of attack based on rather short exposure times and concluded that a layer 5 cm deep resulting from acetic acid seawater attack might protect underlying concrete for up to 10 years provided there was little mechanical abrasion.

The tests were conducted at three temperatures, 20, 40 and 60 °C. It was found that temperature had less effect than might be expected because the layer formation insulated the concrete. It was concluded that temperature effects might have been more pronounced if the concrete sample had been subjected to mechanical abrasion.

It was found that there was no significant difference in properties between the samples that had been preconditioned in seawater before testing and those preconditioned at 100% relative humidity.

As already noted, the pH level had a major effect on the results, with pH 3 having a significant effect. However, even at pH 5 (more typical of actual structures) the losses in properties were small but not negligible with a depth of attack of 2 mm being measured after 90 days in an acetic/sea-water solution at 20 °C. The formula for the depth of attack was developed (Department of Energy 1990a) but based on limited short-term data. This showed that the depth of attack for 1 and 10 years was estimated to be about 20 mm and 50 mm respectively.

It is important to recognise that the exposure time in these tests was relatively short (maximum 90 days). Further work was planned where the DOA testing would be extended for periods of up to a year. However, the authors are not aware that these results have been published.

The authors concluded (Department of Energy 1990a) that oil storage tanks in the North Sea are unlikely to suffer significant damage from the acid produced by bacteria, from preliminary indications at concrete oil storage. One mitigating factor is the thick coating that is expected to exist on the inner walls of the storage tanks due to the presence of oil.

In summary, acidic environments such as might occur in oil storage tanks have the potential to reduce wall thickness depending on the pH value. The loss in strength due to reduced area is also a factor in relation to ageing, as well as the production of softer surface layers, particularly in the presence of acetic acid. These layers will have reduced properties although the depth of the layers in practice is not known.

8.1.10 Thermal Effects

Storing hot oil in the concrete tanks at the base of many installations can induce thermal stresses, potentially causing the concrete to crack. Concrete is particularly vulnerable to significant temperature differences, such as those created by hot oil on one side of the wall and cold seawater on the other. Tests have shown that temperature differences up to 45 °C can be managed with proper design details (Department of Energy 1987). However, if the coolers fail (as the oil is cooled before storage) or unusual conditions occur, oil temperatures can reach up to 90 °C, which may lead to wall cracking. Over extended periods, these thermal effects can accumulate, increasing stresses that cause cracking in the concrete and overstress in the steelwork around the walls and roofs of the storage cells, including the critical junction with the legs.

8.1.11 Erosion

Concrete in offshore environments can suffer erosion due to wave, wind, and ice loading in certain areas. The splash zone is particularly vulnerable because of wave motion and temperature variations.

The degree of erosion depends largely on the quality of the concrete, which is typically high for offshore structures. Cracked sections or areas of low quality are more susceptible to erosion. The primary consequence of local erosion of the concrete cover is the accelerated corrosion of steel reinforcement, which can lead to a loss of strength (see Sect. 3.4.5).

Ice loading, particularly in Arctic regions, can cause significant abrasion and should be considered during the design phase. Ice-concrete friction forces lead to the gradual loss of concrete cover, and in extreme cases, can completely deteriorate the reinforcement cover of marine concrete structures. Special protection may be required to minimize abrasion from ice sheets. Most offshore concrete structures in Arctic waters are not yet in the ageing phase, but ice abrasion will need to be addressed for the life extension of such structures.

Damaged or eroded areas of concrete can be repaired using grout or specially formulated concrete (see Sect. 8.3 and Sharp and Ersdal 2021).

8.2 Detection of Material Changes and Degradation

8.2.1 Introduction

For an assessment for life extension of an ageing concrete structure, various information may be needed. Testing of existing offshore structural materials during operation is generally challenging, as the number of techniques is limited. The methods that are available can broadly be categorised into three types, similar to those for steel structures, which are:

- Destructive
- Semi-destructive
- Non-destructive.

In order to provide the relevant information about a material needed for a life extension assessment, destructive or at least a semi-destructive test is often needed. Destructive and semi-destructive techniques involve removal of materials from the structure being measured, normally followed by reinstatement of new material. As a result, destructive testing is limited in practice for offshore structures. However, the taking of cores is sometime undertaken to establish required information.

Semi-destructive techniques are less invasive than destructive techniques as only a small amount of material is removed for testing. An offshore concrete structure is typically massive and the extraction of small samples of concrete for testing is unlikely to compromise the integrity of the installation. Drilling holes to obtain samples of the concrete is one such method.

There are also a number of non-destructive methods that are further outlined in this section, along with destructive and semi-destructive methods. A good review of inspection methods for concrete is given in Malhotra and Carino (1991), ACI 311.1R-07 (ACI 2007) and ACI 228.2R-98 (ACI 2013). Such information and the appropriate inspection method is shown in Table 8.5.

8.2.2 Visual Inspection for Concrete Materials

General Visual Inspection (GVI) and Close Visual inspection (CVI) are the normal inspection methods, usually from an ROV. Concrete offshore structures are characterised by large areas of concrete, making visual inspection difficult in practice. In addition to external inspection many towers/shafts can be inspected internally manually or using drones and GVI could identify evidence of water ingress or possibly reinforcement corrosion. The Concrete in the Oceans programme reviewed typical features which could be found by visual inspection on legs/towers/shafts by an ROV or manually. These areas are listed in six groups (Department of Energy 1987) as shown in Table 8.6.

Table 8.5 Inspection method and associated parameters

Parameter	Inspection methods
Concrete compressive strength	Core sampling, rebound hammer, ultrasonic pulse velocity,
Cover depth	Magnetic scanner, ground penetrating radar
Concrete constitution	Core sampling, radiometry, permeability testing
Chloride penetration levels	Hole drilling, resistivity testing
Carbonation depth	Hole drilling, core sampling
Moisture content	Core sampling,
Corrosion of reinforcement	Visual inspection, potential measurements,
Presence of voids and cracks	Visual inspection, ultrasonic pulse velocity, infra-red thermography, radiography,
State of cathodic protection	Potential measurements on attached steelwork
Amount and location of reinforcement	Ground penetrating radar, radiography
Amount and location of prestressing tendons	Ground penetrating radar, radiography

Table 8.6 Classification of defects on concrete structures (Department of Energy 1987)

Group	Title	Number of examples
1	Defects from the construction process	34
2	Redundant built-in items	15
3	Concrete materials	12
4	Repairs and surface treatments	13
5	Superficial marks and stains	15
6	Physical damage and deterioration	26

More details of the defects in the more significant groups 4, 5 and 6 are shown in Table 8.7.

The examples in group 6 are the more important and may particularly require follow-up, depending on severity.

In addition, the American Concrete Society has published a guide to visual inspection of concrete (ACI 2008). The descriptions and photographs provided illustrate typical observations encountered during inspections and aid in the preparation of a condition survey report by identifying the characteristics of potential problems and describing their condition. There are several photographs of different forms of cracking with description. In addition, there are examples of delamination, erosion, effervescence, blistering and honeycombing. It is focussed mainly on land-based structures. A checklist of the many details that may be considered in making a report and descriptions for various concrete conditions associated with the durability of concrete is also available.

Table 8.7 Examples of defects on concrete structures (Department of Energy 1987)

Group	Title	Examples
4	Repairs and surface treatments	Making good (4), Resin systems (2), Epoxy/ pitch paints (11), Sprayed concrete (2)
5	Superficial marks and stains	Stains (4), Exudation (4), Surface cleaning marks (1), Marine growth (16)
6	Physical damage and deterioration	Impact damage (9), Abrasion (5), Deterioration of surface treatments (9), Spalling (1), Erosion (4), Deterioration of repairs (2), Deterioration of Jarlan type breakwater wall (4), Corroding reinforcement (2)

NOTE Numbers in brackets indicate number of examples

Visual inspection in the splash zone is more difficult and the main elements of a concrete GBS platform in this zone are the shafts. Inspection of these using GVI requires access, which can be by rope access or more recently by drone. The features of interest are the same as those listed above in Tables 8.6 and 8.7. However, because of the higher oxygen content in the splash zone and that cathodic protection is ineffective, corrosion of the reinforcement is more likely with subsequent rust staining, spalling or delamination of the concrete cover. As already indicated, internal inspection of a shaft could indicate evidence of water ingress and reinforcement corrosion.

Visual inspection has several limitations in that internal defects are generally undetectable by this method. Hence visual inspection is usually supplemented by NDT methods.

8.2.3 Resistivity Testing for Concrete Materials

Electrical resistivity is a method for determining the quality and durability of concrete. This method assesses the ability of concrete to resist a flow of electrical current through the concrete, thus evaluating various parameters including permeability of fluids such as chlorides. When concrete is subject to an electric field, the current starts to move through the solutions within the pores of the material. This current flow is the result of ionic movement within the pore solution. The ease with which this takes place depends on the resistivity. Several standards have been prepared for operation of this technique both in the laboratory and in the field (ASTM 2021). Resistivity is very dependent on the moisture content which needs to be assessed.

In the field the surface electrical resistivity of concrete can be measured using four electrodes—see Fig. 8.3. The two inner electrodes measure the electrical potential V created when the exterior electrodes apply an AC current to the concrete.

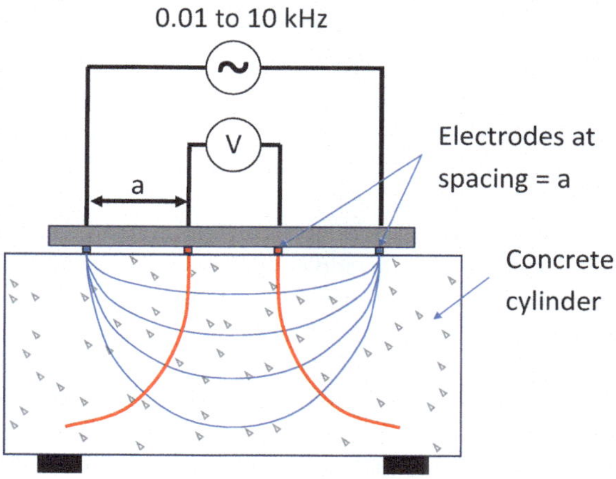

Fig. 8.3 Technique for measuring electrical resistivity of concrete

The presence of cracks and reinforcing bars in concrete can affect the resistivity measurements. In addition, probe measurements are sensitive to the surface condition of the concrete, including the presence of moisture and voids. Making electrical contact with the concrete has proved difficult as surface placed electrodes are sensitive to the surface condition. In the Concrete in the Oceans Programme Ason University devised a method by drilling to a depth of approximately 5 mm and filling the hole with an aqueous suspension of graphite which proved more reliable (Department of Energy 1987). In general, several measurements are needed to obtain a reliable average of the inherent resistivity.

Bulk measurements of resistivity can be made from a sampled core with plate electrodes placed on the top and bottom of the core. This type of measurement provides more relevant information on bulk properties linked to for example chloride ingress, but maintaining the original moisture level in the core is a critical factor.

The correlation of resistivity with chloride diffusion has been investigated. Chloride diffusion is discussed in Sect. 8.1.4. A strong relationship between resistivity and exposure zone was found reflecting the degree of contact with seawater (Department of Energy 1987). The resistivity of concrete in the atmospheric zone was 4–5 times higher than that measured in underwater concrete and concrete containing 20% pfa was found to be three times more resistive than normal concrete (Department of Energy 1987). Resistivity was also found to increase with age.

Concrete resistivity can have a direct influence on the corrosion process affected by the flow of current between anodic and cathodic regions, with an inverse proportional correlation between resistivity and corrosion rate. However, there is no single relationship between the two parameters (Hornbostel et al. 2013) and the authors claim that an upper limit of 1000–2000 Ωm can be identified for which the corrosion rate is likely to be low. In comparison, concrete with a resistivity under 50 Ωm may be vulnerable to significant corrosion, but is dependent on several other parameters.

8.2.4 Hole Drilling Tests for Concrete Materials

During immersion of concrete in seawater, ingress of chlorides occurs over time due to the permeability of the concrete. As discussed in Sect. 8.1.4, the depth of penetration and the chloride content at depths are important in the context of corrosion of the steel reinforcement.

Drilling holes to obtain samples of concrete at different depths is normal practice to establish for example chloride penetration. Samples of concrete powder are collected from various depths which are normally taken to a laboratory for analysis of the chloride ion concentration and cement content. Once results are obtained, depth profiles can be prepared showing the penetration of the chlorides and hence the susceptibility of the reinforcement to corrosion. Such methods are less feasible underwater, but the potential degradation from chlorides is also less important in these locations due to the reduced oxygen level and the presence of cathodic protection. If underwater tests are needed, core sampling can provide such information. Examples of chloride measurements by hole drilling are given in Tables 8.3 and 8.4. Hole drilling tests can also provide information on the degree of carbonation, particularly relevant above water.

After the tests, the holes can (and should) be filled with grout or epoxy to avoid seawater and chloride penetration.

8.2.5 Core Sampling Test for Concrete Materials

Core sampling can be used for more detailed assessments than what is possible with hole drilling. The depth of penetration and related chloride content can be measured by core drilling at sample locations to prescribed depths.

Concrete reinforcement and particularly prestressing cables are avoided during core drilling by controlling the location and depth of the drilling. In special cases, cores can include samples of the reinforcement if required, e.g. to establish the condition and strength of the reinforcement.

Repair after removal is needed to limit chloride ingress to the remaining reinforcement, for example by grouting.

Some results from core sampling are included in Sect. 8.1.4.

8.2.6 Rebound Hammer Testing for Concrete Materials

A Rebound hammer (also called a Schmidt hammer) test is a non-destructive testing method for concrete on-site providing a convenient and rapid measure of concrete material properties. The test involves the plunger of a rebound hammer being pressed against the surface of concrete when a spring-controlled mass with a constant energy strikes the concrete surface. The extent of the rebound is a measure of surface hardness and is measured on a graduated scale. This value is designated as a Rebound Number (rebound index). The objective of a rebound hammer test is to measure the compressive strength and it can be used to give indications of spalling and delamination. Such concrete will absorb more energy providing a lower rebound value. The test has been standardised by ASTM–C805M (ASTM 2018).

8.2.7 Half-Cell Potential Testing for Concrete Materials

The half-cell potential can be used to determine the probability of corrosion within the rebar in reinforced concrete structures. As noted in Sect. 8.1.5, reinforcing steel in concrete is normally protected from corrosion due to the alkaline nature of the concrete. However, in seawater with chlorides penetrating the concrete cover an electrochemical reaction can occur which initiates corrosion in the presence of oxygen and water. The technique utilises a reference electrode which is typically for reinforced concrete applications a copper/copper sulphate electrode ($Cu/CuSO_4$) or silver/silver chloride electrode (Ag/AgCl). By connecting this reference electrode to the embedded reinforcing steel and placing the reference electrode on the surface of the concrete, it is possible to measure the potential difference between the reference electrode and the corroding reinforcement (Fig. 8.4). The half-cell potential test has been standardised by ASTM in its document C876 (ASTM 1999) and provides a guideline on how this measurement can be undertaken. The results of the test can provide a relationship between the measured potential values and the corrosion probability. Interpretation of the result is, however, qualitative and the values given in Table 8.8 are based on the copper/copper sulphate electrode (CSE).

Table 8.8 shows the general interpretation guideline proposed by ASTM, where the measured potential ranges are categorized in three groups. It can be seen that for values less than 200 mV there is less than 10% chance of corrosion, whereas for values > 350mV there is more than 90% chance of corrosion.

The method is relatively simple to operate and it can be very useful in doing a quick assessment in identifying regions where there might be a relatively higher corrosion activity. Mapping a larger area with the half-cell method is the preferred way of localising corrosion. However, it is important to recognise that the information is qualitative and only indicates the chance of corrosion activity. There are also several factors than can influence the result, including cover thickness, concrete resistivity and the state of the concrete surface. It also requires an electrical connection to the rebar, which is not always feasible, unless the concrete cover is locally removed by drilling to allow an electrical connection to the rebar to be made.

Underwater the technique is less useful as the presence of seawater, which is electrically conducting, limits the ability to measure local potentials. Experience in the Concrete in the Oceans programme in using the half-cell method underwater on the Tongue Sands Tower found several problems (Department of Energy 1987). In addition, the reinforcement in the underwater sections of an offshore structure

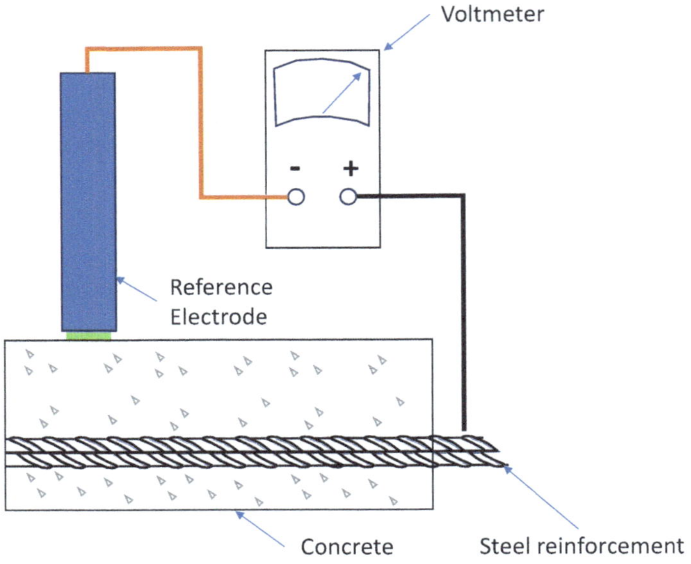

Fig. 8.4 X scheme for measuring potentials for potentially corroding reinforcement in concrete

Table 8.8 Relationship between the potential values and corrosion probability, adapted from ASTM C876 (ASTM 1999)

Measured potential (mV CSE)	Probability of steel corrosion activity
> – 200	Less than 10%
– 200 to – 350	Uncertain
< – 350	More than 90%

Note CSE is Copper/copper sulphate electrode

is usually connected to a CP system provided to protect attached steelwork, which would obscure any variations in potential.

8.2.8 Magnetic Scanners for Concrete Materials (Cover Depth)

A basic easy to operate method for determining cover depth of reinforcing bars has been ferritic scanning. These scanners produce an electromagnetic field during operation and sub-surface steel leads to changes in this field. By evaluating these changes, reinforcing bars can be located and their depth estimated. More recent developments combined with specialist software can produce 2D and 3D plots.

These scanners have some limitations, as they measure cover depth by analysing changes in the magnetic field, they are sensitive to the diameter of the reinforcing bar. As a result, larger amounts of reinforcement can lead to lower than actual cover depths. Hence knowledge of the bar diameter is required which may not be available for older structures. The scanners are also limited in depth to a cover depth of about 100–200 mm (ACI 2013). In most cases this will be sufficient but high cover depths for example in the splash zone of offshore structures may prove more difficult.

8.2.9 Ground Penetrating Radar (GPR) for Testing of Concrete Materials

Ground penetrating radar (GPR) can be applied to examine reinforced concrete structures providing an image of the reinforcement and prestressing, information on concrete resistivity and some information about the condition of the concrete.

This technique uses high-frequency radio waves, usually in the range 10 MHz to 2.6 GHz to examine reinforced and prestressed concrete structures. A transmitter and antenna emit electromagnetic energy into the concrete. When the energy encounters a buried object or a boundary between materials having different permitivities, it may be reflected or refracted or scattered back to the surface. A receiving antenna can then record the variations in the returned signal. This signal contains information on what was reflected, how quickly the signal travelled and how much of the signal was attenuated. These quantities depend on the spatial configuration and electrical properties of the member under investigation. The primary material properties that affect the transmitted and reflected energy are the dielectric constant and the conductivity. The dielectric constant of concrete is affected by the nature of the materials and moisture content. Water is the most significant dielectric contributor to construction materials and explains why radar is highly sensitive to moisture. Metals are strong reflectors of electromagnetic waves and this makes GPR very effective in locating buried metal objects, such as reinforcing bars and metal ducts.

Electrical conductivity, the reciprocal of electrical resistivity, is a measure of the ease with which an electrical current can be made to flow through a material. Concrete conditions such as voids, honeycombing, high moisture and chloride content can be inferred from changes in the dielectric constant and conductivity of the concrete (ACI 2013). The dimensions of concrete e.g. thickness and reinforcing bar depth, can generally be revealed with greater accuracy than properties (such as moisture content, chloride content, delamination) and other forms of deterioration. It has advantages over a cover meter in giving greater depth penetration.

A study was undertaken to investigate the potential to use ground penetrating radar (GPR) to locate voids, measure cover to reinforcement, placement of reinforcement and detect cracks and their depth. (Hagtvedt 2023). Concrete blocks 10 cm thick containing different flaws were manufactured and ground penetrating radar with a frequency of 4.0 GZ was used to inspect these flaws. The inspection findings from the use of ground penetrating radar were also compared to those from more common inspection tools in use at the time (electromagnetic radiation—Profometer). It was concluded that GPR was able to accurately image flaws and irregularities related to the reinforcement. However, the technique failed to provide any advantages for measuring cover-depth when compared with more standard techniques using electromagnetic radiation. Further, GPR was less reliably in identifying cracks and voids in the concrete, compared to echo hammering and visual inspections. In this respect, GPR failed to detect cracks thinner than 5 mm. Hence, the technique failed to be advantageous for the most likely field inspections. However, GPR showed some advantages when used for imaging reinforcement, particularly where the reinforcement was irregular or congested. However, the scanning speed was slower in these tests compared to other methods.

Based on the results of these experiments (Hagtvedt 2023), as well as practical experience in the field with echo hammering and rope access, ground penetrating radar appears to have a limited and specialized area of use during inspection. Overall, it was concluded that the advantage of GPR radar is for it to be used in imaging of reinforcement, or for measuring cover-depth in certain situations.

8.2.10 Ultrasonic Pulse Velocity (UPV) Testing of Concrete Materials

The velocity of an ultrasonic pulse is influenced by those concrete properties which determine its elastic stiffness and mechanical strength. The velocity of ultrasonic waves through concrete is highly dependent on the concrete's porosity and damage level (Kencanawati et al. 2018). The variations obtained in pulse velocity measurements made along different paths in a structure reflect a corresponding variation in the state of the concrete. When a region of low compaction, voids or damaged material is present in the concrete under test, a corresponding reduction in the calculated pulse velocity occurs and this enables the approximate extent of the imperfections to be determined. Hence, good quality concrete is indicated by higher velocities and slower velocities could indicate the presence of cracks and voids.

The UPV method consists of measuring the velocity of an ultrasonic pulse through concrete using a transducer and receiver. A pulse of vibrations is produced by an electro-acoustical transducer, which is held in contact with the surface of the concrete under test. When the pulse generated is transmitted into the concrete from the transducer using a liquid coupling material, it undergoes multiple reflections at the boundaries of the different components in the concrete. A system of stress waves develops, including both longitudinal and shear waves which propagate through the concrete. The longitudinal waves are the first to reach the receiving transducer. The transit time of the pulse can be determined and converted into measures of the compressive strength of the concrete, by calibrating the test results to prepared concrete cubes of different strengths.

The transducers can be operated in several modes but for offshore structures the emitting and receiving transducers are normally placed on the same surface but separated by a distance. With this indirect transmission there is some additional uncertainty regarding the exact length of the transmission path, because of the significant size of the areas of contact between the transducers and the concrete. It is therefore recommended that a series of measurements are made with the transducers at different distances apart to reduce this uncertainty.

Adequate acoustic coupling between the concrete and the transducers is necessary to ensure that the ultrasonic pulses are detected by the receiving transducer. Good acoustical contact can be achieved by the use of a suitable coupling agent and by pressing the transducer against the concrete surface. A standard exists for this test, namely ASTM-C597-16 (ASTM 2016), which covers the determination of the propagation velocity of longitudinal ultrasonic stress wave pulses through concrete.

The degree of saturation of the concrete affects the pulse velocity and this needs to be considered when evaluating test results. Pulse velocity measurements close to reinforcing steel can be higher than plain concrete of a similar composition because the pulse velocity in steel is much higher than that in concrete. This indicates that it is better to avoid measurements close to steel which is parallel to the direction of the pulse propagation.

Research shows that ultrasonic inspection, performed in a laboratory setting, is effective for determining the depth of cracks with a width under 1 mm (Silva et al. 2018).

Equinor has utilised a rolling ultrasound scanner (Elop Insight) to reveal the internal state of concrete structures that are part of one of its offshore platforms on the Norwegian continental shelf (World Oil 2022). The scanner is able to conduct scans quickly and collect data across large areas, which is especially useful for large-scale infrastructure projects, such as offshore platforms. A 3D visualization of the concrete structure on the scanner screen provides a real-time interpretation of the concrete's internal state, including elements and defects such as cracks, air pockets, voids, delamination and rebars. In addition, tendon ducts under the heavy reinforcement were able to be detected at approximately 30–40 cm depth, which is substantially deeper than has been able to be detected when using other inspection tools.

8.2.11 Infrared Thermography Testing (IRT) of Concrete Materials

Infrared thermography (IRT) is a non-destructive testing technique that has proven to be useful in locating voids, delaminations and other defects in concrete structures. The method is based on the fact that heat flows more rapid through solid concrete than through voids and delaminations. These variations in heat flow can lead to localized differences in surface temperature see Fig. 8.5.

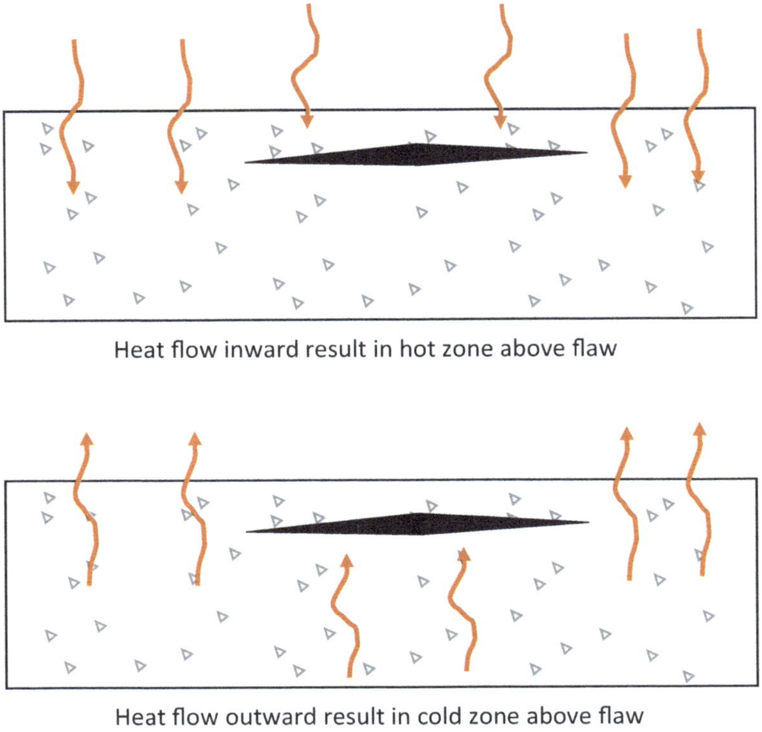

Fig. 8.5 Effect of internal anomaly on surface temperature during heat flow

Infrared thermography can be used to measure these surface temperatures under conditions of heat flow (such as sunlight), allowing the location of damage to be determined. Research has indicated that IRT under low ambient temperature conditions, the temperature difference between delaminated and normal areas in concrete structures could be about 0.2–0.3 °C (Clark 2003). IRT can be used for remote, rapid and accurate imaging of concrete structures. ASTM D 4788 (ASTM 2003) outlines the procedure for undertaking such tests on bridge decks. To measure surface temperatures and for determining delaminated areas, high-resolution infrared thermographic scanners are used, which can detect temperature variations as small as 0.2 °C.

Impulse thermography technique has been used for detection of subsurface voids in concrete structures and studied the effect of concrete age, pore content, aggregate type, reinforcement density and hydration level in detail (Maierhofer et al. 2004).

8.2.12 Permeability Measurements

Many of the degradation mechanisms in concrete involve the penetration of aggressive materials, such as chlorides, sulphates or carbon dioxide. In most cases, water is also required to sustain the degradation mechanism. Thus, concrete that has a surface zone that is highly resistant to the ingress of aggressive materials and water will generally be more durable. The quality of the surface zone is therefore a major factor affecting the rate of degradation of a concrete structure, as further described in Sect. 8.1.6. Various test methods have been devised for assessing the durability potential of a concrete surface.

Absorption tests measure the rate at which water is absorbed into the concrete under a relatively low-pressure head. The absorption rate is a function of the capillary porosity, which is dependent on the water-to-cement ratio and curing history. Several different tests have been developed for this purpose (ACI 2013). One example is the initial surface-absorption test (ISAT) which provides a surface measurement and models water absorption. This is the most common ingress mechanism associated with concrete deterioration. The test has been widely used and a considerable amount of field data and experience has been gained (ACI 2013). The test is non-destructive and can be used on both vertical and horizontal surfaces but cannot easily be used underwater. Another limitation is that this method can only test the outer 10 to 15 mm of the concrete which is not fully representative of the level of offshore cover to the reinforcement.

In addition, a wide variety of tests that measure the flow of air (or other gases) through concrete have been proposed (ACI 2013). There are relationships between air permeability and other durability indicators, such as strength, and curing efficiency (Whiting 1981). Similar to the water-based tests, air-permeability tests involve either drilling a hole into the concrete surface or the application of a chamber to the surface. In most cases, a vacuum is applied and the decrease of the vacuum with time is taken as an indicator of the air permeability. A higher air-permeability index indicates a less permeable concrete. In general, air-permeability tests are simpler to perform than the water-based tests. Air permeability tests appear to be less affected by the moisture content of the concrete than water-permeability tests.

As noted above, some surface tests are affected primarily by the top 15 mm of concrete and do not provide an indication of the penetrability properties of the entire cover zone (ACI 2013). An alternative is the Figg's method which require drilling a hole and exclude the near surface layer in the measurement.

Generally, each test method results in an air-permeability index that can be related to other factors that govern the durability of concrete. However, the index obtained by a specific method is not usually related to the index obtained by another method that uses a different geometry and test pressure.

8.2.13 Nuclear Methods

Nuclear methods for non-destructive evaluation of concrete can be subdivided into two groups, namely radiometric and radiographic methods. Both involve gaining information about a test object due to the interactions between high-energy electromagnetic radiation such as gamma rays and the material in the test object.

Radiometry is used to assess the density of fresh or hardened concrete by measuring the intensity of electromagnetic radiation (gamma rays) that has passed through the concrete. The radiation is emitted by a radioactive isotope and the radiation passing through the concrete is sensed by a detector. One method is based on measurement of gamma rays after transmission directly through the concrete and the other is based on measurement of gamma rays reflected, or backscattered, from within the concrete.

This technique is similar to the ultrasonic through-transmission technique. The most commonly available equipment measures an average density between the immersed source and the surface detector and is not able to identify areas of low compaction at specific depths (ACI 2013).

Radiography involves the use of the radiation passing through the test object to produce an image of the internal structure of the concrete similar to the X-Ray method. Radiographic methods allow the possibility of seeing the internal structure of a concrete member where density variations exist, such as the presence of reinforcement, prestressing and significant voids and cracks. The maximum thickness for gamma-ray radiography is about 500 mm, because thicker members require unacceptably long exposure times. Radiographic procedures are expensive and require evacuation of the structure by persons not involved in the actual testing.

8.3 Mitigation of Concrete, Reinforcement and Tendons as a Material

Damage to concrete structures can arise from numerous causes, including accidental events such as dropped objects, from reinforcement corrosion leading to cracking and delamination and also from excessive loading causing structural cracking.

Typical methods for repair include:

- Injection methods (e.g. cracks or voids).
- Placement methods (damaged concrete material replaced by cementitious material or epoxy resins).
- Re-instating damaged reinforcement and prestressing.

Materials used include concrete, cementitious grouts and resins. For repair in air there are several different products available, while for repair underwater specialised products are required (Sharp and Ersdal 2021).

Injection or placement of a cementitious grout or epoxy material is a versatile and historically widely used repair method. Injection methods are particularly relevant for repairs of cracks and voids. According to OTH 90 318 (Department of Energy 1990b) it has been shown that cementitious grouts have difficulty in penetrating cracks of surface width of 10 mm and impossible below 5 mm. Hence, for cracks less than 10 mm in width epoxy resins would be the repair material of choice.

There are some factors which affect the use of epoxy materials for repair. These include the influence of temperature which can have a significant impact on the material's cure time. This is important underwater with typical North Sea temperatures. The use of epoxy material is not usually employed to fill voids in large volumes due to challenges with heat generation during the curing of the epoxy, creep properties as well as constraints in the material's shelf life.

A study was funded by the Department of Energy to assess the efficiency of different materials for repairing damage to the simulated domes of the roofs of underwater concrete storage tanks (Department of Energy 1990b). The roofs of these storage tanks can suffer damage from accidentally dropped objects; several occurrences of such damage followed by repair have been observed (Sharp and Ersdal 2021).

In this study several concrete slabs and domes (simulating roof members of storage tanks) were subjected to controlled punching shear damage. The level of damage varied from limited cracking on the back of the element to a hole in the element with total loss of concrete from the shear plug. In this worst case the reinforcing steel may still have been intact but damaged.

These were then repaired underwater in a tank of seawater at temperatures of 7 °C, typical of those in the North Sea. After suitable storage time in the tank each repaired member was destructively tested by static loading to assess the performance of the different repair materials.

Materials for undertaking these repairs were identified and included:

- Three cementitious grouts which had properties of high fluidity and non-shrink. One was specially formulated for underwater applications.
- Two epoxy resins, formulated for the repair of water immersed concrete structures by injection.

The tests of the repaired members showed that a regain of a large proportion of the original strength could be achieved. In addition, it was demonstrated that this type of repair can satisfactorily seal cracked elements preventing leakage. This was received as a promising result for the underwater injection repair of damaged concrete members. However, it was noted that severe lack of bond occurred with resin repairs where there had been a problem with surface preparation. The lack of bond was less of a problem for cementitious repairs, but bond strengths where in general less than those best achieved with epoxy resins.

Tests were also carried out on all the repair materials, using slant shear tests to BS 6319 part 4 (British Standards 1984) which tests a resin or grout bonded scarf jointed prism in compression. The compressive strength of the cementitious materials was also undertaken. The specimens were tested after 28 days and for

Table 8.9 Materials used for actual repairs (Sharp and Ersdal 2021)

Cause of damage	Consequence of damage	Repair materials used
Vessel impact	Water seeping though cracked wall	Epoxy resins and caulking
Faulty construction joint	Leak	Initially resins used, later injection of cementitious grouts
Leaky grouted prestressing duct	Water leakage	Resin injection
Dropped object	Damage to cell roof	Prepacked aggregate followed by grout injection
Cracks in external shear walls	Water ingress	Resin
Scour hole at pipeline entry point	Hole extension	Void grouted

periods up to 300 days. The samples were kept under seawater at a typical North Sea temperature. The majority of the concrete compressive strengths after 28 days were in the range 40–54 N/mm^2. Slant shear tests on the resin repaired slabs after 28 days showed changes in bond strengths from 50 to 90% whilst for cementitious grouts the changes in bond strengths were between 50 and 116% of the concrete strength. It was noted that the surface preparation of the parent concrete played a large factor in these results. It was also found that epoxy resin generally bonded well but where there was any contamination the bond was very limited. Actual repairs to offshore concrete structures using both resins and cementitious grouts are shown in Table 8.9.

Repair of damaged reinforcement may be required when significant damage has occurred. In addition, corroded reinforcement may require repair. Removal of the corrosion products can be achieved by high pressure water jets or by abrasive blasting. If the cross section of the reinforcement has reduced significantly new similar reinforcing bars may need to be added, by splicing or welding. However, welding underwater is difficult.

Loss of prestress in tendons can occur and can be reinstated using different methods, as further described by Sharp and Ersdal (2021).

8.4 Assessment of Concrete Material for Life Extension

Concrete material is unusual as the main property, compressive strength, is assumed to increase with age. Concrete shows significant effects of aging caused by the slowly developing micro-structure through the deposition of calcium-silicate-hydrates (CSH). Concrete by itself is a complex composite material consisting of cement paste and aggregate. The addition of reinforcement bars and pre-stressing steels further complicate the situation.

Various chemical substances in seawater have both beneficial and unfavourable impact on the performance of cementitious materials under a marine environment. The primary existence of chloride ions, sulphate ions and magnesium ions in the seawater can cause complex chemical reactions between cementitious material.

As we have seen in this book, the diffusion of chlorides to the steel reinforcement is an important source of degradation of concrete structures. It is recognised that concrete offshore structures have typically been designed for a 20 year lifetime, with high quality concrete (low permeability), adequate cover to the reinforcement and prestressing tendons encased in grouted steel ducts. Within the design life, these mitigations should ensure that chlorides do not reach the reinforcement.

However, as many concrete structures are now in a life extension stage there is more experience indicating that particularly reinforcement corrosion and spalling in the splash and atmospheric zones are major issues. Underwater the concrete structures are showing good performance, although there is less experience of concrete in deep water for long periods. An important factor in life extension will be to determine to what extent the chloride penetration has reached the reinforcement, in addition to the potential beneficial effects of limited oxygen and cathodic protection underwater.

In contrast to steel structures, it has been accepted that concrete structures should be less vulnerable to fatigue degradation.

In general, life extension assessment requires knowledge of the degradation processes that have occurred, in addition to a prediction of their role in the life extension period. Inspection and monitoring are therefore important in providing the necessary information. However, it should be acknowledged that these can only provide a limited understanding of the state of the concrete structure when life extension is being planned.

References

ACI (2007) ACI 311.1R-07 ACI Manual of concrete inspection

ACI (2008) ACI 201.1.R-o8 Guide for conducting a visual survey of concrete in service

ACI (2013) ACI 228.2R-98 Non-destructive test methods for evaluation of concrete in structures

ASTM (1999) ASTM C876 Standard test method for corrosion potentials of uncoated reinforcing steel in concrete

ASTM (2003) ASTM D 4788 Standard test method for detecting laminations in bridge decks using infrared thermography

ASTM (2016) ASTM-C597-16 Standard test method for pulse velocity through concrete

ASTM (2018) ASTM C805/C805M standard test method for rebound number of hardened concrete

ASTM (2021) ASTM C1760 Standard test method for bulk electrical conductivity of hardened concrete

British Standards (1984) BS 6319 Testing of resin and polymer/cement composition for use in construction. Part 4: Method for measurement of bond strength (slant shear method). BSI, London, UK

Clark MR, McCann DM, Forde MC (2003) Application of infrared thermography to the non-destructive testing of concrete and masonry bridges. J NDT&E Int 36(4):265–275. https://doi.org/10.1016/S0963-8695(02)00060-9

Clayton N (1986) International conference on concrete in the marine environment, London, 1986. Concrete Society, London
Department of Energy UK (1987) Concrete in the oceans programme—coordinating report on the whole programme, OTH 87 248, HMSO, London, UK OTH 90 318
Department of Energy (1990a) Concrete offshore in the nineties—COIN, Offshore Technology report OTH 90 320, Prepared by Imperial College for the Department of Energy, HMSO, 1990
Department of Energy (1990b) OTH 90 318 Assessment of materials for the repair of damage concrete underwater, Prepared by Imperial College for the Department of Energy, HMSO, 1990
Department of Energy (1990c) OTH 90 320 Effect of bacterial activity on North Sea concrete
Hagtvedt G (2023) An evaluation of ground penetrating radar for inspection of concrete by experimental method). Thesis, Stavanger University
Haynes HH, Highberg RS (1976) Long term deep ocean test of concrete spherical structures. Civil Engineering Laboratory, Jan 1976, D7.6.5R
Hornbostel K, Larsen CK, Geiker MR (2013) Relation between concrete resistivity and corrosion rate—a literature review. J Cement Concr Compos 39(May 2013):60–72
Hove K, Jacobsen B (1998) Pressure effects on design of deep-water concrete platforms. In: Proceedings of the 2nd international conference on Concrete under severe conditions CONSEC'98, Volume 3, E and FN Spon
HSE (1992) OTH 92 364 Long term durability of grouts in marine environments, prepared by Imperial College for HSE. HSE Books, London, UK
HSE (1997) OTO 97 053 Durability of prestressing components in offshore structures. Gifford and Partners, HMSO London, UK
ISO (2006) ISO 19903 Concrete structures
Jacobsen UH (2011) 14th Euroseminar on microscopy applied to building materials
Kencanawati NN, Akmaluddin A, Anshari B, Paedullah AG, Shigeishi M (2018) The study of ultrasonic pulse velocity on plain and reinforced damaged concrete. MATEC Web of Conf 195:02026, ICRMCE 2018. https://doi.org/10.1051/matecconf/201819502026
Maierhofer C, Wiggenhauser H, Brink A, Röllig M (2004) Quantitative numerical analysis of transient IR-experiments on buildings. J Infrared Phys Technol 46(1–2):173–180. https://doi.org/10.1016/j.infrared.2004.03.022
Malhotra VM, Carino NJ (1991) Handbook on nondestructive testing of concrete. CRC Press Inc.
MATS (1984) Comparative study of the behaviour of reinforced concrete prisms after 2 and 8 years exposure to North sea water at depths of 5 and 100 m. TNO-IBBC report B-84-486-60.5.0400
MATSU (1991), Localised corrosion of the reinforcement in offshore and marine concrete, Prepared by Wilkins NGM and Sharp JV, MaTR87
MTD (1990) Design and operational guidance on cathodic protection of offshore structures, sub-sea installations and pipelines. MTD Publication 90/102
Neville AM (1973) Properties of concrete. Pitman Publishing, Sir Isaac Pitman and Sons LTD, Pitman House, London WC2B 5PB, UK
PSA (2006) Material risk—ageing offshore installations, Prepared by DNV for the Petroleum Safety Authority Norway (now Norwegian Ocean Industry Authority). Stavanger, Norway
Sharp, Ersdal (2021) Underwater inspection and repair for offshore structures
Silva MTA, Rocha JHA, Monteiro ECB, Póvoas YV, Kohlman Rabbani ER (2018) Evaluation of the ultrasound test for estimating the depth of cracks in concrete. https://revistaalconpat.org/index.php/RA/article/view/289/476#info
Whiting D (1981) Rapid determination of the chloride permeability of concrete. Engineering Materials Science
World Oil (2022) Equinor becomes first operator to scan offshore structure with rolling ultrasound scanner (worll.com)
Zhou J, Zhao X, Nie Y, Tian Y (2020) Influence of high water pressure on static and dynamic compressive strength of concrete. Adv Civ Eng 2020

Chapter 9
Other Metallic and Non-metallic Materials

9.1 Introduction

In addition to steel, concrete and grout there are other structural materials which are used in a limited capacity in offshore structures. This includes:

- aluminium (which is primarily used on topside modules like helidecks and living quarters),
- composites (which are mainly used on topsides in addition to being used as a structural repair material),
- coatings for protection of structural and non-structural members (of material such as epoxy, polyurethane, coal-tar and zinc-based),
- adhesives and epoxy resins (which are mostly used for repairs).

A brief overview of these materials, their application, properties and degradation mechanisms are provided in this Chapter.

In addition, titanium is to a very limited extent used for structural purposes, primarily as a cladding where its exceptional corrosion resistance makes it a suitable choice. It helps protect the underlying structural components from corrosion, ensuring the longevity and structural integrity of the offshore platform. Even though titanium is highly reactive, on exposure to the marine environment, it reacts with atmospheric oxygen to form a protective layer. This layer makes it highly corrosion resistant. As a result, titanium requires no corrosion allowance and equipment can be designed to satisfy the minimum requirements for mechanical strength and handling. Titanium alloys are widely used for risers, heat exchangers, downhole tools and subsea production equipment (MTD 1996b), which are outside the scope of this book.

© The Author(s), under exclusive license to Springer Nature Switzerland AG 2025

J. V. Sharp and J. Wintle, *Assessment of Materials in Ageing Offshore Structures*,
https://doi.org/10.1007/978-3-031-84489-8_9

9.2 Aluminium

9.2.1 Introduction

Aluminium is a versatile and widely used structural material with various applications both onshore and offshore. One of the key advantages of aluminium is its low density. Despite its lightweight nature, aluminium exhibits a high strength-to-weight ratio, providing structural strength without excessive weight. This is particularly advantageous in applications where weight savings are critical.

Compared to carbon steels, aluminium alloys have a lower melting point and exhibit a more severe degradation of mechanical properties with increasing temperatures. In almost all cases aluminium alloys need to be protected (insulated) in order to fulfil prescribed fire tolerance requirements. There are a number of different fire protection approaches that may be applied to aluminium alloy structures to enhance their fire resistance (Wang et al. 2022).

Aluminium naturally forms a thin, protective oxide layer on its surface, providing excellent corrosion resistance. This property makes aluminium well-suited for marine applications, where exposure to harsh environmental conditions is of concern. Aluminium is also highly ductile and malleable, allowing it to be easily formed into various shapes and profiles. This makes it adaptable to a wide range of structural designs and manufacturing processes, including extrusion, casting, and rolling.

9.2.2 Applications of Aluminium in Offshore Installations

Aluminium is not commonly used in the primary structure for offshore installations as it has lower strength compared to steel. However, aluminium does find some applications in offshore installations in modules and equipment, where its specific properties are advantageous. Some typical applications for aluminium in offshore installations include:

- Helidecks: The landing area of helidecks is typically constructed from extruded aluminium planks. These have a ‘ribbed’ surface and may be provided with a suitable finish (non-slip paint or grit-blasted finish).
- Accommodation modules: Weight saving may be critical for accommodation modules on offshore installations. As a result, these are often constructed from robust light weight corrosion-resistant materials aluminium alloys to combat the corrosive effects of saltwater.

In addition, aluminium is used in various outfitting and tertiary, such as access platforms, walkways, stairways and handrails.

9.2.3 Microstructure of Aluminium

Various aluminium alloys have been developed to enhance specific properties for structural applications. The microstructure of aluminium alloys commonly used in these applications, plays a crucial role in determining the material's mechanical properties and overall performance. Aluminium alloys are engineered to achieve specific combinations of strength, ductility, and other characteristics suitable for structural use. Aluminium has a face-centred cubic (FCC) crystal structure at the atomic level. This crystal structure contributes to its overall ductility and formability. Structural aluminium alloys often incorporate alloying elements to enhance specific properties (Dwight 1999). Common alloying elements include copper (Cu), magnesium (Mg), silicon (Si), and zinc (Zn). These elements form solid solution strengthening, precipitation hardening phases, or intermetallic compounds within the aluminium matrix. Solution heat treatment involves heating the material to a high temperature to dissolve alloying elements, followed by quenching to trap these elements in a supersaturated solid solution. For example, copper atoms can be dissolved into the aluminium matrix, enhancing the overall strength of the material by impeding the movement of dislocations. Many structural aluminium alloys undergo a precipitation hardening process. This involves the controlled formation of fine precipitates, typically in the form of intermetallic compounds, through heat treatment. These precipitates act as barriers to dislocation movement, contributing to increased strength and hardness (Dwight 1999).

The microstructure of structural aluminium includes grains that form during the solidification process. The size and distribution of these grains influence the material's mechanical properties. Smaller grains, achieved through proper alloying and heat treatment, contribute to improved strength and fatigue resistance. Heat treatment is a critical aspect of shaping the microstructure of structural aluminium (Dwight 1999).

The T6 temper, one of the most common tempers for structural aluminium, involves solution heat treatment followed by artificial ageing. This results in a balanced combination of strength and ductility, making it suitable for various structural applications.

Cold working or deformation during the manufacturing process can lead to the formation of dislocations and distortions in the crystal lattice. Subsequent annealing processes, involving heating and controlled cooling, induce recrystallization, resulting in a refined and more uniform microstructure.

In summary, the microstructure of structural aluminium is carefully engineered through alloying, heat treatment, and other processes to achieve the desired combination of strength, ductility, and other mechanical properties. This tailored microstructure contributes to the effectiveness of aluminium alloys in a wide range of structural applications offshore.

9.2.4 Joining Methods for Aluminium

The joining processes for aluminium includes fusion welding techniques like TIG, MIG and the specialized process of Friction Stir Welding (Dwight 1999). The importance of matching filler metals to parent metals is highlighted for weldability and mechanical properties.

Porosity in the material is often associated with TIG and MIG welding. This is caused by absorption of hydrogen into the weld pool leading to discrete pores in the solidifying weld metal. Several sources of hydrogen exist, including hydrocarbons and moisture from contaminants and from filler wire surfaces and water vapour from the shielding gas atmosphere. Trace levels of hydrogen may also exceed the threshold concentration required to nucleate bubbles in the weld pool. To minimise the likelihood of porosity, rigorous cleaning of material surfaces and filler wire should be carried out. Three cleaning techniques are suitable; mechanical cleaning, solvent degreasing and chemical etch cleaning (TWI 2024).

Air entrainment should be avoided in gas shielded welding, by making sure there is an efficient gas shield and the arc is protected from draughts. Water vapour pickup from gas lines and welding equipment should be avoided. In general, it is recommended that the welding system is purged for about an hour before use (TWI 2024).

Friction stir welding is a unique process that makes it possible to weld extruded aluminium profiles through friction heating without the need for either shielding gas or filler metal. The lower heat required for welding the profiles means less distortion, and the technology produces panels with better mechanical properties than fusion welding.

9.2.5 Degradation Mechanisms for Aluminium

Aluminium, while known for its corrosion resistance in certain environments, can still experience degradation mechanisms in offshore settings due to the harsh and corrosive conditions. While aluminium forms a protective oxide layer that resists corrosion, the presence of chloride ions in seawater can break down this layer, leading to corrosion. In addition, pitting corrosion can occur in offshore environments with high chloride concentrations if the protective oxide layer is breached. Pitting corrosion can be particularly damaging, as it can lead to rapid and localized material loss. Further, stress corrosion cracking (SCC) can be a concern for aluminium components due to the combination of stress and the corrosive environment.

The combination of materials (e.g., aluminium and steel) in offshore structures may lead to galvanic corrosion (occurring when two dissimilar metals are in contact in the presence of an electrolyte). Proper corrosion protection measures, such as the use of sacrificial anodes, are often employed to mitigate galvanic corrosion.

To mitigate these corrosion related degradation mechanisms, proper material selection, corrosion protection strategies (such as coatings and cathodic protection), regular inspection and maintenance practices are essential for offshore installations using aluminium components. Understanding the specific environmental conditions and potential sources of corrosion is crucial for designing and maintaining structures that withstand the challenges posed by the offshore environment.

Aluminium is also sensitive to cyclic loading and fatigue and has a finite fatigue life. Its susceptibility to fatigue depends on factors such as the specific alloy, loading conditions and environmental factors. The fatigue behaviour of aluminium is characterized by the formation and propagation of microcracks, which can eventually lead to macroscopic cracks and structural failure.

9.2.6 Standards for Aluminium as an Offshore Structural Material

As mentioned above, the landing area of helidecks is typically constructed from extruded aluminium planks. Standards for UK offshore helicopter landing areas are covered in UK Civil Aviation Authority's standard CAP 437 (Civil Aviation Authority 2023). API RP 2L (API 2012) provides a guide for planning, designing and construction of helidecks for fixed offshore platforms, including design load criteria and other heliport design recommendations. NORSOK standard C-004 (Standards Norway 2019) also defines the basic requirements for design, arrangement and engineering of helicopter decks on offshore installations. This standard is further supported by the NORSOK N-004 (Standards Norway 2022) for structural issues related to the helideck structure. DNV-OS-E401 (DNV 2001) covers requirements for material and structural strength of the helicopter deck and support structure.

9.3 Composites

9.3.1 Introduction

Composite materials are being used in offshore structures due to their unique combination of properties, such as high strength-to-weight ratio, corrosion resistance and durability. As such these materials offer several advantages over traditional materials in offshore applications. Composite materials typically consist of a matrix (polymer) reinforced with fibres. Common fibres used in offshore structures include glass, carbon and aramid fibres. The choice of fibre depends on the specific requirements of the application.

The high strength-to-weight ratio is particularly beneficial in offshore applications where weight reduction can lead to easier transportation, installation and improved overall structural performance. One of the main advantages of composites is their resistance to corrosion, a significant concern in the harsh marine environment. This advantage extends the service life of offshore structural components and helps to reduce inspection and maintenance costs.

Tailoring the fibre orientation and layup can optimize the structural performance for specific loading conditions. Composite materials can exhibit excellent fatigue resistance, making them suitable for applications where cyclic loading from waves, wind and currents is a factor. However, it's essential to note that the fatigue behaviour of composites is highly dependent on proper design, manufacturing quality, and material selection. Issues such as delamination, fibre breakage or manufacturing defects can impact fatigue performance.

Unlike metals, composite materials are electrically insulating, which can be an advantage in certain offshore applications and reduces the likelihood of corrosion due to stray currents.

Particular concerns in the application of composites offshore include:

- Flammability
- Impact performance
- Durability in the offshore environment
- Ability of use NDT techniques during fabrication and deployment
- Lack of regulations and standards in their early use offshore.

As a result of these concerns research has been undertaken. These programmes included a MTD lead programme entitled "Cost Effective use of Composites Offshore funded by Government and 20 industrial sponsors" (MTD 1996a, b). The programme concentrated on predicting the behaviour of composites subjected to impact, fire, water and other fluids under pressure. In addition, some progress was made in determining buckling and materials failure for composite structures under complex loading. Adhesive methods for fabricating layered panels were also developed. Later work in the programme included wear, erosion, NDT and performance under higher temperatures and pressures.

9.3.2 Applications

Composite materials are used offshore in a number of applications, typically where high strength-to-weight and stiffness-to-weight ratios are required. According to MTD (1996a, b) and Setvati et al (2014) applications include:

- Fire and blast protection panels.
- Equipment enclosures (for example for safety and emergency equipment such as emergency control rooms).
- Caissons (open ended vertical tubes that can house fire water pumps and their risers).
- Protection for wellheads, manifolds and pipelines.

- Walkways and flooring (gratings, handrails etc.).
- Lifeboats and floats (such as TEMPSC).
- Accommodation modules.
- Repair of damaged steelwork.
- Piping systems, tanks and vessels.

The non-conductive and nonmagnetic high-temperature polymer composites are considered to be ideal materials for construction of resistivity and induction logging tools for oil and gas exploration.

Specific examples of the use of composites are (MTD 1996a, b):

- Protection of the helideck and part of the accommodation area of Amerada Hess's Ivanhoe/Rob Roy platform.
- Composite enclosure for the emergency control room on the Claymore A platform.
- Composite fire and blast panels on the Ekofisk Tank control room.
- Composite caissons on the Total Durbar platform.
- Filament-wound carbon fibre/epoxy composite accumulation bottles in tensioning devices for the production risers used on the Shell's Mars production platform in the Gulf of Mexico.

Between 1995 and 2000, over 100,000 m^2 of phenolic gratings were installed on offshore platforms with one installation, the Ursa TLP in the Gulf of Mexico, accounted for over 20,000 m^2 (HSE 2003).

When used in a sandwich configuration to maximize stiffness and fire integrity, composites can achieve a weight advantage of the order of 30% compared with traditional corrugated steel fire and blast wall structures. While composite panels are generally more expensive than carbon steel ones they do not corrode or require painting. They are usually less expensive than stainless steel panels (HSE 2003).

9.3.3 Microstructure of Composites

The microstructure of composites plays a crucial role in determining their mechanical, thermal and electrical properties. The matrix surrounds and binds the reinforcing materials in a composite. It is typically a polymer but can also be a metal or ceramic. A polymer matrix can have an amorphous or crystalline structure. The arrangement of polymer chains affects the overall mechanical and thermal properties of the composite.

The reinforcing phase consists usually of fibres which are usually aligned in a specific direction to maximize strength along the desired load-bearing axis. The orientation and volume fraction of the reinforcing phase significantly influence the composite's anisotropic behaviour. Fibre reinforced composites of glass and carbon are both used as unidirectional strands, woven roving and chopped strands.

Table 9.1 Candidate resins for composites for the offshore industry (MTD 1996a, b)

Resin	Mechanical strength	Corrosion resistance	Fire performance
Polyester	XX	XX	X
Vinyl ester	XXX	XXX	X
Modar (modified acrylic resin)	XX	XX	XXX
Epoxy	XXXX	XXXX	X
Phenolic	XX	XX	XXXX

The interface is the region where the matrix (polymer) and reinforcing materials interact. A strong and well-bonded interface is crucial for transferring loads effectively between the matrix and reinforcing phase. Poor interface bonding can lead to reduced mechanical performance. Surface treatments or coatings on the reinforcing materials are sometimes employed to improve adhesion with the matrix and enhance overall composite properties.

Composite materials may contain fillers or additives to modify specific properties to, for example improving thermal conductivity, flame resistance and reducing cost. The distribution of these elements within the matrix contributes to the composite's overall microstructure.

The selection of resins is important in deterring strength, damage tolerance corrosion resistance and thermal stability. In a study in 1996, MTD (1996a, b) identified the most commonly used resins at the time, as shown in Table 9.1.

Epoxy and vinyl ester resins are the two most commonly used types of resins in composite materials for offshore structures. As can be seen from Table 9.1 these resins have the best mechanical strength and corrosion resistance. However, their fire performance is limited.

The fabrication process has a strong influence in governing the shape, structural integrity and cost. The processes used include contact moulding, resin transfer moulding, hot and cold press moulding, filament winding and pultrusion. Porosity may result from manufacturing processes or inadequate consolidation, and it can weaken the material. Inclusions, if present, may disrupt the uniformity of the matrix and affect overall performance.

Microscopic features such as voids, cracks and fibre/matrix interactions at the microscale play a role in determining the material's overall behaviour. The size and distribution of these features can impact mechanical properties, fatigue resistance and failure modes.

9.3.4 Degradation of Composites

Fiber-Reinforced Polymers (FRP) can undergo degradation over time due to various environmental factors and mechanical stress. The degradation mechanisms depend on the specific polymer matrix, the type of reinforcing fibres used, and the environmental conditions to which the FRP is exposed. Here are some common degradation mechanisms for FRP (MTD 1996a, b):

1. Chemical Degradation leading to polymer matrix attack: Examples include exposure to harsh chemicals (e.g. acids, some solvents) or some environmental conditions which can lead to chemical degradation of the polymer matrix.
2. UV Degradation: Ultraviolet (UV) radiation from sunlight can cause degradation of a polymer matrix. Over time, UV exposure can lead to discoloration, loss of surface gloss, and a decrease in mechanical strength.
3. Fatigue: Repetitive loading or cyclic stresses can lead to mechanical fatigue, especially in dynamic applications. Over time, the repeated stress cycles can cause microcracks to form within the composite, leading to a gradual reduction in strength and stiffness.
4. Microbial Attack: In certain environments, FRP may be susceptible to microbial attack, leading to biodegradation. Microorganisms such as bacteria and fungi can colonize the composite surface, causing chemical changes and deterioration of the polymer matrix.
5. Thermal Degradation: Elevated temperatures can cause thermal degradation of both the polymer matrix and reinforcing fibres. Excessive heat can lead to the breakdown of polymer chains and a reduction in mechanical properties.
6. Abrasion and Wear: In applications where FRP is exposed to abrasive forces, such as in marine environments with sand and debris, abrasion and wear can occur over time. This can lead to a loss of surface finish and, in severe cases, affect the structural integrity of the composite.

9.3.5 NDE/NDT

Composites present particular problems for NDE both during manufacture and in-service inspection. This is limited by their poor electrical conductivity, low thermal conductivity, high acoustic attenuation and significant anisotropy of both the mechanical and physical properties, all which are needed for many of the standard NDE methods. Ultrasonic methods have the greatest potential to detect damage.

The NDT methods with the most potential are ultrasonic, thermal imaging and radiography. The key features of in-service inspection are damage detection and assessment and the assessment of the integrity of bonded joints.

Because of the complexity of composite materials damage identification cannot be achieved via conventional NDE methodology. Only ultrasonic methods are directly sensitive to mechanical changes and can be used to directly assess the mechanical condition and integrity of a composite structure (Djordjevic 2009).

9.3.6 Standards for the Use of Composites

In the early days of the use of composites offshore there were no design codes that readily permitted the use of composites in load-bearing structures. In 1998, the U.S. Coast Guard approved the use of fibre-reinforced plastic (FRP) piping and glass/ phenolic gratings on offshore platforms and drill ships. The International Maritime Organization (IMO) has approved composite piping for use in platform fire-fighting systems carrying water. Limit state codes such as the DNV Design Guideline for Design with Composites (DNV 2000) were also introduced.

The DNV-ST-C501 standard (DNV 2013) provides requirements and recommendations for the structural design and structural analysis procedures for composite components. Aspects related to documentation, verification, inspection, materials, fabrication, testing and quality control are also addressed. It is intended to serve as a common reference for designers, manufacturers and end-users. Failure mechanisms including delamination, matrix cracking, buckling and wear resistance are addressed.

9.4 Coatings for Structural Steel and Concrete

Coatings are widely used to safeguard surfaces against deterioration induced by the marine environment, including salt spray and marine fouling (barnacles). In response to the escalation of fouling associated with elevated water temperatures, contemporary plastic coatings are increasingly utilized for antifouling protection, while specific epoxy coatings find application in corrosion protection. These coatings function as both a physical and chemical barrier, mitigating material degradation. Proper coatings help extend the service life of these structures and ensure their structural integrity. Coatings are used on both steel and concrete offshore platforms.

Typically, protective coatings are applied in multiple layers, resulting in a film with a thickness ranging from 0.25 to 0.5 mm. Common coating materials for corrosion protection include epoxy, coal-tar epoxy, polyurethane and zinc-based coatings. Epoxy Coatings are typically used for corrosion protection and have excellent adhesion, chemical resistance and durability. They are often used as a primer in multi-coat systems. Polyurethane coatings are often used as a topcoat for corrosion resistance. They have resistance to UV and abrasion with good weathering properties. Coal-tar epoxy coatings were earlier used for corrosion protection. As an example, the coating on the West Sole WE platform used this type of coating (see Sect. 12.1). Zinc-based coatings can prove a sacrificial coating as an additional layer of protection.

Coatings must adhere well to the steel surface to ensure effective protection. Proper surface preparation, including cleaning and roughening, is crucial for achieving good adhesion. Additionally, the coating itself should have good cohesion to resist cracking and delamination over time. Proper surface preparation requires a clean, rough and well-adhering substrate for the coating. The specific procedures may vary based on the type of coating, the condition of the substrate and environmental factors.

The surface preparation for offshore coatings would typically include cleaning to remove contaminant using high-pressure water jets. Mechanical abrasion could also be used such as abrasive grit blasting to remove rust, old coatings and mill scale. This process also creates a rough surface profile, enhancing the adhesion of the new coating. After abrasive blasting, solvent cleaning is often used to remove any remaining contaminants, oils or residues. This ensures a clean and dry substrate for coating application. In some cases, a primer or base coat may be applied after surface preparation to provide additional adhesion and corrosion resistance. The choice of primer depends on the coating system being used. Prior to full-scale coating application, adhesion testing may be performed to ensure that the surface preparation has been effective. This helps identify any areas with insufficient adhesion and allows for corrective measures.

It's important to note that the specific requirements for surface preparation are often outlined in coating specifications provided by for example industry standards. Regular inspection and maintenance of coatings are also critical to address any issues that may arise over time.

Coatings on offshore structures are susceptible to degradation. UV exposure from intense sunlight may lead to discoloration and loss of mechanical properties in coatings over time, contributing to the degradation of certain coating materials. In addition, mechanical abrasion due to for example wave actions and vessel collision can cause damage to the coating, reducing its protective properties. Further, temperature fluctuations can damage the integrity of coatings, potentially leading to cracking or delamination. Marine organisms, such as barnacles and algae, can attach to and grow on the coated surface. While antifouling coatings are designed to prevent or minimize this, they may degrade over time, allowing fouling to occur. If the coating is compromised (for example not properly applied) or damaged, water can penetrate the coating and it becomes less effective in protecting against corrosion. As a result, degradation of the underlying steel accelerates.

To mitigate these degradation issues, regular inspection of coatings is essential and damaged areas may need to be recoated having removed the previous coating and any potential corrosion products. When local damage is repaired, feathering of the edges of the repaired area is required to create a smooth transition between the repaired coating and the surrounding intact coating. Periodic inspections can help identify areas of coating damage or degradation early on, allowing for timely repairs or recoating to maintain the protective integrity of the system. Additionally, advancements in coating technologies and the development of more durable and resistant coatings contribute to improving the longevity and performance of coatings on offshore structures.

Epoxy coatings are also widely used on offshore concrete structures for surface protection, particularly in the splash zone and above water. The epoxy coating is providing additional protection against seawater (chloride) ingress. The primary use is to improve the durability of the concrete in areas with low cover, fabrication joints and penetrations as well as protection of completed repairs.

The typical applications and experience with epoxy as surface protection for offshore concrete structures are (DNVGL 2019):

- Full-coverage coating applied as a repair method. There exists good experience with such coatings to minimize chloride penetration. As a result, a complete membrane has been subsequently applied to, for example, the Troll A platform after 10 to 15 years in operation. Similar full-coverage coatings were applied to the Statfjord A platform during construction with good results as discussed earlier (Sharp and Ersdal 2021).
- Local application of a coating is often used in smaller areas to compensate for low concrete cover and to protect the repair material, steel attachments, penetrations and as a membrane for any repairs performed.

Surface protection with an epoxy coating is sensitive to the design and material selection as well as operational temperature and relative humidity. Adequate testing of the method and material should therefore be performed prior to the application. In addition, cleaning and preparation of the concrete surface is essential to achieve good adhesion. The application of epoxy coatings often requires specialist contractors with a good knowledge of both materials and workmanship.

9.5 Adhesives and Epoxy

Adhesives and epoxy are primarily used for repairs. For steel structures this typically includes bonded type repairs and for concrete structures this includes injection of epoxy into cracks.

Bonded types of repairs consist of a fibre-reinforced laminate patch (bonded patch) that is attached to the base material with an adhesive, see Fig. 9.1. Bonded repairs can also involve attaching a metal or composite plate to the base material. The most used bonded repair in the offshore industry is resin repair involving a layer of fibre-reinforced composite material. The fibre-reinforcement is typically glass fibre, carbon fibre or aramid fibre. Such repairs can provide a structural reinforcement and should be designed to allow the transfer of loads from the damaged material into the fibre-reinforced composite repair. The laminate needs to be designed to carry these loads and the adhesive to transfer the forces between the steel structure and the laminate.

Failure modes to be considered in the design are fractures of the laminate, fractures of the adhesive compound and delamination of the laminate. DNVGL-RP-C301 (DNVGL 2015) provides requirements and guidelines for the use of laminating and bonding with composites in repairs and provides guidelines on where such repair can be used and how to obtain a repair of sufficient quality.

The repair method is recommended for non-critical structural elements (DNVGL 2019). The method can also be used on a critical structural element where the damage is considered minor. MSL (2004) describes the use of bonded repairs to caissons but are not clear on whether this was performed underwater or above water.

There is the possibility of using resin techniques for repairs to other parts of an offshore structure such as damaged caissons. These caissons house large pumps for

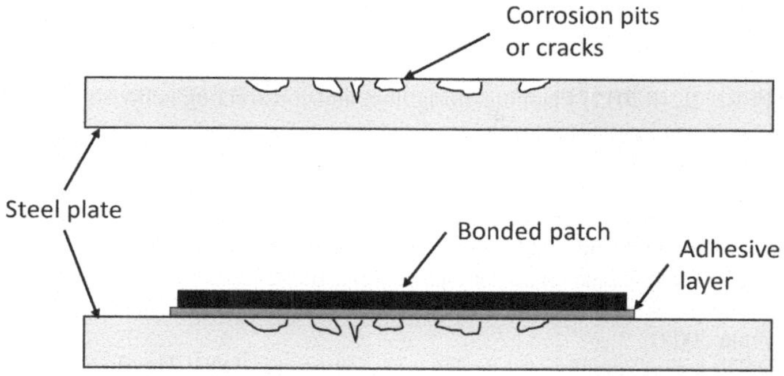

Fig. 9.1 Illustration of bonded patch repair (before and after repair). Based on Sharp and Ersdal (2021)

providing seawater for cooling and fire services. Caissons have suffered damage from galvanic action between the pump and caisson leading to severe localised corrosion and holes in the wall. Another cause of damage is depletion of the original anodes when corrosion can occur if the anodes are not renewed. Repair to a damaged caisson using remote methods is described in Galbraith et al (1994). However, in this case the repair was a stressed elastomer-lined clamp.

Injection of epoxy material is also a versatile and historically widely used repair material on offshore concrete structures. The material is injected into voids such as cracks, delaminations and fabrication joints (to seal a leaking joint). As previously mentioned, according to OTH 90 318 (Department of Energy 1990) for cracks less than 10 mm in width epoxy resins would be the choice repair material. A liquid two-component epoxy may be applied through hoses connected to injection tubes or nipples, which are placed in the concrete to fill voids. The use of epoxy material is not usually used to fill voids in large volumes due to challenges with heat generation during the curing of the epoxy, creep properties as well as constraints in the material's shelf life. Issues concerning the use of epoxy materials include:

- Limited cure life for some of the materials on the market limiting the use to only small volume cavities.
- Influence of temperature can have a significant impact on the material's cure life. This is important underwater with typical North Sea temperatures.
- The influence of relative humidity can have a significant impact on the properties of the final product.
- Injection at too high pump pressures can further damage the concrete.

References

API (2012) API RP 2L (R2012) Planning, designing, and constructing heliports for fixed offshore platforms. American Petroleum Institute

Civil Aviation Authority (2023) CAP 437: standards for offshore helicopter landing areas. CAA, 2023

Department of Energy (1990) OTH 90 318 assessment of materials for repair of damaged concrete underwater. Prepared by Imperial College for the Department of Energy, HMSO London, UK

Djordjevic BB (2009) Ultrasonic characterization of advanced composite materials. In: 10th international conference of the Slovenian society for NDT, September 1–3, 2009, Ljubljana, Slovenia (NDT-Slovenia 2009)

DNV (2000) DNV design guideline for design with composites. DNV, Høvik, Norway

DNV (2001) Offshore standard DNV-OS-E401 helicopter decks, March 2001. DNV, Høvik, Norway

DNV (2013) DNV-ST-C501 composite components. DNV, Høvik, Norway

DNVGL (2015) DNVGL-RP-C301 design, fabrication, operation and qualification of bonded repair of steel structures. DNVGL, Høvik, Norway

DNVGL (2019) Reperasjonsmetoder for bærende konstruksjoner. Work performed for the petroleum safety authority—Norway, Report no.: 2019–1164, Rev. 0, Document no: 524873. DNVGL, Høvik, Norway

Dwight J (1999) Aluminum design and construction. E & FN SPON, An Imprint of Routledge, London and New York

Galbraith D, Lalani M, Sincok P, Geddes T (1994) Beryl Bravo: diverless structural repairs. https://doi.org/10.4043/7501-MS

HSE (2003) HSE RR 039 The cost effective use of fibre reinforced composites offshore. Prepared by A.G. Gibson at the University of Newcastle Upon Tyne for the Health and Safety Executive 2003, HSE Books, UK

MSL (2004) Assessment of repair techniques for ageing or damaged structures. Project 502, MSL Services Cooperation, Egham, Surrey, UK

MTD (1996a) Cost effective use of composites offshore funded by government and 20 industrial sponsors. MTD Publication

MTD (1996b) Titanium alloys in subsea and offshore production systems. Publication 96/100. Published by MTD, London

Setvati MR, Zahiraniza M, Nasir S, Zubair IS (2014) A review on composite materials for offshore structures. In: Proceedings of the 33th international conference on ocean, offshore and arctic engineering OMAE 2014 Jun 8–13, 2014, San Francisco, California, USA OMAE 2014-23542

Sharp JV, nd Ersdal G (2021) Underwater inspection and repair for offshore structures. Wiley, Chichester

Standards Norway (2019) NORSOK C-004:2019+AC:2022 Helicopter decks on offshore installations. Standard Norge, Lysaker, Norway

Standards Norway (2022) NORSOK N-004:2022 Design of offshore structures. Standard Norge, Lysaker, Norway

TWI (2024) Aluminium alloys—Job Knowledge 21. https://www.twi-global.com/technical-knowledge/job-knowledge/weldability-of-materials-aluminium-alloys-021, visited September 2024

Wang Z, Li M, Han Q, Yun X, Zhou K, Gardner L, Mazzolani FL (2022) Structural fire behaviour of aluminium alloy structures: review and outlook. Engineering structures, vol 268, p 114746. ISSN 0141-0296. https://doi.org/10.1016/j.engstruct.2022.114746

Chapter 10
Assessment of Materials Integrity During Life Extension

10.1 Introduction

As we have seen in this book, materials used in offshore structures contain defects from the fabrication process and are susceptible to degradation processes in the operational phase. Hence, a significant part of the assessment of materials in an ageing structure is to establish an understanding of the condition of the material after many years of service, taking into account the state of the material after fabrication and operation.

In addition, we have in this book also shown that there have been technological developments over the years in the quality of materials and their joining methods, which have in many cases been followed by stricter requirements for the materials. The result is that the material in an ageing structure may not be of the quality that would be provided at the present time.

Furthermore, we have in this book shown that it may be challenging to determine the condition of the material in an ageing structure. This can be due both to practical and physical limitations in testing and inspection of the materials for many degradation mechanisms and possibly also as a result of information about the original material being limited. As a result, assessment of materials in ageing structures is to a larger extent an evaluation whether what we know about the condition of the material in the structure is adequate, taking into account present requirements for materials.

However, the challenge is not only the condition of the materials at present, but the material also has to be proven acceptable for the life extension period. This requires an estimate and prediction of the expected degradation in this period. In such a task it is important to use information from design, fabrication, inspection and the present condition to predict how materials can be expected to degrade in the life extension period. Figure 10.1 is illustrating some common degradation histories, which can be helpful in making predictions about the future degradation.

In Fig. 10.1, the different development of degradation may be interpreted as:

© The Author(s), under exclusive license to Springer Nature Switzerland AG 2025
J. V. Sharp and J. Wintle, *Assessment of Materials in Ageing Offshore Structures*,
https://doi.org/10.1007/978-3-031-84489-8_10

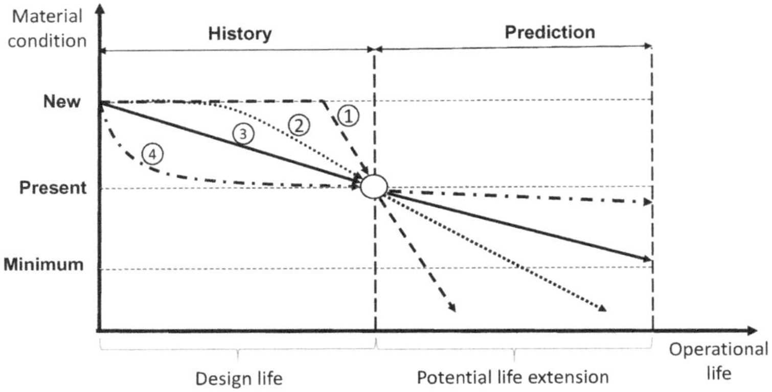

Fig. 10.1 Prediction of degradation in potential life extension based on history of degradation

1. Slow starter then fast, which can be expected to continue fast. Fatigue cracking is typically of this type and mitigation is normally needed prior to life extension.
2. Slow starter then linear, which can be expected to continue linearly. Corrosion is typically of this type and also here mitigation is normally needed prior to life extension.
3. Linear from the start, which is expected to continue linearly. Wear and tear are typically of this type and the need for mitigation during the life extension period is often possible to estimate based on such linear behaviour.
4. Initially rapid and then slow, which is expected to continue slowly. Free corrosion in sea water can be of this type and mitigation may not be needed if the degradation level is acceptable now (with some margin).

These issues of assessment of materials in ageing structures are illustrated in Fig. 10.2, where it is highlighted that the assessment needs to determine if the material is adequate at present and that the material will remain adequate in the life extension period.

Evaluation of the robustness of the material is one possible method for life extension assessment. As previously mentioned in Sect. 1.3, robust materials are those that can tolerate to a reasonable extent the environment, operational use and loading to which they are exposed while still being able to perform their full function (fit for purpose). However, assessment of the robustness of materials for life extension has specific issues:

- The assessment needs to take the present condition of the materials as a starting point, rather than the material as fabricated.
- The materials are already in-place with little or no opportunity for replacement (you have to live with what you got).

The task is therefore to determine if the materials that are there can be accepted as sufficiently robust.

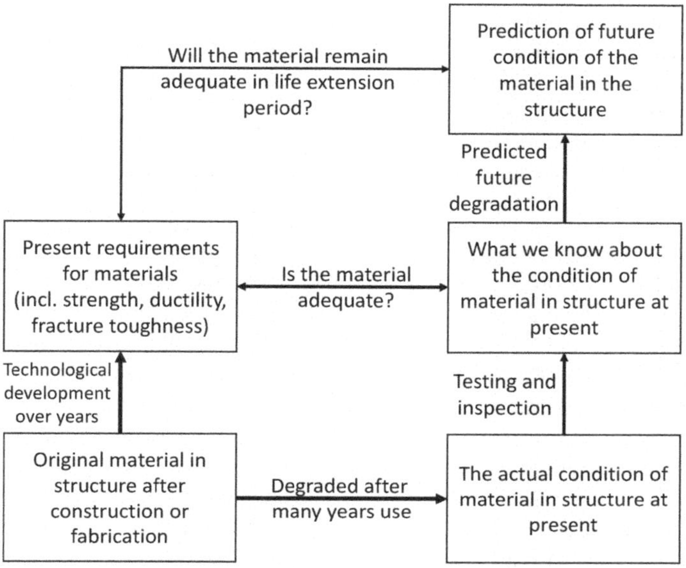

Fig. 10.2 The process of assessment of materials in ageing structures

When reviewing the materials in a structure for the purposes of life extension, a first stage (Drugli et al 2003; Lange et al 2004) in the assessment would be to evaluate both:

- whether the original materials selection process was sufficiently rigorous,
- the extent to which the original materials have proved to be robust, based on the experience gained during operation so far.

These issues should be used as the basis for evaluating the capability of the materials to be fit for life extension.

Understanding the original material selection should include establishing what materials were used and an evaluation of the materials selection process. Further, it may also include an evaluation of the condition after construction or fabrication. These investigations should establish the baseline for subsequent changes. Further, the material can be shown to be sufficiently robust by information available from experience, inspection and testing. The evaluation of the robustness of the materials can be based on the level of degradation (within expected limits) and to what extent experience and testing is proving that the key properties of the material remain intact.

In addition, the following issues need to be included in assessing the robustness of the materials:

- What are the present requirements for materials (production, construction and fabrication), to know if the original material meets these current requirements (for

example strength, ductility, defects, fabrication tolerances and fracture toughness that are required in present structural standards).
- What the condition of the material in the structure is after many years of service (the as-present condition).
- Whether the material can continue to be used in the environment and with the loading it will be exposed to in the life extension period, which includes proving the materials continued robustness.

However, even when all the usual failure mechanisms have been considered and degradation has been properly modelled, there is still the possibility of a premature component failure by unexpected hazards or an uncommon combination of mechanisms (Molyneaux and Speck 1997). As a result, one inadequacy of these original design codes and material specification is that they do not include the experience gathered from later operations. In addition, design codes and material specifications from the early days of offshore structures for oil and gas exploration are unlikely to have had in mind the life extensions we are now getting used to and, hence, care is needed if these are used as a basis for an assessment. It should be mentioned that the materials specified at the design stage are typically designed to last for the planned design life, with allowances for anticipated degradation and deterioration. However, in an assessment for life extension, the effects of deterioration and other ageing factors are likely to exceed these.

The approach to the assessment of all materials can be viewed in a general sense, taking into account knowledge gaps and uncertainties. This has been developed by Lange et al (2004) and further developed by Hörnlund et al. (2008, 2011). This approach is presented in this chapter.

10.2 Model for Robust Material Selection

Lange et al (2004) divided the risk related to selection of materials into four categories, depending on how much is known about the materials (proven or un-proven) and how much is known about the environment (known or new). This is illustrated in Fig. 10.3.

An example of an unproven material is a material that lacks or has little field history, whereas a proven material applied in a known environment implies that there is sufficient field data with respect to materials and design. An example of unproven materials is the earliest use of hight strength steels in jack-up structures, where later experience showed that these were susceptible to hydrogen embrittlement and hydrogen related cracking (see Sect. 6.2).

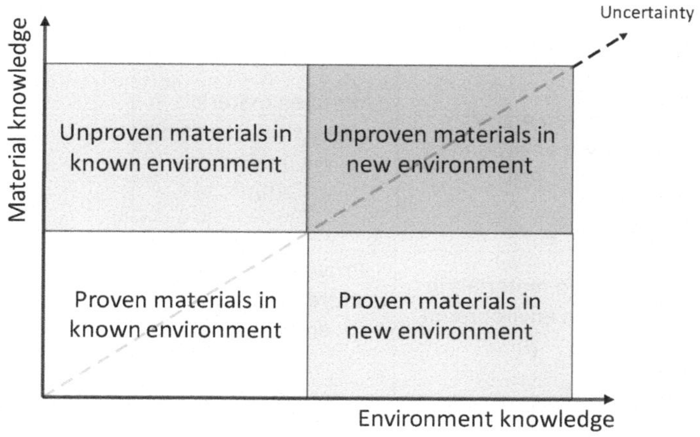

Fig. 10.3 Risk and uncertainty in relation to robust materials selection

10.3 Model for Evaluating Robustness of Materials in Existing Structures

Hörnlund et al. (2011) developed a similar set of categories for the assessment of materials in existing structures. The process suggested by Hörnlund et al. (2011) includes activities to establish the degree of uncertainty associated with the materials and the environment to which they are subjected, both retrospective, present and forward-looking. This set also included four categories related to knowledge gaps and uncertainties, depending on the operational history. This approach adapts DNV's recommended practice for qualification procedures for new technology (DNV 2019) where there is a similar, but more detailed categorization of classes of uncertainty.

The retrospective assessment should gather knowledge about material performance and degradation during operation, as well as information about the environment that the materials have been exposed to. Further, it should establish the degree of uncertainties in the operational phase. Essentially the environment is normally well known for materials in offshore structures. However, there may be situations where the level of cathodic protection is uncertain, which may have caused degradation (corrosion and fatigue) beyond what is expected. In addition, overprotection may have caused hydrogen embrittlement. Furthermore, subsidence may have exposed materials to a more corrosive environment. The evaluation of the present condition will add to this information and may confirm or otherwise these uncertainties.

Finally, a prediction about future development should be undertaken and the uncertainties linked to these predictions should be identified, as indicated in Fig. 10.4.

In Fig. 10.4, materials that are proven and known in well tested designs in an established stable location without change to the operating environment may be categorized as "proven materials applied in known environments". Defining the materials to be within this categorisation implies that the operator has sufficient field test data

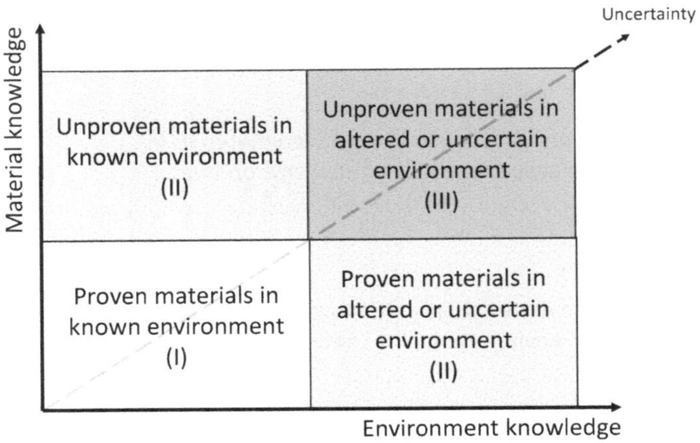

Fig. 10.4 Uncertainty evaluation on environmental and material knowledge from operational history

with respect to materials and environment to affirm the selection of robust materials within the normal design life.

Proven materials that have experienced changes in the environment during service can be categorized as "proven materials applied in new environments" because the design presumptions are changed. Example of a changed environment may be the before mentioned examples of unknown cathodic protection and extended splash zone exposing new structural members. Under these circumstances an evaluation of the materials for the new environment and loading is necessary. A further and more critical example of proven materials applied in new environments would be risers and pipelines exposed to new well streams such as sour conditions. In such cases laboratory and field testing under the new conditions should be considered.

In life extension, unforeseen degradation mechanisms or unmitigated degradation over the long term could eventually make the material unproven. This statement results from the fact that data are in practice not, or at best only sparsely, available for materials used in such long-term timeframes. Hence, using a known structure with known materials beyond the original design life in an unchanging environment will in time represent "unproven materials applied in known environments". Generally, the use of "unproven materials in known environments" is more challenging than "proven materials in a changing environment". This is because knowledge about an aging material has to be built up with respect to possible degradation processes and failure modes that may occur.

The category representing the most demanding challenge is "unproven materials applied in new environments". This is because there is limited knowledge about the material and of the performance of materials in this new environment. To demonstrate that materials continue to be robust for such purposes, the various possible degradation processes must be mapped and closely monitored to detect any early changes in performance. It cannot be assumed that the information from laboratory testing and research work are fully representative of the actual conditions. Application of known structures or systems with proven materials beyond the design life but with altered functions, will fall into this category. Such modified functions can possibly be exemplified by change of structures from maintained production and operation to a situation where production has ceased (sometime referred to as "lighthouse mode") and normal maintenance has been reduced. In this mode, materials originally protected by coatings or cathodic protection may be exposed to higher corrosion rates and other deterioration processes when these safeguards are degraded. The assumptions made at the time of manufacture and fabrication of the oldest existing facilities were probably not those to be encountered in this situation. Consequently, there is likely to be more uncertainty in the material properties and performance.

In order to deal with the uncertainties and gaps in knowledge with the use of old materials in structures intended for life extension, there is a need for continued testing of such materials accelerated over a longer timescale to map the various degradation processes. Simultaneously, there is an opportunity available for such testing by using structures after they have been decommissioned. Many of these structures represent the same type and age as those who are now presented for life extension.

10.4 Categorisation of Materials

An assessment of possible life extension should determine the state of knowledge about the specific materials used and their interaction with their environment. In addition, it should identify gaps in knowledge and uncertainty (see for example Table 10.1). There are several key questions to be asked about the materials covering the entire lifetime of the facility. The answers to these questions will determine in which one of the four categories of knowledge and uncertainty a material lies corresponding to the boxes in the diagram shown in Fig. 10.4.

An important conclusion from this assessment may be that the as-is-condition is within the original design parameters (for example within the corrosion allowance),

Table 10.1 Categorisation of materials

Original design and fabrication	Operational history	As-is condition
• The conditions that determined the choice of materials in the original design • Material certificates and joining procedures • Repairs or material replacement during fabrication	• Inspection results from operation and trends • Occurrence of any accidental damage, degradation mechanisms and failure modes • Environment during operation • Repairs undertaken during operation	• Identify any sign of detrimental changes • Investigations of materials lacking material certificates and materials that are impossible to inspect • Evaluation of knowledge related to challenges with similar materials and welding procedures

which would in most cases make the assessment for life extension relatively simple. If so, the task would mostly be to document that the as-is condition can be shown to be sufficiently robust. Such documentation can, for example, be based on a fitness-for-service assessment for metallic structures or a durability assessment for concrete structures.

The assessment of the material for life extension should be based on the criteria and properties which are needed for materials in this phase. A key question is if there any evidence of potential changes in the materials which could develop in the life extension period affecting the functionality of the materials. In addition, it can be questioned if all materials in older offshore structures can be successfully welded in a repair situation with modern consumables and welding procedures. This raises the issue of compatibility, which is part of the "technological changes" shown in Fig. 1.1.

Future degradation and performance of the materials should be assessed. This should include limitations of the material introduced by the degradation mechanism and the prediction of the future environment. Models for future degradation should be established where possible. In addition, it should be noted that structures designed to good standards are in most cases robust and able to tolerate single failures, but in a life extension period possible multiple site degradation should be considered.

Based on these assessments, the material and corrosion engineers should give guidance to integrity assessments, taking into account the effect of the degradation mechanism on the robustness of the facility.

Typically, one could say that a material is "known" if the following are present:

- Design and fabrication:
  - Good knowledge of original materials and fabrication. This includes having material certificates, NDT-documentation, welding procedures, repairs, change orders etc. available.
- Operation and as-is-condition:
  - All significant items have been inspected regularly throughout the operational phase and show little sign of accidental damage or material degradation.
  - Inspection records and trends are well documented (and used in the assessment).
  - Repairs involving new welds, material replacement and modifications have not caused any significant material issues and are well documented.
  - Well documented as-is-condition (which could be based on the above mentioned as-built condition and inspection records and trends).
- Future performance:
  - Good knowledge on relevant future degradation mechanisms leading to predictions of future changes of material properties.

On the other hand, a material is "unknown" if significant amount of this above-mentioned information is missing. This could include:

- Design and fabrication:
  - Lacking knowledge of original materials and fabrication.
- Operation and as-is-condition:
  - Inspection records and trends are poorly documented.
  - Undocumented repairs or modifications.
  - Significant accidental damages or material degradation has occurred (for example corrosion beyond design limitations).
  - Significant amount of structural members are non-inspectable or have not been inspected during the operation phase. Such lack of information creates considerable uncertainties that need to be taken into account in the assessment.
  - Unexpected changes of material properties.
  - Poorly documented as-is-condition.
- Future performance:
  - Lacking knowledge on future degradation mechanisms.
  - Limited prediction of material degradation in life extension period is performed.

Similarly, a known environment can be described by these conditions:

- Good knowledge about the environment that determined the choice of materials in the original design.
- Well documented environmental data during operation, including for example cathodic protection levels, changes in seawater levels and degree of inundation, the presence of harmful bacteria.
- Evidence that the environment in the extended life will be the same as that in the original design life.

While an unknown environment can be described by the lack of such knowledge and information, for example that increased temperatures or more hostile loading are expected in the life extension period.

Based on the answers to these questions, a material is allocated to one of the four categories of uncertainty shown in Table 10.2.

Table 10.2 Categorisation of a material corresponding according to the state of knowledge

Level 3 Unknown material condition in known environment	Level 4 Unknown material condition in unknown environment
Level 1 Known material condition in known environment	Level 2 Known material condition in unknown environment

10.5 Actions Following Categorisation of a Material for Life Extension

For a level 1 material (known material condition in known environment), essentially no actions would be needed, as it would be regarded as fit for purpose for life extension.

For a level 2 material (known material condition in unknown environment), actions should be taken to better establish the past and predict the future environment to reduce uncertainty. The actual environment and its influences on the materials should as a result be monitored throughout the life extension period. Better methods and models for predicting the future degradation and the associated failure modes for the actual material in the predicted environment may need to be established. The involvement of qualified corrosion engineers may be needed to meet these requirements.

For a level 3 material (unknown material condition in known environment), inspection, testing and other actions are needed to gain better knowledge on the present condition. Information from similar facilities may be used to add to this knowledge, accounting for the assumed correlation between these facilities. Methods and models for determining the future degradation need to be established. The involvement from qualified materials- and corrosion engineers is needed in the life extension assessment and possibly in the extended operational phase.

For a Level 4 material (unknown material condition in unknown environment), life extension will in most cases not be feasible or will be limited to short periods, unless significant efforts are carried out. Hence, inspection, testing and better knowledge of the present condition and the environment is important to be established so that either the material condition or the environment can be considered as known.

Based on the information and data gathered, new estimates of characteristic values should be established for the various material properties. Essentially, this type of statistical information is only expected to be fully available for level 1 materials. A part of the further investigation of level 2–4 materials would be to gain such information and identify knowledge gaps. However, suitable safety factors may need to be applied in the structural analyses for level 2–3 materials. Furthermore, additional or special inspections or other measures might need to be included in a structural integrity management plan.

Level 4 materials should, as mentioned above, not be used in life extensions unless sufficient information can be obtained to bring the material to level 2–3.

10.6 Evaluation of Materials for Life Extension

Several roles in a life extension assessment may have an interest in the material evaluation. These range from the life extension manager, who mostly should be informed early in case of any issues, to the materials and corrosion engineers. In addition, the mechanical and structural engineers will also need to be informed on any findings, due to their need for material properties relevant to their analyses.

A suggestion for tasks likely to be involved are listed by Hörnlund et al. (2011). The tasks mentioned by Hörnlund et al. (2011) for the material and corrosion engineer includes tasks that are essential to this book. These include:

- Determining the original material used including material certificates and, if possible, the criteria that determined this selection, which are discussed in Chap. 3 for steel materials and in Chap. 7 for concrete materials.
- Determine the fabrication and construction procedures and possible issues with these, which are discussed in Chaps. 4 and 5 for steel materials and Chap. 7 for concrete structures.
- Evaluate the expected and actual behaviour of the materials (degradation mechanisms and failure modes) from inspection results and trends, which is discussed in Chap. 6 for steel materials and Chap. 8 for concrete materials.

In addition, Hörnlund et al. (2011) raises questions related to the uncertainty about the materials, as discussed earlier in this chapter. These include (Hörnlund et al. 2011):

- Materials that lack relevant certificates and that are impossible to inspect.
- Fabrication and construction methods and procedures that have later been shown to have limitations.
- Limitations of the material introduced by the degradation mechanism. In practical terms this implies an evaluation of uncertainty related to:
  - Unpredicted degradation and deterioration (existence, extent and trends).
  - Unknown phenomena.
  - Unexpected failure modes.
- Uncertainty related to the material when degraded.

It can be seen that these questions raise significant issues about suitability for life extension. Particularly, materials lacking relevant certificates and without inspection data are a possible show-stopper for life extension. In addition, the material and corrosion engineer need to provide estimates of future degradation and performance of the materials.

10.7 Degradation of Materials

Degradation of materials can both severely affect their properties (intrinsic changes) and change the form of the component (extrinsic changes). Extrinsic changes include mechanisms which in different ways can remove or change the material at the surfaces, in addition to cracking which creates new surfaces. Extrinsic changes are relevant for the integrity and capacity of structures as they typically include thickness loss, geometric changes and introduction of discontinuities, such as macro-scale cracks and sharp transitions in the structure.

Intrinsic ageing of materials can be seen as any change in material properties from when it was new and after 20, 30 or 40 years in service. If tests were performed when the material was new and after years of service, any difference that is seen is an indication of intrinsic ageing of the material. Typical intrinsic changes to material properties are hydrogen embrittlement and hardening for ferritic materials and chloride ingress, carbonation and sulphate attack to cementitious materials like concrete. Another very important change is propagation and aggregation of microcrack and other impurities (inclusions, grain boundaries, second phase particles and voids) from the manufacturing and welding process due to e.g. cyclic and static loading.

References

DNV (2019) DNV-RP-A203 technology qualification. DNV Høvik, Norway

Drugli JM, Thaulow C, Ødegård J, Rogne T, Stokke R, Berge S, Berget J (2003) Pre-project: robust material selection in the offshore industry. SINTEF, Trondheim, Norway

Hörnlund E, Sævik Ø, Hinderaker RH, Ersdal G (2008) Ageing of materials. In: Proceedings of the ASME 2008 27th international conference on ocean, offshore and arctic engineering, OMAE 2008, June 15–20, Estoril, Portugal

Hörnlund E, Ersdal G, Hinderaker RH, Johnsen R, Sharp J (2011) Material issues in ageing and life extension. In: Proceedings of the ASME 2011 30th international conference on ocean, offshore and arctic engineering, OMAE 2011, June 19–24, Rotterdam, The Netherlands

Lange H, Berge S, Rogne T, Glomsaker T (2004) Robust material selection. SINTEF, Trondheim, Norway

Molyneaux MJ, Speck JB (1997) Lessons to be learned from five unusual cases of high temperature material degradation. In: Penny PK (ed) Ageing of materials and methods for the assessment of lifetimes of engineering plants. A.A. Balkema

Chapter 11
Conclusions and Future Possibilities

Ascertaining the continuing good performance of the materials used in offshore structures is essential for their safety and reliability, particularly during the period of life extension. This book has examined in depth the performance of the two widely used materials, namely steel and concrete. Offshore structures constructed of both steel and concrete have now been in use for several decades and a wealth of experience has been gained on their performance. However, in recent years life extension has been required for these structures bringing more significant longer term material issues which have been addressed in this book.

The early North Sea offshore steel structures were designed and constructed based on US Gulf of Mexico practice. Early component failures showed that the North Sea environment was more demanding than that in the Gulf of Mexico and more robust designs and materials were needed for this location. Research has played an important part in improving the knowledge of material's performance and this book has highlighted and presented the results of relevant areas of research, the results of which are not readily available in all cases.

The book has highlighted the quality of materials and joining methods typically used at different periods of the North Sea offshore development, in addition to the factors that determined this quality. Further, the book has also examined the different stages of a material's life from selection and design to construction and further to operation in a harsh environment. In addition, the experience gained from offshore oil and gas installations is proving to be valuable for the substantial development of offshore wind energy.

Degradation from immersion in seawater, together with the effect of wind and waves creates many problems which need to be assessed, particularly in life extension. The different ways in which the two main materials can degrade is an important part of this book. The book has also highlighted the importance of the microstructure of these materials which affects their performance in operation and the possibility for their continued use during life extension.

© The Author(s), under exclusive license to Springer Nature Switzerland AG 2025
J. V. Sharp and J. Wintle, *Assessment of Materials in Ageing Offshore Structures*,
https://doi.org/10.1007/978-3-031-84489-8_11

Testing of materials in offshore platforms to identify their condition is difficult which has several disadvantages when considering life extension. However, some very useful information has been obtained from a few decommissioned structures which is presented in this book. If further decommissioned structures could be made available for post-service examination and testing, then greater insights into their performance and effects of long-term ageing could be gained for wider benefit.

Rehabilitation of degraded materials itself is difficult (if not impossible) but components containing degraded material can be repaired or replaced. Improvement of repair and replacement techniques would allow more structures to remain in-service for longer. Whereas some repair methods for degraded materials are described in the book, further details on typical repair methods for material and components are described in Sharp and Ersdal (2021).

This book is a part of a series of three books on different aspects of ageing of the types of offshore structures used for oil and gas production and wind energy generation (Ersdal et al. 2018; Sharp and Ersdal 2021). The three books together cover topics that are important for the long-term performance and life extension of offshore structures, including structural integrity, inspection, monitoring, repair and the assessment of original materials for continued service. They highlight the changes in technology and knowledge that have occurred during the last 60 years from service experience and extensive research programmes.

It is hoped that readers of the book will benefit from its content, both students and practicing material engineers who have the task of assessing materials in ageing structures for life extension.

References

Ersdal G, Sharp JV, Stacey A (2018) Life extension of offshore structures. Wiley

Sharp JV, Ersdal G (2021) Underwater inspection and repair for offshore structures. Wiley

Chapter 12
Appendix—Case Studies

12.1 Case Study 1—West Sole WE Platform

The West Sole WE platform was fabricated for BP in 1965 and installed in July 1966 as an unmanned single well satellite unit from which gas was exported to the nearby manned platform WB. It was one of the first platforms to be installed in the UK sector of the North Sea and was designed for a life of 25 years.

In 1978, BP decided to remove the platform and abandon the well due to severe corrosion in the well tubing, which would have required a major workover (repair and mitigation). This was the first decommissioning in UK waters. BP recognised that there was an important opportunity to learn from recovery of this platform and a series of tests was performed together with a detailed examination of the structure once removed (British Petroleum 1979). The information from these tests provided some very useful data on the performance of materials in the 13-year period of exposure to North Sea condition.

The report indicates that the platform was designed and fabricated according to Gulf of Mexico practice, regarded as the best standards available at the time, as was common for all platforms in European waters at that time. As part of the testing after the removal of the platform, circumferential weld cracks were found on three of the four leg nodes at elevation + 28.5 feet (8.7 m) above sea level. Below sea level only two nodes had minor crack indications associated with welds. One of these cracks was believed to be from the removal operation.

These experiences and other similar experiences with early platforms led to the recognition of more demanding requirements being required in North Sea conditions, as the particular cyclic loading conditions were significantly different from those in the Gulf of Mexico. For many of the early structures this required strengthening and several of the other West Sole platforms were strengthened in 1974/75. However, due to the well tubing problems and the potential need for strengthening, the West Sole WE platform was decided to be removed for economic reasons.

© The Author(s), under exclusive license to Springer Nature Switzerland AG 2025

J. V. Sharp and J. Wintle, *Assessment of Materials in Ageing Offshore Structures*,
https://doi.org/10.1007/978-3-031-84489-8_12

The platform comprised four 33″ diameter raked legs, cross braced and piled though the legs using 30″ diameter piles with grout in the annuli between the leg and the pile, see Fig. 12.1. The steel in the splash zone was protected with coal-tar epoxy. Below water the platform was provided with a sacrificial anode cathodic protection system (Table 12.1).

Prior to decommissioning two different vibration analysis tests were conducted on the platform to detect damage both offshore and also on the recovered platform at Nigg. The platform was deliberately damaged prior to removal by cutting members. Vibration measurements were made before the damage and after each cut. These are briefly reviewed in Sharp and Ersdal (2021).

Offshore surveys were also carried out prior to decommissioning. These showed:

Fig. 12.1 West sole WE platform (courtesy of John V Sharp)

Table 12.1 Specifications for the West Sole WE platform (British Petroleum 1979)

Steel	The main tubular members were specified as ASTM 252 grade 2 steel. The piles were a German steel grade St. 52, grade 3 steel. The bracing was specified as refinery piping API grade 5LB steel
Welding	Construction was stated to be to an accepted standard such as BS 1856 (British Standards 1952). Welds should have generated a tensile strength not less than that of the members to be joined. Pre-heating to a minimum of 200 °C was required for all weld connecting or adjacent to steel of ¾ "(19 mm) thickness or greater. All welded joints in material of a thickness greater than ¾" (19 mm) should have been stress relieved after welding by heating to between 580 and 620 °C and maintaining the temperature for an hour per inch of the material thickness
Sacrificial system	This design consisted of 12 kg aluminium alloy anodes with an estimated life of 10 years without need for replacement
Painting	Below water all welded joints were to be given a single coat of Wailes Dove no. 50 coal tar preparation. In the splash zone after surface preparation surfaces were to be coated with a zinc primer to a thickness of 25 microns followed by a layer of epicoat tar to a minimum thickness of 300 microns. Above the splash zone one coat of red lead, followed by one coat of an approved marine paint
Cladding	In the splash zone the main members to be reinforced with mild steel wrappers of 8 mm welded around the tubular thickness

- Cathodic potentials—at level − 73.5 the potentials ranged from − 0.88 to − 0.92 (w.r.t Ag/AgCl). At level − 37.5′ the potentials were from − 0.87 to − 0.93 and at level − 3.5′ the range was − 0.87 to − 0.91.
- Crack monitoring: CVI was used on a weld on one of the legs. No defects were reported. MPI was also undertaken on the same weld. Two indications were reported but were deemed to more like scratches than cracking.

The West Sole WE platform was removed in two sections, in September 1978, the first being the superstructure. The second lift was the remaining 230 tonne jacket after the piles had been cut. The jacket was towed to the Humber and then transferred onto a barge to be taken to the Highland Fabricators yard at Nigg in Scotland. A photo of the jacket at Nigg is shown in Fig. 12.2.

The condition of the platform after recovery was reported as:

- Marine growth: This was considerably less than expected, possibly due to some cleaning during annual surveys offshore. The growth on the anodes was variable and of a different type to that observed on the structure. Marine fouling on the recovered platform can be seen in Fig. 12.3.
- Anodes: All the anodes were intact with less wastage than anticipated with an average weight loss of 40%—see Fig. 12.3.
- Coatings and paint: Despite the specification for painting the jacket had not been painted below level − 3.5′ (− 1 m). Considerable breakdown of the coating had occurred up to the underside of the cellar deck due to weathering. The splash zone area had been painted with coal tar epoxy over a zinc rich epoxy primer. The protection of the cross bracings in this zone had suffered serious breakdown

Fig. 12.2 West Sole platform and topside after recovery at Nigg (courtesy of John V Sharp)

from wind and spray action. It was noted that this could have been minimised by maintenance painting during the life of the platform.

- Corrosion: Pitting was observed in the splash zone in the form of corrosion pitting and metal wastage with depths of 1/8″ (3 mm) in places. It was concluded that this would not have affected the strength of the platform significantly. Corrosion on the underwater parts of the jacket was negligible with little evidence of metal wastage thus indicating that the CP system had been effective. The topsides equipment and structure were found to be in generally good condition, with little corrosion and metal loss.
- Cracking: Circumferential weld cracks were found on three of the four leg nodes at elevation + 28.5′ (+ 8.7 m). Below sea level only two nodes had minor crack indications associated with welds. One of these cracks was believed to be from the removal operation.
- Structural damage: Minor damage was observed partly due to boat collisions and partly due to the removal process.
- Pile connections: Good quality grout in sound condition was found in the leg/pile annuli of three of the legs. No strength tests of the grout were made. The remaining leg was un-grouted although the original specification called for all four leg-pile annuli to be grouted. Pile eccentricity was noted in the legs with touching of the leg walls in some cases.
- Dimensions: Wall thicknesses and dimensional surveys after recovery showed significant differences from the original specifications. In practice thicker and stronger members had been used which it was noted could have contributed to the overall good condition of the structure.

Fig. 12.3 Aluminium anode recovered from West Sole WE platform (courtesy of John V Sharp)

12.2 Viking AD Clamp

The Viking AD platform was decommissioned in 1998, and one of the reinforcement clamps introduced post-installation was acquired by the UK Health and Safety Executive for assessment. The Health and Safety Laboratory in Buxton conducted a thorough examination of this clamp after its twelve years in service. The first phase of the project (HSE 2000a) involved the following tasks, performed approximately one year after the clamp had been out of the water and had been transported between sites and lifted multiple times:

- Creating a visual record of the clamp and its dimensions at the end of service

- Assessing marine growth in terms of thickness, density, and distribution
- Conducting a corrosion survey
- Performing a non-destructive assessment of the grout layer between the clamp shell and the jacket tubulars
- Measuring residual stress around the welds in the tubulars
- Disassembling one half of the clamp and measuring stress in the stud bolts

The clamp measured approximately 8 m long, including the tubular ends. It was about 1.7 m wide across the strong-backs, with each clamp box being 0.75 m deep. The clamp box flanges were 1.0 m across (see Fig. 4.7). The two halves of the clamp were secured by 12 stud bolts made from medium carbon, low alloy steel, likely grade L7 in AISI 4140, with a tensile strength of 930 MPa, equivalent to grade 8.8 bolts.

A sample for metallographic examination was taken from a longitudinal weld in one of the tubulars. Microscopic examination showed that the tubulars were made from rolled low carbon steel plate (estimated at approximately 0.2% carbon) in the normalized condition. The double-sided butt weld was well-made, with the weld metal being of similar composition and condition to the parent plate, and no apparent defects were found in the examined section of the weld.

Fracture toughness specimens (three-point bend) were created from the parent material, heat-affected zone (HAZ), and weld metal of the tubulars and tested according to BS 7448 (British Standards 1991) at approximately 0 °C. Charpy impact test data at around 0 °C showed values between 99 and 116 J for the parent plate, 110 and 120 J for the HAZ, and 74 and 86 J for the weld. It was concluded that the impact toughness of the tubular material was within the expected range for 50D steel.

The threshold value of the stress intensity factor range (ΔK_{th}) and the fatigue crack growth rate (da/dN) in air were determined using three point bend fracture toughness specimens. The results indicated that ΔK_{th} was approximately 3.5 MPa $m^{1/2}$ at an R ratio of 0.35. The Paris Law constants m and C were found to be 3.78 and 5.89×10^{-13} respectively. Published fatigue threshold data on grade 50D show a wide range of values and the present result falls into the lower end of this range.

After cleaning, a visual examination of the strengthening clamp was conducted to estimate the degree of corrosion and assess the effectiveness of the impressed current protection. This examination revealed no significant localized pitting or general corrosion on the plate material used to fabricate the clamp boxes or the strong backs. The only components with noticeable corrosion were the threaded stud-bolts, which showed minor general corrosive attack where they passed through the pockets on the upper surface. This corrosion was confined to areas covered by sand, grit, and molluscs, while other parts of the threaded stud-bolts were in excellent condition. Ultrasonic measurements showed the grout between the clamp shell and the jacket tubulars was uniformly thick, indicating the clamp and tube were concentric. Strain gauging confirmed all the stud-bolts were correctly tensioned as per the drawing requirements, indicating satisfactory installation of the clamp. The node weld in the centre of the clamp was inspected and found free of defects. The residual stress pattern near the node welds was complex, suggesting incomplete stress relief, and installation of the clamp appeared to have induced compressive stress around the node welds.

The second phase of the project (HSE 2000b) involved the following tasks:

- Dismantling one half of the clamp
- Conducting compression tests on grout samples
- Assessing the residual fatigue life of the stud-bolts
- Measuring fracture toughness and impact toughness of parent plate, weld metal, and heat-affected zone (tubulars)
- Measuring the fatigue crack growth rate and threshold stress intensity in the parent plate (tubulars)

One half of the clamp was dismantled successfully with no significant deterioration observed. Visual and mechanical testing of the grouted annulus between the clamp body and the tubular showed a complete and relatively uniform thickness with no large voids. The bond between the clamp body and the grout material was very good. Compression tests indicated the grout's compressive strength met design requirements, but shear properties were anisotropic due to the layered structure, with shear strength along a layer being about one-third of that across a layer, possibly due to early age cycling, see Sect. 7.6.4. These results could impact the clamp's strength, which relies on the bonding and mechanical properties of the grout material.

Test samples, each approximately 0.5 m in length, were cut from the stud-bolts removed from the upper half of the clamp. Fatigue tests conducted in air at around 2 Hz produced an S/N curve for the axial tensile loading of the stud-bolts, which was compared with BS 7608 standards (British Standards 1994). The residual fatigue life of the stud-bolts was found to be consistent with new bolts, indicating no significant reduction due to service. Further testing showed that the fatigue crack growth rate and threshold stress intensity matched previously reported values for this material. Additionally, the fracture toughness and impact toughness of the tubular material were within the expected range for 50D steel.

Overall, it was concluded that twelve years in service had not led to significant corrosion, and the measured material properties were similar to those of new materials.

References

British Petroleum (1979) North Sea—West Sole field platform WE; post removal condition report and lifetime performance survey

British Standards (1952) BS 1856:1952 General requirements for the metal-arc welding of mild steel

British Standards (1991) BS 7448—fracture mechanics toughness tests

British Standards (1994) BS 7608:1994 Guide to fatigue design and assessment of steel products British Standards Institute

HSE (2000a) Assessment of strengthening clamp from the Viking offshore platform: phase I, offshore technology report 2000/057. HSE Books, London

HSE (2000b) Assessment of strengthening clamp from the Viking offshore platform: phase II, offshore technology report 2000/058. HSE Books, London
Sharp JV, Ersdal G (2021) Underwater inspection and repair for offshore structures. Wiley

The manufacturer's authorised representative in the EU is Springer Nature Customer Service Centre GmbH, Europaplatz 3, 69115 Heidelberg, Germany. If you have any concerns regarding our products, please contact ProductSafety@springernature.com

Printed and bound by CPI Group (UK) Ltd, Croydon, CR0 4YY

01/10/2026

02225928-0010